Design of Crystal and Other Harmonic Oscillators

Design of Crystal and Other Harmonic Oscillators

BENJAMIN PARZEN
Consulting Engineer

With a contribution by
ARTHUR BALLATO
U.S. Army Electronics Technology & Devices Laboratory
Fort Monmouth, New Jersey

A Wiley-Interscience Publication
JOHN WILEY & SONS
New York • Chichester • Brisbane • Toronto • Singapore

Copyright © 1983 by John Wiley & Sons, Inc.

All rights reserved. Published simultaneously in Canada.

Reproduction or translation of any part of this work beyond that permitted by Section 107 or 108 of the 1976 United States Copyright Act without the permission of the copyright owner is unlawful. Requests for permission or further information should be addressed to the Permissions Department, John Wiley & Sons, Inc.

Library of Congress Cataloging in Publication Data:

Parzen, Benjamin, 1913-
   Design of crystal and other harmonic oscillators.
   "A Wiley-Interscience publication."
   Bibliography: p.
   Includes index.
   1. Oscillators, Electric.   2. Oscillators, Crystal.
I. Ballato, Arthur.   II. Title.
TK7872.07P26 1983   621.3815′33   82-13620
ISBN 0-471-08819-6

Printed in the United States of America

10 9 8 7 6 5 4 3 2

To the Memory of
My Father and Mother

Preface

Oscillators are an integral part of electronic equipment in many fields—communication, navigation, data processing, and so on. The harmonic oscillator is the type of oscillator with the greatest stability. Probably millions or billions of this type of circuit are in existence and more will be needed in the future. It is therefore important to gain as much knowledge as possible about harmonic oscillators, and this book has been written to help this.

As the title signifies, the book treats oscillators producing quasi-sine-waves whose frequencies are determined by networks of two types of energy storage elements—such as inductors and capacitors, or the equivalent, for example, crystal resonators.

Despite the long history of harmonic oscillators, including crystal oscillators, their successful design still depends largely on the skill and experience of the designer, and usually proceeds in a "cut and try" fashion. This has been greatly influenced by the well-known fact that it is not at all difficult to assemble a configuration that will oscillate somehow, since useful oscillations are produced over a very wide range of parameters. However, the procedure required to make the oscillator satisfy prespecified conditions is not at all easy and has not been successfully developed. This book attempts to correct this situation.

The literature on harmonic oscillators is voluminous, but has made very little progress toward achieving a practical, systematic, and quantitative design procedure over a wide frequency range. Examination of the literature shows that it generally falls into the following categories:

1. General qualitative and quantitative exposition of oscillator theory with little attention paid to its practical application to produce oscillators. The exposition may be at a relatively elementary or a very advanced level.
2. Qualitative and perhaps some quantitative descriptions of oscillator theory including specific circuit diagrams for particular designs, with some sketchy statements about the performance of these designs but with no clue as to how the circuit diagrams were derived.

3 Exploration of the effect on frequency variation by the variation of some circuit parameter, without corresponding attention to the effect of the same variation upon the remaining characteristics of the oscillator, for example, amplitude.

The literature thus far lacks a relatively simple and direct design procedure which bridges the gap between theory and practice, and which transforms oscillator performance requirements into circuit component values. This book attempts to significantly contribute to the realization of such a procedure.

In this book the literature and the writer's theoretical work and practical experience are gathered together and modified by suitable assumptions and approximations to synthesize a straightforward, unified, and direct design technique for the most popular configurations. In achieving the result, the following criteria were followed:

1 The technique should involve only direct manipulation of equations and relatively easily determinable values for the approximate values of the parameters of these equations. Graphical procedures are therefore excluded, and the technique can be used with simple computers or with the more powerful hand-held calculators. This includes the necessary algorithms for such programming.
2 Only a minimum of information about the various components, particularly the transistor, should be necessary for an approximate design, provided that the resonator properties are reasonably well known.
3 The design technique should be useable by people having only moderate skill and knowledge and lacking the understanding of the derivation of the technique, such as medium-grade electronics technicians. On the other hand, personnel of much greater skill, such as engineers, should be able to study the derivation, including the assumptions and approximations, learn much about the behavior of oscillators, and, hopefully, go on to advance the state of the art and improve the technique.

The basis of the technique is the judicious combination of fields of knowledge, which have long been available, and some recently formulated unpublished work by the writer. These fields of knowledge are:

1 Linear circuit theory.
2 The nonlinear behavior, outlined in Chapters 2 and 6, of practically all silicon bipolar transistors under the idealized conditions described.

These subjects are combined while making the important and simplifying assumption that the circuit operates independently on each component of the current and/or voltage, fundamental, and harmonics, without interference from each other. The purist can rightly object to this assumption because of the obvious nonlinearity, but, since only an approximate solution is acceptable, it is heuristically permissible.

The main reason for the lack of satisfactorily simple conversion of theory into practice has been the great complexity of the reasonably accurately detailed oscillator model, due to its highly nonlinear nature and the resulting insufficient knowledge concerning the magnitude of the parameters in the model. It was, therefore, necessary to simplify the models by suitable assumptions and approximations, self-evident in theory or justifiable by experience. It is obvious that such assumptions and approximations can only result in approximately accurate designs. However, it is questionable whether a more accurate and therefore more complicated procedure will yield a more usefully accurate design in view of the tolerances which exist for all the components of the oscillator and particularly the wide variations of the transistor parameters. In addition, the calculated design will always require modification to accommodate the standard values of available components, and to compensate for the unknown stray elements inherent in the layout which cannot be included in the calculations.

The traditional design technique in industry often consists of the following steps:

1 The designer receives the set of specifications outlining the oscillator requirements.
2 The designer examines the firm's past experience and locates the previous design closest to the design now requested.
3 Depending upon his or her skill and knowledge, the designer will, on paper, alter some of the components of this previous design to more closely suit the new requirements. Very often this step is omitted and step 4 is done immediately.
4 The unit of step 3, called the "prototype," is then built.
5 The designer proceeds to adjust the component values until the desired performance is achieved. The time duration and degree of success of this step depends largely on the skill, intuition, and experience of the designer. Often enough, this is done in more or less a random fashion, and is therefore very costly. Catastrophe results when a large number of identical units based upon this design are desired, as each unit will then require more or less extensive individual adjustment.

The design technique presented in this book is the following:

1a The designer receives the set of specifications outlining the oscillator requirements.
2a The designer selects the oscillator circuit type suitable for the application.
3a Using the algorithm, which presumably has been programmed into a computer or calculator, for that circuit type, provided in this book, the

designer calculates the prototype which is deliberately tailored to satisfy the specifications.

4a The prototype is built.

5a The prototype is adjusted for the desired performance. The adjustment procedure, called "trimming," is performed systematically using oscillator theory and trimming instructions developed in this book. The person performing the trimming must understand the instructions or be under the direction of such a person. After a short period of trimming under the direction of skilled personnel, the unskilled person will be able to trim without direction in a more or less cookbook fashion.

It follows that the latter design technique, because of its systematic nature, will lead to more uniform units and fewer production problems. It also has the great advantage that it can be formally taught to people relatively new to the field, since it does not depend on experience and intuition.

Since the crystal is the heart of any crystal oscillator, the book includes a chapter on piezoelectric resonators, written by Arthur Ballato under my editorial guidance. The discussion of the bulk acoustic wave resonator is quite intensive, and considerable information is also provided for the surface-wave acoustic resonator. The treatment is strongly slanted toward resonator application but, of necessity, includes substantial material on resonator design and fabrication.

A common difficulty in crystal oscillator design is the interfacing of the crystal to the circuitry so that optimum performance and the exact desired operating frequency is achieved. Often it has even been found necessary to supply a sample circuit to the crystal supplier to ensure the correct operation and frequency of the complete oscillator. Accordingly, extensive quantitative treatment of the crystal specifications and characteristics, clearly indicating their influence upon the over-all oscillator performance, is given throughout the book. The treatment also supplies much information on the crystal drive power.

The quality of the oscillators treated in this book ranges from moderate to the best available at the present state of the art. I hope that some readers will use the material to progress beyond the state of the art.

Full advantage is taken of the present availability of low-cost high-performance transistors and the recent development of new crystal cuts such as the SC. The material has been prepared to include the use of these components.

The book pays much attention to the practical problems of oscillator design. To that end, chapters are included on special problems encountered in oscillators and measurements of components and circuitry during trimming. Also, many references and recommendations are made as to what is practical during the course of the development.

This book, of necessity, must use mathematics extensively, but every effort has been made to keep it simple even at the expense of slight losses in accuracy. The trimming will compensate for these additional inaccuracies. The analytic parts of the book are for readers who already have considerable knowledge of electronics and some knowledge of oscillators. For convenience, background material in circuit concepts and transistors is briefly presented, but only as final results and references are made to the literature where more extensive treatment may be obtained. Several different methods of analysis of oscillators are developed and the advantages of each explained. But in every case of application, except for didactic purposes, the tendency is to give preference to that method which entails the least work in solving that particular problem.

The goal of this book is not to present typical circuits with component values, but to develop a logical procedure for calculating the component values. For this reason, the component values are omitted in circuit diagrams except in design examples.

The book is concerned with the steady-state conditions and only incidentally mentions the transient state. Other subjects omitted are temperature compensation, oven design, and VCOs. RC and relaxation oscillators are excluded, as they are not harmonic oscillators.

I should like to express my sincere appreciation to the staff, and particularly to Martin Bloch, of Frequency Electronics, Inc., where I developed much of the original material and conceived the idea of writing the book; to Arthur Ballato, for his contribution of the chapter on piezoelectric resonators and for his invaluable assistance in obtaining background material from the more inaccessible domestic and foreign sources; to the faculty and staff of the Department of Electrical Engineering and Computer Sciences at the University of California at San Diego, where I wrote most of the book; and to my brother Emanuel for his advice and encouragement.

BENJAMIN PARZEN

San Diego, California
November 1982

Contents

Chapter 1   Basic Concepts of Oscillators   1

    1.1   Introduction, 1
    1.2   Summary of Network Theory, 3
    1.3   Basic Configurations of Oscillators, 12
    1.4   Oscillator Description and Specifications, 17
    1.5   Oscillator Circuit Classification, 19
    1.6   References, 20

Chapter 2   Transistor Properties Applicable to Oscillator Design   22

    2.1   Introduction, 22
    2.2   Two-Port Small-Signal Transistor Models, 22
    2.3   The Small-Signal Common Emitter Hybrid-PI Model of the Bipolar Transistor, 37
    2.4   Large-Signal Common Emitter Bipolar Transistor Characteristics, 47
    2.5   The Common Base Bipolar Transistor, 52
    2.6   Biasing of Bipolar Transistors, 57
    2.7   The Junction Field Effect Transistor, 59

Chapter 3   Piezoelectric Resonators   66

    By Arthur Ballato

    3.1   Introduction, 66
    3.2   Equivalent Circuits, 69
    3.3   Doubly Rotated Cuts, 96
    3.4   Surface Acoustic Wave (SAW) Devices, 100
    3.5   Miniature Resonators, 102
    3.6   Undesired Responses, 104

3.7 Static Temperature Effects, 105
3.8 Aging Effects, 109
3.9 Environmental Effects (Excluding Static Temperature Effects), 110
3.10 Drive Level Effects, 114
3.11 Activity Dips, 118
3.12 Radiation Effects, 118
3.13 Noise in Quartz Crystals, 119
3.14 General Specifications, 120
3.15 Military Specifications, 121
3.16 National and International Specifications, 122

Chapter 4 Other Resonators   123

4.1 Inductance Capacitance Circuits (LC Resonators), 123
4.2 Electromechanical Resonators, 124
4.3 The Magnetostriction Resonator, 124

Chapter 5 The Pierce, Colpitts, Clapp Oscillator Family   125

5.1 Introduction, 125
5.2 The Idealized Oscillator, 128
5.3 Examples of Idealized Oscillator Applications, 130
5.4 Effect of the Phase Angle of g_m in the Idealized Oscillator, 136
5.5 The Real Oscillator, 139
5.6 The Theory Applied to Crystal Oscillators, 151
5.7 Load Impedance Transformation, 160
5.8 f_T and β_o Requirements, 163
5.9 Frequency Stability Analysis, 164

Chapter 6 Limiting   169

6.1 Introduction, 169
6.2 The Self-Limiting Single Bipolar Transistor Oscillator, Sinusoidal v_{BE} or $v_{BE'}$, 170
6.3 The Self-Limiting Single Bipolar Transistor Oscillator, Sinusoidal i_E, 178
6.4 The Self-Limiting Single Junction Field Effect Transistor Oscillator, Sinusoidal v_{GS}, 179
6.5 Diode Limiter, 179
6.6 Principal Reference, 180

Chapter 7 The Normal Pierce Oscillator   181

7.1 Introduction, 181
7.2 Oscillator Circuit Analysis, 181

7.3 Frequency Variations Due to External Factors, 192
7.4 The Design Procedure, 194
7.5 The Pierce Oscillator, Collector Base Limiting, 195
7.6 The Pierce Oscillator, *be* Cutoff Limiting, 199
7.7 Measurement Techniques for the Pierce Oscillator, 205

Chapter 8 The Isolated Pierce Oscillator — 208

8.1 Introduction, 208
8.2 Oscillator Circuit Analysis, 208
8.3 Limiting, 215
8.4 The Design Procedure, 215
8.5 Algorithm and Design Examples, 215

Chapter 9 The Normal Colpitts Oscillator — 220

9.1 Introduction, 220
9.2 Oscillator Circuit Analysis, 221
9.3 Frequency Variations Due to External Factors, 223
9.4 The Design Procedure, 223
9.5 The Colpitts Oscillator, Collector Base Limiting, 224
9.6 The Colpitts Oscillator *be* Cutoff Limiting, 227

Chapter 10 The Semiisolated Colpitts Oscillator — 233

10.1 Introduction, 233
10.2 Oscillator Circuit Analysis, 233
10.3 Frequency Variations Due to External Factors, 241
10.4 The Design Procedure, 241
10.5 The Design Algorithm and Design Examples, 241

Chapter 11 The Butler Oscillator — 250

11.1 Introduction, 250
11.2 Oscillator Circuit Analysis, 251
11.3 The $X_A = 0$ Design Procedure, 264
11.4 Theory and Design of the Stable Butler Oscillator, 268

Chapter 12 Algorithms — 280

12.1 Introduction, 280
12.2 General Discussion of the Algorithms, 282

  Algorithm 12.1 Pierce Oscillator, Collector Limiting, 286
  Algorithm 12.2 Pierce Oscillator, be *Cutoff Limiting*, 294
  Algorithm 12.3 Isolated Pierce Oscillator,
                  Collector Limiting, 302

Algorithm 12.4 Colpitts Oscillator, Collector Limiting, 310
Algorithm 12.5 Colpitts Oscillator, be Cutoff Limiting, 319
Algorithm 12.6 Semiisolated Colpitts Oscillator,
                be Cutoff Limiting, 329
Algorithm 12.7 $X_A = 0$ Butler Oscillator, R_{IN} Limiting, 339
Algorithm 12.8 Stable Butler Oscillator, R_{IN} Limiting, 346

Chapter 13   Other Oscillator Configurations   353

13.1 Introduction, 353
13.2 Single Transistor Oscillators, 353
13.3 The Darlington Transistor Pair, 358
13.4 The Cascode Amplifier, 362
13.5 Other Oscillator Circuits, 364

Chapter 14   ALC Oscillators   369

14.1 Introduction, 369
14.2 General Description of the ALC Oscillator, 370
14.3 Description of the Components of the ALC Oscillator, 374
14.4 The Prediction of the Approximate Phase Noise
      Performance of the Oscillator, 382

Chapter 15   Gate Oscillators   383

15.1 Introduction, 383
15.2 Classification of Gate Elements, 384
15.3 Application of the Various Gate Types, 385
15.4 Biasing, 386
15.5 Conversion of Gates Having Different Functions into
      Equivalent Inverters, 389
15.6 The Inverter Approximate Equivalent Circuit, 390
15.7 Type of Limiting Used in Gate Oscillators, 391
15.8 Gate Oscillators with Crystals Operating in the
      Inductive Region, 392
15.9 Gate Oscillators with Crystals Operating Near
      Series Resonance, 394
15.10 Closing Remarks, 395

Chapter 16   Integrated Circuit Oscillators   397

16.1 Introduction, 397
16.2 Gate Oscillators with Digital Output, 398
16.3 Linear Integrated Circuit Oscillators, 399
16.4 Miniature Packaged Crystal *Oscillators*, 400

Contents xvii

Chapter 17 Circuitry Frequency Stability Requirements in Crystal Oscillators, Excluding the Crystal   **402**

17.1 Introduction, 402
17.2 The Osci and Llator Concepts, 403
17.3 Oscillator Relationships Using the Osci Llator Concept, 404
17.4 Crystal Oscillators Wherein the Crystal Network Operates in the Inductive Region, 405
17.5 Experimental Procedure for the Oscillator of Section 17.4, 406
17.6 Crystal Oscillators Wherein the Crystal Network Operates Near Series Resonance, 407
17.7 Experimental Determination of the Oscillating Point in the Oscillator Using the Osci and Llator Concepts, 408
17.8 Conditions for the Applicability of Sections 17.1 to 17.7, 409
17.9 Short-Term Performance, 409

Chapter 18 Special Problems in Oscillators   **411**

18.1 Introduction, 411
18.2 Spurious Oscillations, 411
18.3 Spurious Signals Due to Crystal Responses, 413
18.4 Squegging, 414
18.5 Crystal Physical Location and Connections, 415
18.6 Oscillator Starting, 415
18.7 Isolation, 416
18.8 Aging of Components 417
18.9 *Oscillator* Testing, 417

Chapter 19 Component and Circuitry Measurements   **418**

19.1 Introduction, 418
19.2 General Characteristics of the Measurement Procedures, 419
19.3 dc Measurements, 419
19.4 ac Voltage and Current Measurements, 420
19.5 The Oscilloscope as a Measurement Tool, 422
19.6 Linear and Small-Signal Immittance Measurements, 422
19.7 Large-Signal Immittance and Phase Measurements, 428
19.8 Z_{LL} Measurement Procedure, 429

References and Bibliography   **431**

List of Symbols   **441**

Index   **447**

Design of Crystal and Other Harmonic Oscillators

1

Basic Concepts of Oscillators

1.1 INTRODUCTION

1.1.1 Definition of an Oscillator

An oscillator is a device for producing ac power, the output frequency of which is determined by the characteristics of the device. In Fig. 1.1, the input power P_i at frequency f_i is fed to the device which delivers output power P_o at frequency f_o. If f_o is not correlated to f_i, then the device is an oscillator. Usually P_i is high-quality dc power, but that is not a necessary condition. A *harmonic oscillator* is an oscillator producing quasi-sine-wave oscillation, the frequency of which is mainly determined by two types of energy storage elements, such as inductors or capacitors or equivalent; for example, crystal resonators. It is interesting to note that the sine wave does not necessarily have to be present at the oscillator output terminals, but it does exist somewhere within the oscillator and may be either a voltage or current.

In order to realize harmonic oscillations, the following are required:

1 An active element producing amplification.
2 Positive feedback leading to negative resistance.
3 A frequency selective network which mainly determines the oscillation frequency.
4 A nonlinearity, which is hereafter called "limiting" to maintain the oscillation amplitude in stable equilibrium.

Very often items 2 and 3 are performed by the same components.

2 Basic Concepts of Oscillators

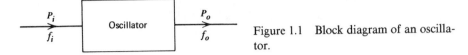

Figure 1.1 Block diagram of an oscillator.

1.1.2 Classification of Oscillators

The oscillators treated in this book have one common characteristic: good frequency stability; but otherwise there are many ways in which they can be classified, for example

1. By frequency range.
2. By power output range.
3. By function; for example, the frequency can be readily modulated or shifted by an externally applied voltage.
4. By the number of active devices; for example, single transistor where the same transistor provides the amplification and the limiting.
5. By the manner its frequency is stabilized for the changes in environment; for example, oven controlled.
6. By the manner of limiting; for example, self-limiting, automatic level control.
7. By the degree of frequency stability; for example, moderate, high.

There are additional classifications, too numerous to mention. For further classifications, see Section 1.5.

To facilitate the classification process, a system of abbreviations has been gradually devised. Some of these abbreviations are:

O	oscillator
X	crystal
LC	inductor capacitor
VC	voltage controlled
TC	temperature-compensated
OC	oven controlled
ALC	automatic level controlled

These basic abbreviations may be combined to form a new abbreviation, for example, ALCTCVCXO would be an automatic-level-controlled temperature-compensated voltage-controlled crystal oscillator.

1.1.3 Crystal Oscillators Discussed in This Book

Detailed discussions are given of oscillators in the frequency range from 0.8 to 200 MHz. Crystal oscillators below this frequency range are rarely used because of the availability of inexpensive, low-power, miniature frequency dividers. However, Chapter 13 will include schematic diagrams of lower-frequency crystal oscillators.

1.1.4 Names of Oscillator Circuits

During the history of oscillator circuit development, names have been given to various circuits; the names were usually the name of the person who originated the circuit, such a Pierce oscillator, Hartley oscillator, and so on. At other times, the name is a part description of the circuit; for example, tuned plate, tuned grid oscillator. In this book, names have also been given to the various circuits described, and great effort has been expended to keep these names consistent with history. However, during the course of time, the names have become somewhat obscured and varied, so that there may be readers who may dispute the choice of name. Of course, the writer makes no claims as to the accuracy of the names and apologizes in advance for any errors he may have made either by commission or omission.

1.2 SUMMARY OF NETWORK THEORY

1.2.1 Two-Terminal Immittance

This subsection states without proof the various forms in which immittance may be expressed and the relationships existing between the different forms. For further information and proofs, one may consult any book on circuit theory of which the number is legion. The notation is standard and universal and will therefore only be explained briefly in the List of Symbols at the end of the book.

1.2.1.1 Basic Definitions and Relations
Rectangular notation:

$$Z = R + jX \qquad (1.1)$$

Polar notation:

$$Z = |Z|e^{j\theta} \quad \text{or} \quad |Z|\angle\theta \qquad (1.2)$$

where
$$R = |Z|\cos\theta, \qquad X = |Z|\sin\theta \qquad (1.3)$$

$$|Z| = \sqrt{X^2 + R^2} \qquad (1.4)$$

$$Y = \frac{1}{Z} = G + jB \qquad (1.5)$$

4 Basic Concepts of Oscillators

1.2.1.2 Derived Relationships

From Eqs. (1.2), (1.3), and (1.5),

$$Y = \frac{e^{-j\theta}}{|Z|} = |Y|e^{-j\theta} \tag{1.6}$$

$$G = \frac{\cos\theta}{|Z|}, \quad B = \frac{-\sin\theta}{|Z|} \tag{1.7}$$

and from Eq. (1.3)

$$G = \frac{R}{|Z|^2}, \quad B = \frac{-X}{|Z|^2} \tag{1.8}$$

From Eq. (1.8), and by definition,

$$Q_Z = \left|\frac{X}{R}\right| = \left|\frac{B}{G}\right| \tag{1.9}$$

Figure 1.2 shows the schematic diagram of Z and Y in the rectangular form. The schematic diagram for the polar form does not exist, as it is a mathematical concept.

1.2.1.3 Approximations of Eq. (1.8)

When $Q_Z \geq 5$,

$$|Z|^2 \approx X^2 \tag{1.4a}$$

and Eq. (1.8) becomes

$$G \approx \frac{R}{X^2}, \quad B \approx -\frac{1}{X} \tag{1.8a}$$

1.2.1.4 Some Alternate Specialized Forms

Some calculations are more easily made in alternate representations of Z and Y and some immittance-measuring instruments present their data in these forms.

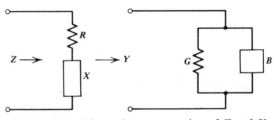

Figure 1.2 Schematic representation of Z and Y.

$Z(f) = \dfrac{1}{Y(f)}$

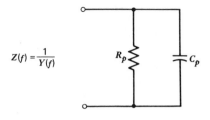

Figure 1.3 Alternate representation of $Z(f)$ for RX meter.

Figure 1.3 shows the data form of the H.P. 250 RX meter. In this figure,

$$R_p = \frac{1}{G} \tag{1.10}$$

$$C_p = \frac{B(10)^6}{2\pi f} \tag{1.11}$$

Figure 1.4 shows the data form of the Wayne–Kerr 100 admittance bridge. The symbols in this figure have already been defined in Eqs. (1.7) and (1.11).

In addition, the very popular Q meter presents its data in Q_Z and C_p.

The H.P. 4815 and 4193 Vector Impedance Meters present their data in $|Z|$ and θ.

The General Radio 1606 Radio Frequency bridge presents its data as R and X.

The General Radio 1402 UHF Admittance Meter, which is useful down to 40 MHz, presents its data in G and B.

1.2.1.5 The $Z = R + \underline{X}$ Representation

The writer in using the $Z = R + jX$ form in long complicated calculations involving multiplication and division of complex immittances found that he made many errors of sign caused by the relationship $j^2 = -1$, and extensive periods of time were expended in correcting the errors. The $Z = R + \underline{X}$ representation was therefore formulated.

All calculations are performed using the $Z = R + \underline{X}$ form and all signs are $+$. Thus no error of sign can be made. At the very end, the $Z = R + \underline{X}$ form is

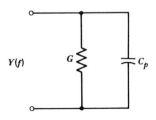

Figure 1.4 Alternate representation of $Y(f)$ for Wayne–Kerr admittance meter.

6   Basic Concepts of Oscillators

transformed into the $R + jX$ form, using the following operational rules:

$$\underline{X} = jX \tag{1.12}$$

$$\underline{X}_A \underline{X}_B = -X_A X_B \tag{1.13}$$

$$\frac{\underline{X}_A}{\underline{X}_B} = \frac{X_A}{X_B} \tag{1.14}$$

It should be noted that X_n includes its sign. Thus,

$$3 - jA = 3 + \underline{A}$$

as a special case,

$$\underline{X}^2 = -|X|^2 \tag{1.15}$$

In performing the calculations, care must be taken that values of R and $\underline{X}$ are not added in the sense that apples cannot be added to pears. Also, the sign of X_n must be taken into account.

1.2.1.6   Practical Realizable Tuned Load Values

In many ideal oscillators, the power output increases directly with the value of the transistor-tuned load resistance R_L. Obviously, as higher and higher values of R_L are approached, they become physically unrealizable, primarily due to the frequency of operation, so upper limits to this value, depending upon the frequency of operation, must be set.

The relationship that was developed over long experience is

$$R_L = \frac{10,000}{\sqrt{f}} \tag{1.16}$$

(R_L in Ω, f in MHz). This is based on the fact that at 100 MHz, a reasonable circuit capacity is 15 pF and for an operating Q of 10, $R_L = 1000\ \Omega$. Similarly for 1 MHz, a reasonable minimum circuit capacity is 150 pF and for $Q = 10$, $R_L = 10,000$. Equation (1.16) is thus derived.

1.2.1.7   Maximum Realizable Resistor Values

Occasionally, it is required to use a resistor which is as purely resistive as possible and as high a value as possible. See Fig. 1.5.

The relationship developed is

$$R_{max} = 32,000/f \tag{1.17}$$

(R in Ω, f in MHz). This is based on the fact that, in a small good noninductive

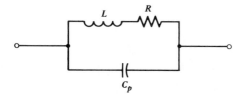

Figure 1.5 Equivalent circuit of a resistor.

high-value resistor, L can be neglected and $C_p \approx \frac{1}{2}$ pF. If a good resistor is defined as $X_{C_p} \geq 10R$, then Eq. (1.17) results.

1.2.2 Two-Port Linear Network Parameters

Any linear network possessing input and output ports can be represented as in Fig. 1.6. By a linear network is meant:

1. A network composed of passive elements whose values are independent of signal amplitude; or
2. A network composed of active and passive components whose values may be considered independent of small excursions about the mean currents and voltages but are strong functions of the total mean voltages and currents. In this case, the linear relationships apply to the excursions only. This is often called the small-signal network model. Obviously, the model changes when the mean values change.

In oscillators, condition 2 is not valid because the steady-state excursions are not small, but the theory is useful because it will predict whether the oscillator will start, for the excursions approach zero at the starting point.

1.2.2.1 y Parameters

If the relationships between the ports' currents and voltages are written as

$$I_1 = y_{11}V_1 + y_{12}V_2 \qquad (1.18)$$

$$I_2 = y_{21}V_1 + y_{22}V_2 \qquad (1.19)$$

Figure 1.6 Two-port network.

8 Basic Concepts of Oscillators

then

$$y_{11} = \left.\frac{I_1}{V_1}\right|_{V_2=0}$$

$$y_{12} = \left.\frac{I_1}{V_2}\right|_{V_1=0}$$

$$y_{21} = \left.\frac{I_2}{V_1}\right|_{V_2=0}$$

$$y_{22} = \left.\frac{I_2}{V_2}\right|_{V_1=0} \quad (1.20)$$

where $V_2 = 0$ means that the output port is short-circuited and $V_1 = 0$ means that the input port is short-circuited.

1.2.2.2 z Parameters

The network relationships are

$$V_1 = z_{11}I_1 + z_{12}I_2 \quad (1.21)$$

$$V_2 = z_{21}I_1 + z_{22}I_2 \quad (1.22)$$

and

$$z_{11} = \left.\frac{V_1}{I_1}\right|_{I_2=0}$$

$$z_{12} = \left.\frac{V_1}{I_2}\right|_{I_1=0}$$

$$z_{21} = \left.\frac{V_2}{I_1}\right|_{I_2=0}$$

$$z_{22} = \left.\frac{V_2}{I_2}\right|_{I_1=0} \quad (1.23)$$

where $I_2 = 0$ means that the output port is open-circuited and $I_1 = 0$ means that the input port is open-circuited.

1.2.2.3 h Parameters

The network relationships are

$$V_1 = h_{11}I_1 + h_{12}V_2 \tag{1.24}$$

$$I_2 = h_{21}I_1 + h_{22}V_2 \tag{1.25}$$

where

$$h_{11} = \left.\frac{V_1}{I_1}\right|_{V_2=0}$$

$$h_{21} = \left.\frac{I_2}{I_1}\right|_{V_2=0}$$

$$h_{12} = \left.\frac{V_1}{V_2}\right|_{I_1=0}$$

$$h_{22} = \left.\frac{I_2}{V_2}\right|_{I_1=0} \tag{1.26}$$

It will be noted that h_{11} is an impedance, h_{21} and h_{12} are dimensionless, and h_{22} is an admittance. Also, some of the parameters are obtained under short-circuited conditions, while others are obtained under open-circuited conditions. For these reasons, the parameters are called hybrid.

1.2.2.4 Some Relationships in Network Parameters

1. In all three types, $(\ )_{21} = (\ )_{12}$ for passive networks.
2. For y parameters, as shown in Fig. 1.7, the parameters of composite network Y_c created when connecting networks Y_a and Y_b in parallel are

$$y_{mnc} = y_{mna} + y_{mnb} \tag{1.27}$$

where m and n are 1 or 2

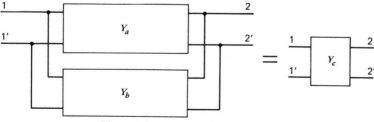

Figure 1.7 Y networks in parallel.

10   Basic Concepts of Oscillators

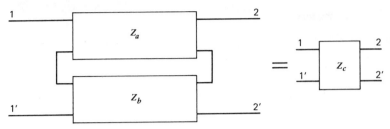

Figure 1.8   Z networks in series.

3 For z parameters, as shown in Fig. 1.8, the parameters of composite network Z_c created when connecting network Z_a and Z_b in series are

$$z_{mnc} = z_{mna} + z_{mnb} \tag{1.28}$$

1.2.2.5 Some Important Relationships in Y Networks

1 For two networks connected in cascade, as shown in Fig. 1.9,

$$y_{11c} = y_{11a} - \frac{y_{12a} y_{21a}}{y_{22a} + y_{11b}} \tag{1.29}$$

$$y_{22c} = y_{22b} - \frac{y_{12b} y_{21b}}{y_{22b} + y_{11b}} \tag{1.30}$$

$$y_{21c} = -\frac{y_{21a} y_{21b}}{y_{22a} + y_{11b}} \tag{1.31}$$

$$y_{12c} = -\frac{y_{12a} y_{12b}}{y_{22a} + y_{11b}} \tag{1.32}$$

2 For a Y network terminated as shown in Fig. 1.10

$$Y_{in} = y_{11} - \frac{y_{12} y_{21}}{y_{22} + Y_L} \tag{1.33}$$

$$Z_{in} = \frac{y_{22} + Y_L}{\Delta y + y_{11} Y_L} \tag{1.34}$$

$$Y_{out} = y_{22} - \frac{y_{12} y_{21}}{y_{11} + Y_S} \tag{1.35}$$

$$Z_{out} = \frac{y_{11} + Y_S}{\Delta y + y_{22} Y_S} \tag{1.36}$$

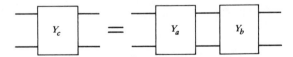

Figure 1.9   Y networks in cascade.

1.2 Summary of Network Theory   11

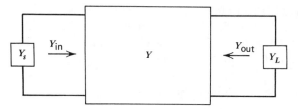

Figure 1.10   Terminated Y network.

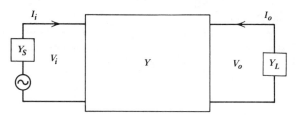

Figure 1.11 Y network fed by generator.

3   For a Y network connected to a generator as shown in Fig. 1.11

$$\frac{V_o}{V_i} = A_V = \frac{-y_{21}}{y_{22} + Y_L} \qquad (1.37)$$

$$\frac{I_o}{I_i} = A_I = \frac{-y_{21}Y_L}{\Delta y + y_{11}y_{21}} \qquad (1.38)$$

where

$$\Delta y = y_{11}y_{22} - y_{12}y_{21} \qquad (1.39)$$

4   Incorporation of the termination into the network, as shown in Fig. 1.12, yields

$$y_{12a} = y_{12b}, \qquad y_{21a} = y_{21b} \qquad (1.40)$$

$$y_{11b} = y_{11a} + Y_S \qquad (1.41)$$

$$y_{22b} = y_{11a} + Y_L \qquad (1.42)$$

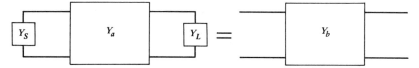

Figure 1.12   Incorporation of the terminations into the network.

12 Basic Concepts of Oscillators

Equations (1.40) through (1.42) are important in that they apply equally well to nonlinear networks since the amplitudes of the signals are not changed when the terminations are incorporated into the network. This incorporation will be useful in the analysis of real oscillators.

1.2.2.6 Nonlinear Two-Port Networks

It is possible to configure similar networks for nonlinear two-ports, in which the I, V, Y, Z terms are the effective fundamental components. However, none of the theory presented for linear networks applies, with the exception of item 4 in Section 1.2.2.5.

An interesting and very practical case often found in oscillators occurs when Y_a in Fig. 1.7 is a passive network and Y_b is an active network. Then, Y_c can be considered equal to $Y_a + Y_b$, provided that care is taken to ensure that the voltages and currents in Y_b are equal to those in the active components of Y_c.

1.2.2.7 Transformations between Y and Z Networks

1.2.2.7.1 z to y

$$y_{11} = \frac{z_{22}}{\Delta z}, \quad y_{12} = \frac{-z_{12}}{\Delta z}, \quad y_{21} = \frac{-z_{21}}{\Delta z}, \quad y_{22} = \frac{z_{11}}{\Delta z} \quad (1.43)$$

where

$$\Delta z = z_{11} z_{22} - z_{11} z_{21} \quad (1.44)$$

1.2.2.7.2 y to z

$$z_{11} = \frac{y_{22}}{\Delta y}, \quad z_{12} = \frac{-y_{12}}{\Delta y}, \quad z_{21} = \frac{-y_{21}}{\Delta y}, \quad z_{22} = \frac{y_{11}}{\Delta y} \quad (1.45)$$

where

$$\Delta y = y_{11} y_{22} - y_{12} y_{21} \quad (1.46)$$

1.2.2.8 Final Remarks on Two-Port Networks

The y parameters will prove to be the most useful type in oscillator analysis and design. For an excellent exposition on how these parameters are applied in oscillator analysis, see Ref. 1.1. For further information on the derivation and application of two-port networks in general circuit design, see Refs. 1.2 and 1.3.

1.3 BASIC CONFIGURATIONS OF OSCILLATORS

In considering basic configurations of oscillators, it is convenient to set up oscillator models. There are two different popular models, which upon deeper

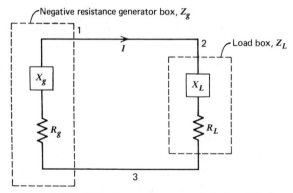

Figure 1.13   Negative resistance model of oscillator.

examination, turn out to be the identical model for 3-terminal active devices. Both models are valid for nonlinear circuits (if it is kept in mind that the model component values change when the signal amplitudes are changed).

1.3.1   The Negative Resistance / Conductance Models

Figure 1.13 shows an oscillator negative resistance model. R_g is a negative resistance, the absolute value of which decreases as the signal amplitude increases. Initially at small amplitudes, $R_g > R_L$. As the amplitude increases, R_g decreases until $R_g = R_L$. The oscillator is then at equilibrium and assumes the frequency of operation at which $X_g + X_L = 0$. Thus, the following is true at equilibrium:

$$R_g = -R_L \qquad (1.47)$$

$$X_g = -X_L \qquad (1.48)$$

Figure 1.14 shows the dual version of the model of Fig. 1.13. In this figure, G_g is a negative conductance, the absolute value of which decreases as the

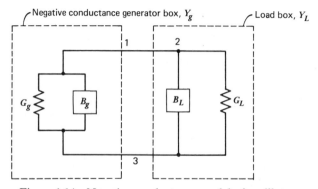

Figure 1.14   Negative conductance model of oscillator.

signal amplitude increases. When G_g reaches G_L, the oscillator is at equilibrium and assumes the frequency at which $B_g + B_L = 0$ so that the equations at equilibrium become

$$G_g = -B_L \qquad (1.49)$$

$$B_g = -B_L \qquad (1.50)$$

Both models are equally correct and the choice of models depends only upon which model is more suitable in the calculations for the particular application.

A very interesting and important point is that in a given oscillator, no matter what components are allocated to the g and L boxes, the values of the real and imaginary immittance components change but Eqs. (1.47) through (1.50) will always be satisfied by those values.

The models are remarkably simple and they directly provide sufficient information to design oscillators including the amplitude values, provided that the behavior of Z_g or Y_g is known. They are very useful in analysis, experimental verification, and in analysis combined with experimental verification.

At this point, only the negative conductance model will be used but it is to be remembered that anything done with this model can be duplicated in a dual sense with the negative resistance model.

Some interesting applications of the model are presented:

Consider a particular operating oscillator. The g and L boxes are configured in any manner except that all the active devices are in the g box. The operating frequency f is measured. The two boxes are then separated (disconnected) and Y_L is measured at f. Y_g can be computed from Eqs. (1.49) and (1.50).

Furthermore, when f is measured, the current I going into Y_L is also measured. G_L is then varied and for each value of G_L, f and I are measured. From these data, curves of Y_g versus I, G_L, and frequency, which can be used in oscillator design and analysis, can be plotted. The same can be done for variations in B_L.

Some special configurations of g and L boxes are now considered.

1. The passive components which have maximum values of dB/df are placed in the L box and the remaining components in the g box. Again Eqs. (1.49) and (1.50) are valid. By definition, in this arrangement, the g box is called the "llator" and the L box the "osci." Obviously, when the two boxes are connected an "oscillator" is created. This particular arrangement will be used for important analyses in Chapter 17.

2. Another configuration which appears trivial, but in reality is very important, is where all the components are put into the g box. Then from Eqs. (1.49) and (1.50), $Y_g = 0$ (see Fig. 1.15a).

3. Now consider the two-port Y network in Fig. 1.15b. If a voltage exists across terminals 1 and 1', then by definition, the network is an oscillator since there is no input signal but only output signals. The oscillator load

1.3 Basic Configurations of Oscillators

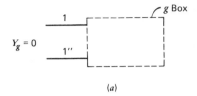

(a)

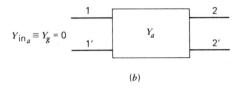

(b)

Figure 1.15 Two-port oscillator representation. (a) Negative conductance model where all components are in g box. (b) Two-port network equivalent of the g box.

normally connected to port 2 has been incorporated into the network as demonstrated in Fig. 1.12 and Eq. (1.42). (Note that as explained in Section 1.2.2.6, this is valid for nonlinear circuits.)

Obviously the network Y_a is the same as the g box in Fig. 1.15a and, therefore,

$$Y_{in_a} = Y_g = 0 \tag{1.51}$$

Since $Y_{in_a} = 0$ and $Y_{L_a} = 0$, it follows from Eq. (1.33) that

$$0 = y_{11a} - \frac{y_{12a} y_{21a}}{y_{22a}} \tag{1.52}$$

or

$$0 = y_{11a} y_{22a} - y_{12a} y_{21a} \tag{1.53}$$

or from Eq. (1.39)

$$0 = \Delta y_a \tag{1.54}$$

where Δy_a is called the determinant of network Y_a.

Equation (1.54) is important because it states the conditions necessary for oscillation for a Y_a type of network. This equation will be used later in the section on oscillator design.

1.3.2 The Feedback Oscillator Model

Figure 1.16a shows one form of a feedback oscillator model called the y model. In this model, A is an amplifier of variable gain, $|A|\angle\theta_A$ and β is the feedback network and has the transfer function $|\beta|\angle\theta_\beta$. Both A and β are strong

16 Basic Concepts of Oscillators

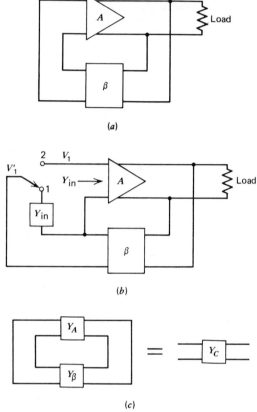

Figure 1.16 Oscillator feedback model. (*a*) Basic model. (*b*) Model for calculating A_L. (*c*) Transformation of (*a*) into two-port network Y_c.

functions of the operating frequency f. In general $\partial \theta_\beta / \partial f$ is very large in stable oscillators.

Figure 1.16*b* shows the setup for calculating the open loop gain

$$A_L = \frac{V_1'}{V_1} = A\beta$$

when the switch is in position 1.

When $|A_{L_0}| > 1$ and $\theta_{A_{L_0}} \approx 0$ or $2\pi n$, where n is an integer, the system will oscillate when the switch is transferred to position 2 since V_1' is then the same as V_1. The loop is closed and

$$A\beta = 1 \tag{1.55}$$

$$\theta_A + \theta_\beta = 0 \quad \text{or} \quad 2\pi n \tag{1.56}$$

where n is an integer or in rectangular coordinates at oscillation,

$$\text{Re}(A\beta) = 1 \qquad (1.57)$$

$$\text{Im}(A\beta) = 0 \qquad (1.58)$$

Equation (1.55) and its derived forms are very often known as the Barkhausen or Nyquist criterion. The reduction of $(A\beta)$ that occurs is called the limiting process.

It has been found convenient to make A_{L_0} the small-signal value which, of course, is always larger, in practical oscillators, than the large-signal value.

If the load is incorporated into the amplifier and the combined load and amplifier shown as Y_A (see Fig. 1.16c), the β network now becomes Y_β. Y_β and Y_A then combine into Y_C as given by Eq. (1.27) which is now the same as Y_a in Fig. 1.15b. Thus, this model has been converted into the negative conductance model.

Two models which appear different (but are really equivalent) have been developed. The choice of model depends only upon which is more suitable for the calculations and the particular application.

1.4 OSCILLATOR DESCRIPTION AND SPECIFICATIONS

1.4.1 Introduction

The word "oscillator" is ambiguous since it has two recognized meanings:

1. The complete device into which is fed the dc power and out of which comes the desired frequency at the desired power and impedance level. This device is denoted by *oscillator*.
2. That part of the *oscillator*, denoted by oscillator, which generates the desired frequency. It is usually followed by buffer and/or final amplifiers, means of temperature stabilization or compensation, voltage regulation, and so on.

1.4.2 Characteristics of the *Oscillator* versus the Oscillator

The additional circuitry may improve the overall performance in some characteristics and deteriorate the performance in other characteristics. Examples are:

1. Power level increased.
2. Impedance level transformed from the oscillator optimum load impedance to the desired load impedance.

18 Basic Concepts of Oscillators

3 The temperature stabilization means will improve the performance under variable ambient temperatures.

4 The short-term stability of the *oscillator* will be deteriorated but the long-term stability will not be affected.

5 The isolation, meaning the effect of change of load impedance upon the frequency, will be improved.

1.4.3 *Oscillator* **Specifications**

This book is mainly concerned with the design of the oscillator but the user rarely orders an oscillator. He does order *oscillators*!! In doing so, he must prepare specifications which must be satisfied by the *oscillator*. In this connection, it should be noted that the tendency is to overspecify since usually it is much easier to play safe by overspecifying than to exert the efforts required to determine the minimum performance required for the application. For most components, overspecification is often employed to compensate for the gradual deterioration in performance that takes place with time. However, the *oscillator* performance usually improves with operating time, except for catastrophic failures, so the overspecification is not required.

In writing specifications, the following must be considered and included if necessary for the particular application:

1 Output frequency.
2 Short-term stability in the frequency and/or time domain (see Section 1.7).
3 Long-term aging performance.
4 Variation of frequency with load (isolation).
5 Variation of frequency with power supply voltage.
6 Variation of frequency with temperature
7 Frequency settability, coarse and fine.
8 Setting resolution.
9 Minimum number of years for which the *oscillator* may be reset to f, the specification frequency.
10 Frequency stabilization time; for example, 10^{-8} 15 min after turn-on; 10^{-9} 2 hr after turn-on, and so on.
11 Frequency repeatability; for example, 10^{-9} in 8 hr after turn-on, after a turn-off for 48 hr with respect to the frequency before turn-off.
12 Output power into the specified load impedance.
13 Output wave shape and harmonic distortion characteristics. In this connection, it should be noted that the distortion requirement is very often overspecified as many systems fed by the *oscillator* will generate the harmonics anyway.

14 Output spurious signals. (Note: a spurious signal is a signal of frequency not related to f.) Typical spurious signals are power line frequency sidebands. (Also see Sections 18.2 and 18.3.)

15 Signal-to-noise ratio in a specified bandwidth, usually 30 kHz, excluding spurious signals.

16 Power supply power and voltage available for the *oscillator* stating the characteristics of the voltage; for example, 12 V ± 1%, 10 mV peak-to-peak ripple.

17 Maximum power consumption of the *oscillator* under different environmental and/or operating conditions.

18 Environmental conditions such as temperature, humidity, vibration, shock, acceleration, and nuclear radiation. These conditions must be specified for operating, storage, and/or survival.

19 Dimensions and configuration.

The above demonstrates that the task of writing a proper set of specifications for the *oscillator* requires much effort and substantial analysis of the system requirements.

1.5 OSCILLATOR CIRCUIT CLASSIFICATION

Because the field of oscillator design has such a long history, there are many circuit configurations and many different forms of classification. The choice of which configuration to use in a particular application can only be made after these configurations have been discussed in detail and, therefore, is treated in Chapter 12.

However, an attempt will now be made to create a system of classification as the process of classification will give some insight into the ramifications of oscillator design.

1 Classification by circuit name. This is only important historically and serves the function of identifying the particular circuit; for example, Colpitts oscillator, tuned plate, tuned base oscillator, and so on.

2 Classification by order of stability (e.g., high-, medium-, and low-stability oscillators).

3 Classification by power output (e.g., high, medium and low power). Usually, the higher the power the lower the stability, but there are some exceptions as will be shown later in the discussions on particular circuit configurations.

4 Classification by frequency region (e.g., low, medium, and high frequency).

5 Classification by the type of circuit element(s) which largely determine the frequency (e.g., LC, Crystal).
6 If the element is a crystal—is it used near series resonance or far from series resonance in its inductive region?
7 The type of limiting used (e.g., self-limiting, diode-limiting, ALC limiting, auxiliary transistor(s) limiting). In general, the circuits which have auxiliary devices or circuits for limiting purposes have higher stability.
8 By the ratio, η, of the power output, P_o, to the power dissipated in the main frequency determining element(s), usually called the *drive power*, P_3 or P_x in a crystal oscillator. In general, for a given circuit and a given power output, the stability is lower as η increases. In many circuits, $\eta \ll 1$.
9 Classification by the magnitude of the drive power. In general, the lower the drive power the higher the long-term stability. However, the optimum short-term stability exists at a considerably higher P_x. Thus, one has to compromise on the values of P_x in accordance with the desired performance.
10 Classification by the number of active devices in the circuit (e.g., single transistor, where the one transistor performs both the limiting and signal generating functions. This type of circuit is by far the most popular by reason of its economy. Because of the transistor's dual function, it is also the most difficult to design on a scientific basis. However, when properly designed, it is capable of remarkably good performance. The design algorithms and much of the detailed circuit analysis presented in this book are for this type of circuit.)
11 Classification by which element of the signal generating transistor is at "ground" potential (e.g., grounded emitter, grounded base, etc.).

It is certain that the reader after completing this book will be able to state additional categories of classification but the above is a representative sampling of the many different types of classification.

1.6 REFERENCES

This book could not have been written without the predecessor literature on oscillators, especially Ref. 1.1 and Refs. 1.4 to 1.10.

It is mandatory to have a working knowledge of the subjects listed below in order to fully understand the material in Chapters 13 through 17. These subjects are not included in this book because there is already readily available a very extensive and satisfactory literature. References 1.11 through 1.15 are cited to assist those readers who wish to increase their expertise in these subjects, which are:

1. Characterizations of frequency stability in both the time and frequency domains.
2. Translation of frequency domain stability data into time domain stability data and vice versa.
3. Oscillator noise models.
4. The theory and practice of the measurement of frequency and frequency stability in both the time and frequency domains.

2
Transistor Properties Applicable to Oscillator Design

2.1 INTRODUCTION

This chapter presents a brief review of transistor models suitable for oscillator design and describes the approximations and simplifications necessary to render the models amenable to relatively simple mathematical manipulation. In doing so, full advantage is taken of the present availability of high-performance low-cost transistors which incidently also help to simplify the application theory.

The models are very correct at low frequency of operation. As the frequency increases, they become less and less accurate until they become almost useless.

An additional major operating parameter which strongly influences the accuracy of the models is the emitter dc current, I_E. Again, as the current increases, the accuracy of the models decreases.

In order to ensure that the transistor characteristics are such as to result in good accuracy of the oscillator circuit algorithms, the input data for each algorithm includes a relationship between the f_T, f, and some direct function of I_E. If this relationship is satisfied, the algorithm will be reasonably accurate.

It is almost impossible to develop a satisfactory transistor model with fixed parameters or parameters that vary in a simple fashion with frequency over a wide frequency range, particularly at high frequencies. At the large signals which exist in real oscillators, the problem is rendered much more difficult.

2.2 TWO-PORT SMALL-SIGNAL TRANSISTOR MODELS

The two-port network theory stated in Section 1.2.2 applies equally well to transistors at small-signal levels. This theory results in transistor parameters

2.2 Two-Port Small-Signal Transistor Models

which, because they are only for small signals, are useful in determining whether the oscillator small-signal gain is sufficiently large to start oscillations in a given oscillator configuration. However, that is not a trivial achievement and for that reason, the small-signal theory is presented. The small-signal theory will also serve as a starting point for the large-signal theory.

As pointed out in Section 1.2.2, network models can be developed for transistors operating at large-signal levels but the mathematical operations using these models are very limited.

The usual transistor has three terminals and one of these terminals can be considered a common terminal, as shown in Fig. 2.1. The transistors shown are bipolar but they can just as well be FET's, in which case *gate* is substituted for *base*, *drain* for *collector*, and *source* for *emitter*.

To indicate clearly that the device, within the box, is an active device, a different type of notation may be used. For example:

$$y_{11e} \text{ is written } y_{ie}$$

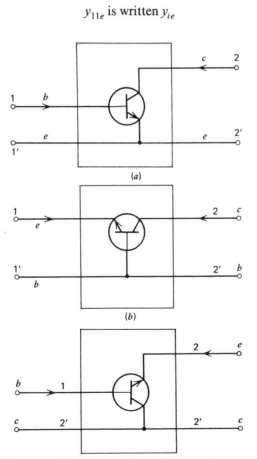

Figure 2.1 The three possible representations of a transistor as a two-port network. (*a*) Common emitter. (*b*) Common base. (*c*) Common collector.

where i signifies 11 and e indicates that the emitter is common. Similarly,

y_{fe} is equivalent to y_{21e}

y_{re} is equivalent to y_{12e}

y_{oe} is equivalent to y_{22e}

where $i \equiv$ input, $f \equiv$ forward, $r \equiv$ reverse, and $o \equiv$ output. In this book, both forms of notation are used.

Three types of parameters have already been discussed in Section 1.2.2, the y, z, and h parameters. For transistors, an additional type of parameter, the s parameter, or scattering coefficients, is also used and is gradually becoming more and more popular. More transistor data are being presented in s parameter form. The advantage of the s parameter is that the measurements can be made easily since the terminations are always 50 Ω or some other medium impedance standard termination which has been highly developed over many years. The y parameters are short-circuit parameters; the z parameters are open-circuit parameters; and the h parameters are combination short- and open-circuit parameters. It is very difficult to obtain good open circuits, and shorts at high frequencies while 50-Ω accurate terminations are widely available and the s parameter measurements can be easily made with conventional network analyzers. Also, the transistor may oscillate with an open- or short-circuit termination.

The user of the transistor has to have the performance data available. Unfortunately, some manufacturers may supply the data in only one parameter type and one common element configuration. The user has to be able to transform this information into some other parameter type and other configuration. To perform this transformation, the relationships given below can be used. These relationships are in addition to those in Section 1.2.2.7.

2.2.1 Conversions among h, y, and z Parameters

2.2.1.1 h to y

$$y_{11} = \frac{1}{h_{11}}, \quad y_{12} = \frac{-h_{12}}{h_{11}}, \quad y_{21} = \frac{h_{21}}{h_{11}}, \quad y_{22} = \frac{\Delta h}{h_{11}} \quad (2.1)$$

where $\Delta h = h_{11} h_{22} - h_{12} h_{21}$.

2.2.1.2 y to h

$$h_{11} = \frac{1}{y_{11}}, \quad h_{12} = \frac{-y_{12}}{y_{11}}, \quad h_{21} = \frac{y_{21}}{y_{11}}, \quad h_{22} = \frac{\Delta y}{y_{11}} \quad (2.2)$$

where $\Delta y = y_{11} y_{22} - y_{12} y_{21}$.

2.2.1.3 h to z

$$z_{11} = \frac{\Delta h}{h_{22}}, \quad z_{12} = \frac{h_{12}}{h_{22}}, \quad z_{21} = \frac{-h_{21}}{h_{22}}, \quad z_{22} = \frac{1}{h_{22}} \quad (2.3)$$

2.2.1.4 z to h

$$h_{11} = \frac{\Delta z}{z_{22}}, \quad h_{12} = \frac{z_{12}}{z_{22}}, \quad h_{21} = \frac{-z_{21}}{z_{22}}, \quad h_{22} = \frac{1}{z_{22}} \quad (2.4)$$

where $\Delta z = z_{11} z_{22} - z_{12} z_{21}$.

2.2.2 Conversion between y and s Parameters

$$s_{11} = \frac{(1 - y_{11})(1 + y_{22}) + y_{12} y_{21}}{(1 + y_{11})(1 + y_{22}) - y_{12} y_{21}} \quad (2.5)^\dagger$$

$$s_{12} = \frac{-2 y_{12}}{(1 + y_{11})(1 + y_{22}) - y_{12} y_{21}} \quad (2.6)^\dagger$$

$$s_{21} = \frac{-2 y_{21}}{(1 + y_{11})(1 + y_{22}) - y_{12} y_{21}} \quad (2.7)^\dagger$$

$$s_{22} = \frac{(1 + y_{11})(1 - y_{22}) - y_{21} y_{12}}{(1 + y_{11})(1 + y_{22}) - y_{12} y_{21}} \quad (2.8)^\dagger$$

$$y_{11} = \left[\frac{(1 + s_{22})(1 - s_{11}) + s_{12} s_{21}}{(1 + s_{11})(1 + s_{22}) - s_{12} s_{21}} \right] \frac{1}{Z_0} \quad (2.9)$$

$$y_{12} = \left[\frac{-2 s_{12}}{(1 + s_{11})(1 + s_{22}) - s_{12} s_{21}} \right] \frac{1}{Z_0} \quad (2.10)$$

$$y_{21} = \left[\frac{-2 s_{21}}{(1 + s_{11})(1 + s_{22}) - s_{12} s_{21}} \right] \frac{1}{Z_0} \quad (2.11)$$

$$y_{22} = \left[\frac{(1 + s_{11})(1 - s_{22}) + s_{12} s_{21}}{(1 + s_{22})(1 + s_{11}) - s_{12} s_{21}} \right] \frac{1}{Z_0} \quad (2.12)$$

where Z_0 is the characteristic impedance of the transmission lines used in the scattering parameter system, usually 50 Ω.†

---

† In converting from y to s parameters, the y parameters must first be multiplied by Z_0, and then substituted in the equations for conversion to s parameters.

26 Transistor Properties Applicable to Oscillator Design

2.2.3 Conversions among Common Emitter, Common Base, and Common Collector Parameters for y and h Parameters

2.2.3.1 Common Emitter y Parameters in Terms of Common Base and Common Collector y Parameters

$$y_{11e} = y_{11b} + y_{12b} + y_{21b} + y_{22b} = y_{11c} \tag{2.13}$$

$$y_{12e} = -(y_{12b} + y_{22b}) = -(y_{11c} + y_{12c}) \tag{2.14}$$

$$y_{21e} = -(y_{21b} + y_{22b}) = -(y_{11c} + y_{21c}) \tag{2.15}$$

$$y_{22e} = y_{22b} = y_{11c} + y_{12c} + y_{21c} + y_{22c} \tag{2.16}$$

2.2.3.2 Common Base y Parameters in Terms of Common Emitter and Common Collector y Parameters

$$y_{11b} = y_{11e} + y_{12e} + y_{21e} + y_{22e} = y_{22c} \tag{2.17}$$

$$y_{12b} = -(y_{12e} + y_{22e}) = -(y_{21c} + y_{22c}) \tag{2.18}$$

$$y_{21b} = -(y_{21e} + y_{22e}) = -(y_{12c} + y_{22c}) \tag{2.19}$$

$$y_{22b} = y_{22e} = y_{11c} + y_{12c} + y_{21c} + y_{22c} \tag{2.20}$$

2.2.3.3 Common Collector y Parameters in Terms of Common Emitter and Common Base y Parameters

$$y_{11c} = y_{11e} = y_{11b} + y_{12b} + y_{21b} + y_{22b} \tag{2.21}$$

$$y_{12c} = -(y_{11e} + y_{12e}) = -(y_{11b} + y_{21b}) \tag{2.22}$$

$$y_{21c} = -(y_{11e} + y_{21e}) = -(y_{11b} + y_{12b}) \tag{2.23}$$

$$y_{22c} = y_{11e} + y_{12e} + y_{21e} + y_{22e} = y_{11b} \tag{2.24}$$

2.2.3.4 Common Emitter h Parameters in Terms of Common Base and Common Collector h Parameters

$$h_{11e} = \frac{h_{11b}}{(1+h_{21b})(1-h_{12b}) + h_{22b}h_{11b}} \approx \frac{h_{11b}}{1+h_{21b}} = h_{11c} \tag{2.25}$$

$$h_{12e} = \frac{h_{11b}h_{22b} - h_{12b}(1+h_{21b})}{(1+h_{21b})(1-h_{12b}) + h_{22b}h_{11b}} \approx \frac{h_{11b}h_{22b}}{1+h_{21b}} - h_{12b}$$

$$= 1 - h_{12c} \tag{2.26}$$

$$h_{21e} = \frac{-h_{21b}(1 - h_{12b}) - h_{22b}h_{11b}}{(1 + h_{21b})(1 - h_{12b}) + h_{22b}h_{11b}} \approx \frac{-h_{21b}}{1 + h_{21b}}$$

$$= -(1 + h_{21c}) \qquad (2.27)$$

$$h_{22e} = \frac{h_{22b}}{(1 + h_{21b})(1 - h_{12b}) + h_{22b}h_{11b}} \approx \frac{h_{22b}}{1 + h_{21b}} = h_{22c} \qquad (2.28)$$

2.2.3.5 Common Base h Parameters in Terms of Common Emitter and Common Collector h Parameters

$$h_{11b} = \frac{h_{11e}}{(1 + h_{21e})(1 - h_{12e}) + h_{11e}h_{22e}} \approx \frac{h_{11e}}{1 + h_{21e}}$$

$$= \frac{h_{11c}}{h_{11c}h_{22c} - h_{21c}h_{12c}} \approx \frac{-h_{11c}}{h_{21c}} \qquad (2.29)$$

$$h_{12b} = \frac{h_{11e}h_{22e} - h_{12}(1 + h_{21e})}{(1 + h_{21e})(1 - h_{12e}) + h_{11e}h_{22e}} \approx \frac{h_{11e}h_{22e}}{1 + h_{21e}} - h_{12e}$$

$$= \frac{h_{21c}(1 - h_{12c}) + h_{11c}h_{22c}}{h_{11c}h_{22c} - h_{21c}h_{12c}} \approx (h_{12c} - 1) - \frac{h_{11c}h_{22c}}{h_{21c}} \qquad (2.30)$$

$$h_{21b} = \frac{-h_{21e}(1 - h_{12e}) - h_{11e}h_{22e}}{(1 + h_{21e})(1 - h_{12e}) + h_{11e}h_{22e}} \approx \frac{-h_{21e}}{1 + h_{21e}}$$

$$= \frac{h_{12c}(1 + h_{21c}) - h_{11c}h_{22c}}{h_{11c}h_{22c} - h_{21c}h_{12c}} \approx \frac{-(1 + h_{21c})}{h_{21c}} \qquad (2.31)$$

$$h_{22b} = \frac{h_{22e}}{(1 + h_{21e})(1 - h_{12e}) + h_{11e}h_{22e}} \approx \frac{h_{22e}}{1 + h_{21e}}$$

$$= \frac{h_{22c}}{h_{11c}h_{22c} - h_{21c}h_{12c}} \approx \frac{h_{22c}}{h_{21c}} \qquad (2.32)$$

2.2.3.6 Common Collector h Parameters in Terms of Common Base and Common Emitter h Parameters

$$h_{11c} = \frac{h_{11b}}{(1 + h_{21b})(1 - h_{12b}) + h_{22b}h_{11b}} \approx \frac{h_{11b}}{1 + h_{21b}} - h_{11e} \qquad (2.33)$$

$$h_{12c} = \frac{1 - h_{21b}}{(1 + h_{21b})(1 - h_{12b}) + h_{22b}h_{11b}} \approx 1 = 1 - h_{12e} \qquad (2.34)$$

28 Transistor Properties Applicable to Oscillator Design

$$h_{21c} = \frac{h_{12b} - 1}{(1 + h_{21b})(1 - h_{12b}) + h_{22b}h_{11b}} \approx \frac{1}{1 + h_{21b}}$$
$$= -(1 + h_{21e}) \tag{2.35}$$

$$h_{22c} = \frac{h_{22b}}{(1 + h_{21b})(1 - h_{12b}) + h_{22b}h_{11b}} \approx \frac{h_{22b}}{1 + h_{21b}}$$
$$= h_{22e} \tag{2.36}$$

2.2.4 The Application of Two-Port Parameter Types

The z parameters are used principally in network analysis. Very often the feedback network is most conveniently expressed in z parameters.

The h parameters are used for describing transistor characteristics.

The y and s parameters are used for describing network and transistor performance. Except for h_{21e}, the h parameters are gradually becoming obsolete. The s parameters, the scattering coefficients, were originally used at microwave frequencies but are becoming more popular at much lower frequencies. Very often because of their ease of measurement, the s parameters are measured and the y and h parameters are then calculated using the conversion formulas.

2.2.5 Important Relationships for Transistor Parameters of Different Types

Obviously, the parameters of the same type are independent quantities. Some parameters have had long histories and often are used in other transistor models by other names. In addition, some parameters of one type have particularly important relationships to parameters of other types. Some examples of both kinds are

$$\beta \equiv h_{fe} \equiv h_{21e} \tag{2.37}$$

$$\beta_o = h_{FE} \equiv h_{fe} \quad \text{at low frequencies} \tag{2.38}$$

$$\alpha \equiv h_{fb} \equiv h_{21b} \tag{2.39}$$

$$\beta = \frac{\alpha}{1 - \alpha} \approx \frac{1}{1 - \alpha} \quad \text{up to reasonably high frequencies} \tag{2.40}$$

$\alpha \approx 1$, while β has a large range of values especially when $\alpha \to 1$. (2.41)

$$g_{m_0} \equiv y_{21e} \equiv y_{fe} \tag{2.42}$$

$$g_{m_0} = \frac{\beta}{h_{ie}} = \beta y_{ie} \tag{2.43}$$

2.3 The Small-Signal Common Emitter Hybrid-PI Model of the Bipolar Transistor

where g_{m_0} denotes the value of g_m at small-signal conditions or

$$\beta = \frac{g_{m_0}}{y_{ie}} \qquad (2.43a)$$

Often the symbol β is used for $|\beta|$ and great care must be exercised in determining what is meant.

Several important characteristics of the above quantities are:

All except h_{FE} are complex quantities and are strong functions of the dc emitter current and the operating frequency. (Some are also weak or strong functions of the dc voltages as indicated in Table 2.1.)

Even though these quantities were derived for a specific transistor configuration, for example, common emitter, they are valid for all circuit configurations using the appropriate transformations.

2.2.6 Typical Two-Port Parameter Data Sheets

Figures 2.2, 2.3, and 2.4 have been included to give the reader some quantitative knowledge of bipolar transistor behavior in terms of y and h parameters. These figures also demonstrate the forms in which the data are presented.

A study of the data of the latter figures, which is substantial, shows that it does not describe the complete range of operation and tremendous amounts of additional data are required to do so.

The parameter concepts are very useful for the analysis of the total circuitry, but the amount of data required to fully present the transistor characteristics is enormous, particularly if the large-signal performance is to be considered. It is obviously desirable to have a transistor model from which one can calculate with reasonable accuracy the important transistor characteristics under varying conditions. Such a model is the well-known hybrid-PI.

2.3 THE SMALL-SIGNAL COMMON EMITTER HYBRID-PI MODEL OF THE BIPOLAR TRANSISTOR

2.3.1 General Discussion

Figure 2.5 shows the small-signal hybrid-PI common emitter model of the intrinsic transistor which is that part of the complete transistor which performs the amplification function.

g'_{m_0} theoretically has the remarkable property quantitatively common to all bipolar transistors in that its value is a function only of the emitter current and the temperature and is given by

$$g'_{m_0} \approx \frac{1}{r_{e_0}} \qquad (2.44)$$

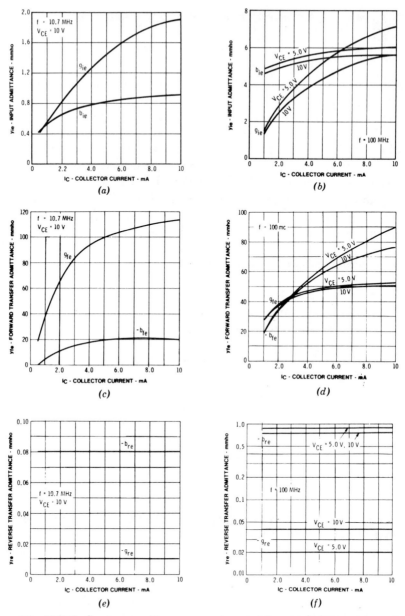

Figure 2.2 2N918. Common emitter y parameters. (a) Input admittance versus I_C at 10.7 MHz. (b) Input admittance versus I_C at 100 MHz. (c) Forward admittance versus I_C at 10.7 MHz. (d) Forward admittance versus I_C at 100 MHz. (e) Reverse admittance versus I_C at 10.7 MHz. (f) Reverse admittance versus I_C at 100 MHz. (g) Output admittance versus I_C at 10.7 MHz. (h) Output admittance versus I_C at 100 MHz. (i) Input admittance versus frequency. (j) Forward admittance versus frequency. (k) Reverse admittance versus frequency. (l) Output admittance versus frequency. (Courtesy of Fairchild Semiconductor.)

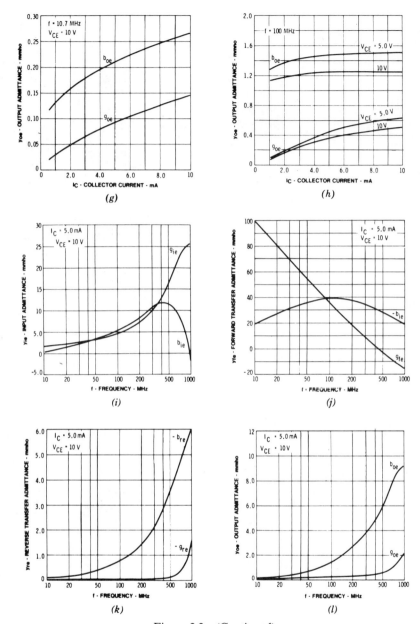

Figure 2.2 (Continued).

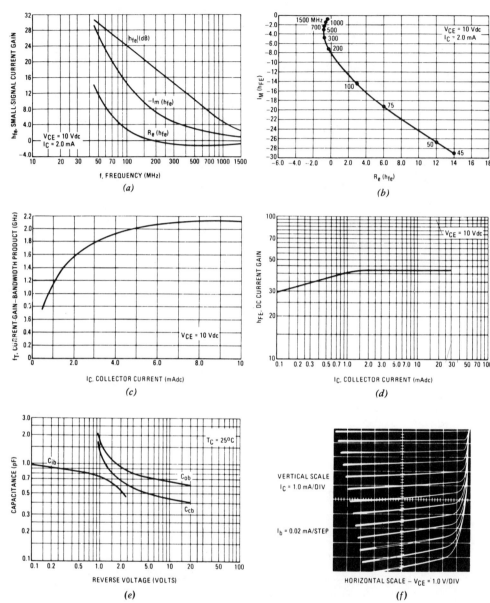

Figure 2.3 2N4957 characteristics. (a) h_{fe} versus frequency. (b) Polar h_{fe} versus frequency. (c) f_T versus I_C. (d) h_{FE} versus I_C. (e) Capacitance versus reverse voltage. (f) Collector characteristics. (g) y_{ie} versus frequency. (h) Polar y_{ie} versus frequency. (i) y_{fe} versus frequency. (j) Polar y_{fe} versus frequency. (k) y_{oe} versus frequency. (l) Polar y_{oe} versus frequency. (m) y_{re} versus frequency. (n) Polar y_{re} versus frequency. (o) y_{ib} versus frequency. (p) Polar y_{ib} versus frequency. (q) y_{fb} versus frequency. (r) Polar y_{fb} versus frequency. (s) y_{ob} versus frequency. (t) Polar y_{ob} versus frequency. (u) y_{rb} versus frequency. (v) Polar y_{rb} versus frequency. (w) y_{ie} versus I_C. (x) y_{ib} versus I_C. (y) y_{fe} versus I_C. (z) y_{fb} versus I_C. (aa) y_{oe} versus I_C. (bb) y_{ob} versus I_C. (cc) y_{re} versus I_C. (dd) y_{rb} versus I_C. (Courtesy of Motorola Inc.)

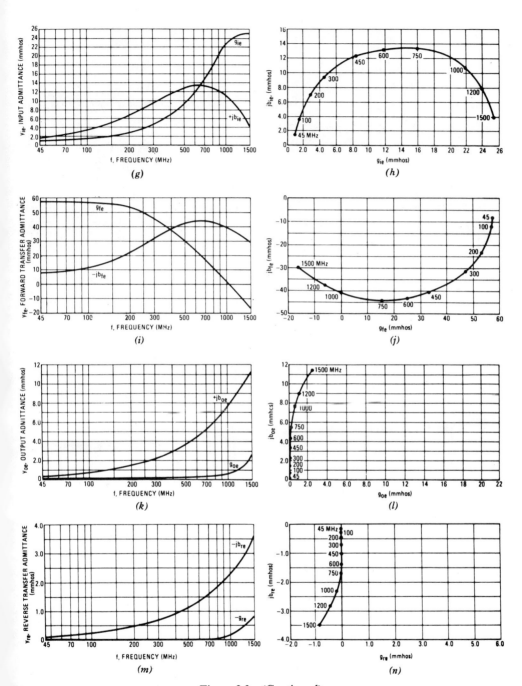

Figure 2.3 (Continued).

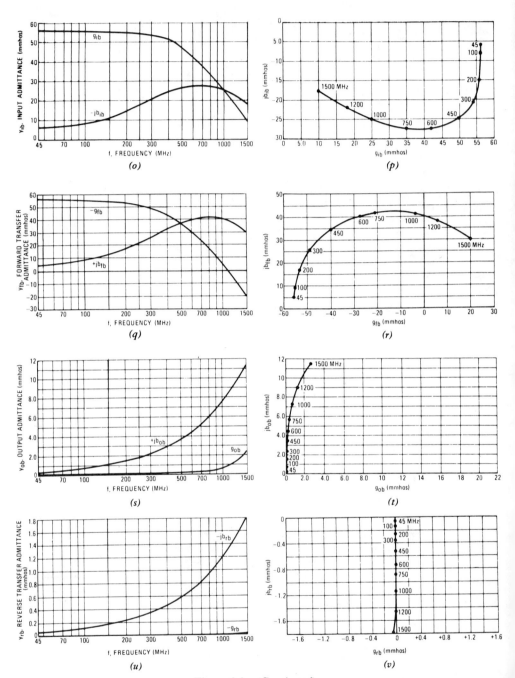

Figure 2.3 (Continued).

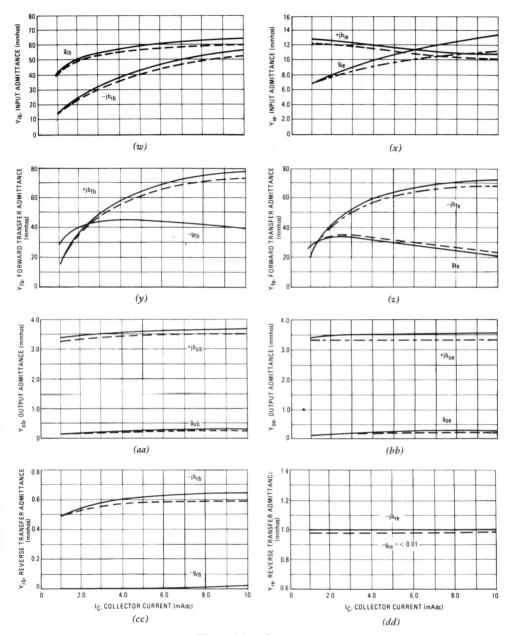

Figure 2.3 (Continued).

2N2857 JAN
(a) TYPICAL COMMON EMITTER Y-PARAMETERS AT 25°C LEAD TEMPERATURE
V_{CE} = 6.0 VOLTS, MAGNITUDE IN MILLIMHOS

FREQUENCY (MHz)	I_c (mA)	INPUT ADMITTANCE Y_{ie}	ϕ_{yie}	FORWARD TRANSFER ADMITTANCE Y_{fe}	ϕ_{ie}	REVERSE TRANSFER ADMITTANCE Y_{re}	ϕ_{yre}	OUTPUT ADMITTANCE Y_{oe}	ϕ_{yoe}
100	1.0	3.9	78°	34.7	−12°	0.48	−92°	1.2	84°
	1.5	5.1	76°	43.8	−12°	0.48	−90°	1.2	84°
	3.0	7.2	67°	82.1	−20°	0.50	−86°	1.1	78°
	5.0	8.1	63°	110.9	−27°	0.46	−87°	1.2	73°
200	1.0	6.9	74°	35.9	−17°	0.72	−90°	1.8	84°
	1.5	7.7	72°	44.6	−18°	0.72	−92°	1.8	83°
	3.0	10.6	61°	81.0	−29°	0.72	−90°	1.8	80°
	5.0	12.9	53°	108.4	−38°	0.73	−91°	1.7	80°
300	1.0	11.0	69°	35.1	−26°	1.1	−92°	2.7	89°
	1.5	12.0	66°	44.8	−29°	1.1	−94°	2.9	84°
	3.0	15.5	51°	74.3	−43°	1.1	−93°	2.8	83°
	5.0	17.4	43°	91.5	−52°	1.1	−92°	2.8	82°
400	1.0	13.5	66°	35.7	−31°	1.4	−91°	3.4	87°
	1.5	14.6	61°	44.1	−35°	1.4	−93°	3.5	85°
	3.0	17.9	46°	69.1	−50°	1.4	−91°	3.4	83°
	5.0	19.4	38°	82.6	−60°	1.3	−91°	3.4	82°
450	1.0	14.9	60°	35.7	−35°	1.4	−93°	3.8	85°
	1.5	16.2	60°	39.5	−39°	1.4	−90°	3.8	81°
	3.0	19.7	45°	61.0	−55°	1.4	−92°	3.9	78°
	5.0	21.2	37°	78.3	−64°	1.4	−91°	3.7	80°
500	1.0	16.4	59°	35.3	−40°	1.7	−92°	4.1	85°
	1.5	17.6	56°	42.3	−44°	1.7	−91°	4.2	82°
	3.0	20.4	40°	62.4	−61°	1.7	−92°	4.2	82°
	5.0	21.7	33°	71.2	−70°	1.7	−92°	4.1	81°
600	1.0	20.1	53°	35.9	−48°	2.2	−93°	5.0	82°
	1.5	21.0	48°	42.2	−54°	2.1	−93°	5.1	81°
	3.0	22.8	34°	57.3	−70°	2.1	−93°	4.9	82°
	5.0	23.4	27°	63.4	−79°	2.1	−92°	4.8	81°
700	1.0	22.3	50°	37.2	−54°	2.5	−92°	5.7	79°
	1.5	22.8	44°	42.5	−59°	2.5	−92°	5.7	81°
	3.0	24.0	30°	55.1	−75°	2.5	−91°	5.4	81°
	5.0	24.3	24°	59.9	−85°	2.5	−91°	5.4	81°

Figure 2.4 2N2857 JAN. Common emitter characteristics. (Courtesy of Microwave Associates.) (a) y parameters. These parameters have been calculated from S-parameter data measured on a computer controlled network analyzer. (b) h parameters. These parameters have been calculated from S-parameter data measured on a computer controlled network analyzer. The magnitude of h_{ie} is in Ω, and the magnitude of h_{oe} is in $m\Omega^{-1}$.

2N2857 JAN
(b) TYPICAL COMMON EMITTER h-PARAMETERS AT 25°C LEAD TEMPERATURE $V_{CE} = 6.0$ VOLTS

FREQUENCY (MHz)	I_c (mA)	SHORT CIRCUIT INPUT IMPEDANCE		FORWARD CURRENT TRANSFER RATIO		REVERSE VOLTAGE TRANSFER RATIO		OPEN CIRCUIT OUTPUT ADMITTANCE	
		h_{ie}^2	ϕ_{hie}	h_{fe}	ϕ_{hfe}	h_{re}	ϕ_{hre}	h_{oe}	ϕ_{oe}
100	1.0	254	−78°	8.8	−91°	0.122	10°	4.5	14°
	1.5	195	−76°	8.6	−88°	0.094	13°	4.4	17°
	3.0	137	−67°	11.3	−88°	0.069	25°	6.0	15°
	5.0	122	−63°	13.5	−90°	0.057	29°	6.8	12°
200	1.0	144	−74°	5.2	−92°	0.103	14°	4.2	23°
	1.5	129	−72°	5.8	−91°	0.093	15°	4.6	20°
	3.0	93	−61°	7.6	−91°	0.067	27°	5.9	15°
	5.0	77	−53°	8.3	−91°	0.056	35°	6.5	12°
300	1.0	90	−69°	3.2	−95°	0.107	18°	4.3	30°
	1.5	82	−66°	3.7	−95°	0.092	19°	4.9	28°
	3.0	64	−51°	4.8	−94°	0.075	35°	6.2	19°
	5.0	57	−43°	5.2	−95°	0.067	43°	6.6	16°
400	1.0	73	−66°	2.6	−98°	0.108	22°	4.8	34°
	1.5	68	−61°	3.0	−97°	0.098	25°	5.3	31°
	3.0	55	−46°	3.8	−97°	0.080	41°	6.3	23°
	5.0	51	−38°	4.2	−98°	0.072	49°	6.7	20°
450	1.0	67	−60°	2.4	−96°	0.097	25°	4.9	41°
	1.5	61	−60°	2.4	−99°	0.092	29°	5.2	37°
	3.0	50	−45°	3.0	−100°	0.072	42°	5.8	29°
	5.0	47	−37°	3.6	−101°	0.066	51°	6.2	23°
500	1.0	60	−59°	2.1	−99°	0.106	28°	5.2	40°
	1.5	56	−56°	2.4	−101°	0.098	31°	5.6	35°
	3.0	48	−40°	3.0	−102°	0.084	47°	6.3	27°
	5.0	45	−33°	3.3	−103°	0.079	54°	8.5	23°
600	1.0	49	−53°	1.8	101°	0.110	33°	6.0	42°
	1.5	47	−48°	2.0	−102°	0.103	38°	6.3	38°
	3.0	43	−34°	2.5	−105°	0.096	52°	6.6	28°
	5.0	42	−27°	2.7	−106°	0.093	60°	6.9	25°
700	1.0	44	−50°	1.7	−104°	0.114	38°	6.7	40°
	1.5	43	−44°	1.8	−104°	0.111	43°	6.9	38°
	3.0	41	−30°	2.2	−106°	0.105	57°	7.2	29°
	5.0	41	−24°	2.4	−109°	0.103	64°	7.3	26°

Figure 2.4 (Continued).

38 Transistor Properties Applicable to Oscillator Design

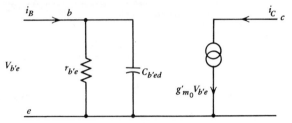

Figure 2.5 Small-signal hybrid-PI intrinsic bipolar common emitter transistor model.

where r_e is the dynamic emitter resistance and, in general, the subscript 0 denotes small signal.

$$r_{e_0} \approx \frac{kT}{qI_E} \tag{2.45}$$

where k is Boltzmann's constant, T is the temperature in degrees Kelvin, q is the charge on the electron, and I_E is the dc emitter current;

$$r_{e_0} = \frac{26\ \Omega}{I_E} \tag{2.45a}$$

at $T = 300°K$ and I_E in mA.

Other important properties for this model are

$$r_{b'e_0} = \frac{\beta_o}{g'_{m_0}} \tag{2.46}$$

where $\beta_o \equiv h_{fe}$ at low frequencies $\equiv h_{FE}$.

$C_{b'ed}$ is called the base emitter diffusion capacitance and

$$C_{b'ed_0} = \frac{g'_{m_0}(10^6)}{2\pi f_T} \tag{2.47}$$

where f_T is the gain bandwidth product which will be discussed later. C is in pF and f_T is in MHz.

Both β_o and f_T are strong functions of the construction of the transistor and therefore have to be determined for each individual transistor; but it is remarkable that once these two quantities are known, the rest of the model can be calculated.

Experimental investigations of g'_{m_0} have shown that Eqs. (2.44) and (2.45) are not completely true and the model must be modified to agree with experiment.

If the model is analyzed at low frequencies and at small signals, agreement can be obtained by adding a constant r'_e to r_e, called the extrinsic emitter

2.3 The Small-Signal Common Emitter Hybrid-PI Model of the Bipolar Transistor

dynamic resistance, and/or adding a resistance, $r_{bb'}$, called the base spreading resistance in series with the base. Then Eq. (2.45) becomes

$$r_{e_0} = \frac{kT}{qI_E} + r_e' \qquad (2.48)$$

For small power transistors, r_e' has been found to vary between 1 and 3 Ω. In this work, for simplicity,

$$r_e' = 1\,\Omega \qquad (2.49)$$

so that

$$r_{e_0} = \frac{26T}{I_E T_0} + 1 \qquad (2.50)$$

where

$$T_0 = 300°\mathrm{K} \qquad (2.50)$$

Figure 2.5 has been modified to become Fig. 2.6a. This figure now includes $r_{bb'}$ and capacitors C_{bet}, called the base emitter transition capacitance, $C_{b'c}$, the

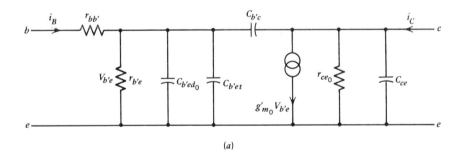

(a)

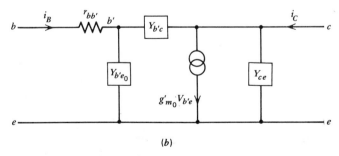

(b)

Figure 2.6 Small-signal models of the common emitter bipolar transistor. (a) Small-signal model. (b) y equivalent of (a).

capacitance between base and collector, and C_{ce}, the capacitance between collector and emitter. All the latter capacitances are functions of the voltage across them as shown in Fig. 2.3e, but they will be considered constant. Also it should be noted that additional capacitances exist because of the transistor packaging but these have been either lumped into the capacitors shown or disregarded as being negligible. Figure 2.6a also includes r_{ce} which is necessary to account for the slope of the collector voltage versus collector current characteristic as shown in Fig. 2.3f.

Again considering the model of Fig. 2.6a at low frequencies and small signals, it is obvious that

$$g_m \equiv \frac{I_e}{V_{be}} = \frac{V_{b'e}}{V_{be}} g'_m \tag{2.51}$$

and from Eq. (2.46)

$$\frac{V_{b'e}}{V_{be}} = \frac{r_{be'}}{r_{b'e} + r_{bb'}} = \frac{\beta_o/g'_{m_0}}{\beta_o/g'_{m_0} + r_{bb'}} \tag{2.52}$$

so that

$$g_{m_0} = \left(\frac{1}{g'_{m_0}} + \frac{r_{bb'}}{\beta_o}\right)^{-1} \tag{2.53}$$

$$= \left(\frac{26T}{I_E T_o} + \frac{r_{bb'}}{\beta_o} + 1\right)^{-1} \tag{2.54}$$

from Eqs. (2.44) and (2.50).

As noted in Section 2.1, the algorithms specify conditions which make Eq. (2.54) valid for the entire frequency range for which the algorithms will be used.

It should be noted that while $r_{bb'}$ is treated as a constant, it actually varies with frequency and the collector current, being smaller at higher collector currents. Also, while $r_{bb'}$ is considered resistive at some frequencies, it may have a reactive component. It is normally difficult to obtain data for the value of $r_{bb'}$ and it is therefore recommended that for small power transistors, $r_{bb'}$ be assumed to be 60 Ω, which is somewhat pessimistic. A more accurate value of $r_{bb'}$ may be obtained from curves such as in Fig. 2.3h, which shows that at the maximum frequency, $y_{i_e} \equiv g_{i_e} \approx 25$ m℧ so that $r_{bb'} = 40$ Ω.

Table 2.1 shows the variation of hybrid-PI parameters with voltage, current, and temperature.

2.3 The Small-Signal Common Emitter Hybrid-PI Model of the Bipolar Transistor

Table 2.1 Variation of Hybrid-PI Parameters with Current, Voltage, and Temperature

Parameter	Variation with Increasing...		
	I_C	V_{CE}	T
g_m	Linear	Independent	Decreases
r_{be}	$1/I_C$	Increases	Increases
C_{be}	Linear	Decreases	Varies with transistor
$r_{bb'}$	Decreases with high current	?	Increases
C_{cb}	Independent	Decreases	Independent
β_o	Independent	Increases slowly	Increases

2.3.2 The Calculation of β from the Model of Fig. 2.6a

$$\beta \equiv \left. \frac{I_c}{I_b} \right|_{V_{ce}=0}$$

If $C_{b'e}$ and $C_{b'c}$ are neglected, which is permissible when I_E is reasonably large, then

$$V_{b'e} = I_b \left(\frac{1}{r_{b'e}} + \frac{j}{X_{C_{b e d_0}}} \right)^{-1}$$

$$= I_b \left(\frac{g'_{m_0}}{\beta} + \frac{jg'_{m_0}f}{f_T} \right)^{-1} \tag{2.55}$$

from Eqs. (2.46) and (2.47). But $I_c = g'_{m_0} V_{b'e}$, so that from Eq. (2.55)

$$\beta = \left(\frac{1}{\beta_o} + \frac{jg'_{m_0}f}{f_T} \right)^{-1}$$

$$= \frac{\beta_o [1 - j\beta_o f/f_T]}{1 + (\beta_o f/f_T)^2} \tag{2.56}$$

In polar coordinates

$$\beta = \frac{\beta_o}{\sqrt{1 + (\beta_o f/f_T)^2}} \angle \tan^{-1}\left(-\beta_o \frac{f}{f_T}\right) \tag{2.56a}$$

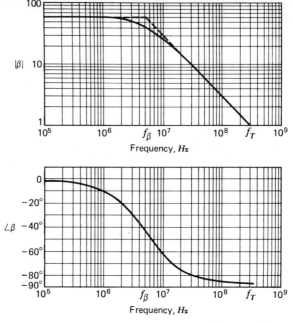

Figure 2.7 Typical magnitude and phase of β versus frequency.

It is seen from Eq. (2.56a) that

$$|\beta| = \beta_o \qquad \text{at } f \approx 0 \tag{2.57a}$$

$$|\beta| = 0.707\beta_o \qquad \text{at } f = \frac{f_T}{\beta_o} \equiv f_\beta \tag{2.57b}$$

$$|\beta| = 1 \qquad \text{at } f = f_T \tag{2.57c}$$

Figure 2.7 shows a plot of Eq. (2.56a) for a transistor having a β_o of 60 and an f_T of 300 MHz. This plot is idealized since if measurements were taken, the plot would be valid up to some frequency between f_β and f_T and would be completely inaccurate near f_T because of stray elements in the transistor. Thus f_T should be considered as an extrapolated value from data taken at some frequency above f_β where the ideal curve of $|\beta|$ versus f is a straight line of slope -1.

For example, the plot gives $\beta = 10$ at $f = 30$ MHz. Therefore, f_T would extrapolate to $30 \times 10 = 300$ MHz. The same procedure would be followed in determining f_T from a transistor data sheet where $|\beta| \equiv |h_{fe}|$ is specified at some arbitrary high frequency.

It should be noted that f_T is a function of both the currents and voltages in the transistor, so that efforts should be made to obtain f_T at the correct

2.3 The Small-Signal Common Emitter Hybrid-PI Model of the Bipolar Transistor

operating conditions. However, if the assumed operating conditions are somewhat inaccurate, the resulting errors will not be terribly significant.

Figures 2.3a and 2.3b show plots of $|\beta|$ versus frequency and Fig. 2.3c shows the plot of f_T versus collector current. These figures are presented to give some idea of how these parameters vary in a high-frequency transistor suitable for oscillators.

Figure 2.4b shows the variation of h_{fe} for different values of current and frequency in polar form. Equation (2.56a) states that the maximum absolute value of the phase angle is 90°, while Fig. 2.4b shows larger angles at the very high frequencies, which means that the model is insufficiently accurate at these frequencies. But, this transistor is not used at a frequency higher than 100 MHz where $f_T \approx 1100$ MHz. If it is assumed that $\beta_o = 100$, then $\tan^{-1}\theta = 10$ and θ is 85°, which is not far from the indicated value of $\approx 88°$.

2.3.3 The Calculation of y_{ie} from the Model of Fig. 2.6a

Neglecting C_{bet} and $C_{b'c}$, it is seen by inspection that

$$y_{ie} = \frac{(1/r_{bb'})\left[1/r_{b'e} + j/X_{C_{b'ed_0}}\right]}{1/r_{bb'} + 1/r'_{be} + 1/X_{C_{b'ed_0}}}$$

$$= \frac{(1/r_{bb'})\left[g'_{m_0}/\beta_o + jg'_{m_0}f/F_T\right]}{(1/r_{bb'} + g'_{m_0}/\beta_o) + jg'_{m_0}f/f_T} \tag{2.58}$$

Equation (2.58) is rather complicated and it is better to put it into polar form:

$$y_{ie} = \frac{(g'_{m_0}/r_{bb'}\beta_o)\sqrt{1 + (\beta_o f/f_T)^2}}{\sqrt{(1/r_{bb} + g'_{m_0}/\beta_o)^2 + (g'_{m_0}f/f_T)^2}} \tag{2.58a}$$

at an angle

$$\tan^{-1}\frac{\beta_o f}{f_T} - \tan^{-1}\frac{\beta_o f/f_T}{(1 + \beta/r_{bb'}g'_{m_0})}$$

Equation (2.58a) is also complicated, but is is evident from the angle term that the angle is always positive but its absolute magnitude first increases with frequency and then decreases until it approaches 0, as shown in Figs. 2.3g, 2.3h, and 2.4a.

2.3.4 The Calculation of y_{21e}, y_{12e}, and y_{22e} from the Model of Fig. 2.6a

The calculation of $y_{21e} \equiv y_{fe}$ can most easily be performed with the aid of Eq. (2.43) which states

$$y_{fe} \equiv g_{m_0} = \beta y_{ie} \tag{2.59}$$

This can be done by multiplying the results of Eqs. (2.56a) and (2.58a). The form of the absolute magnitude is rather complicated and therefore difficult to analyze, but the angle is

$$\tan^{-1}\left(-\frac{\beta_o f}{f_T}\right) + \tan^{-1}\left(\frac{\beta_o f}{f_T}\right) - \tan^{-1}\left(\frac{\beta_o f/f_T}{1 + \beta_o/r_{bb'}g'_{m_0}}\right)$$

$$= -\tan^{-1}\frac{f/f_T}{1/\beta_o + 1/r_{bb'}g'_{m_0}} \tag{2.59a}$$

which shows that the angle is lagging and its magnitude increases with frequency and with β_o, g'_{m_0}, and $r_{bb'}$, and hence with the dc emitter current.

Figure 2.4a shows that the angle at 100 MHz is 12° for a current of 1.5 mA, which is fairly large for a precision oscillator. The 12° angle is not particularly harmful and since the great majority of oscillators in the common emitter transistor configuration wil not exceed 50 MHz, it is seen that for all practical purposes, the angle can be considered to be 0° for this particular transistor type.

From inspection,

$$y_{re} \equiv y_{12e} = j2\pi f C_{bc} \tag{2.60}$$

$$y_{oe} \equiv y_{22e} = j2\pi f(C_{bc} + C_{ce}) + \frac{1}{r_{ce}} \tag{2.61}$$

2.3.5 The Effect of Local Feedback upon the Transistor Performance in the Common Emitter Configuration

Local emitter degenerative feedback is often used to stabilize the performance of transistors in the common emitter configuration. The effects of such feedback is studied in this section.

Figure 2.8 shows the schematic of a transistor, the emitter of which is fed to a resistor R_E and the end of the resistor serves as the common connection point.

2.3 The Small-Signal Common Emitter Hybrid-PI Model of the Bipolar Transistor

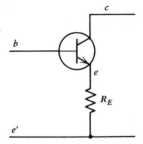

Figure 2.8 Transistor local emitter degeneration.

Figure 2.9 shows the two-port equivalent of Fig. 2.8 and the problem is to calculate the equivalent network Y_c which is composed of networks $Y_A + Y_B$ in series. This can be done in a straightforward manner as follows:

1. Transform the Y networks into Z networks.
2. Add the corresponding parameters to obtain Z_c.
3. Transform the Z_c network into the equivalent Y_c network.

The above procedure is very straightforward but is tedious and laborious and gives little insight into what is happening to the various elements in the hybrid-PI model.

A better procedure is to deal directly with the hybrid-PI model. In this case, Fig. 2.6b becomes Fig. 2.10.

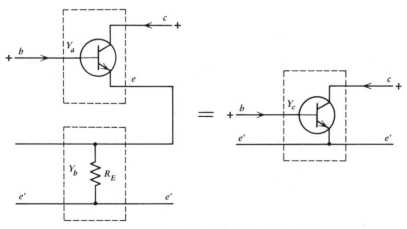

Figure 2.9 Two-port equivalent of Fig. 2.8.

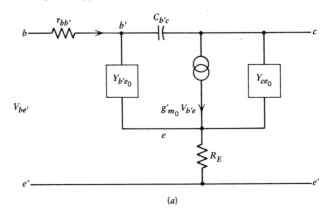

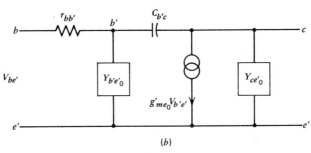

Figure 2.10 Hybrid-PI model of emitter degeneration. (*a*) PI model. (*b*) Equivalent of (*a*).

Analysis of the circuit in Fig. 2.10*a* using mesh analysis results in the following at *frequencies well below* f_T (See Fig. 2.10b):

1. $\beta, r_{bb'}$, and $C_{b'c}$ are unchanged. (2.62)
2. $Y_{be'} = Y_{b'e}\left(1 + \dfrac{R_E}{r_e}\right)^{-1}$. (2.63)
3. Similarly, $Y_{ce'} = Y_{ce}\left(1 + \dfrac{R_E}{r_e}\right)^{-1}$. (2.64)
4. $g'_{me_0} = g'_{m_0}\left(1 + \dfrac{R_E}{r_e}\right)^{-1}$. (2.65)
5. Equation (2.58a) shows that $y_{ie'}$ is decreased to a degree strongly dependent upon $r_{bb'}$.
6. Equation (2.59) shows that $y_{fe'}$ is decreased both in magnitude and phase.

It is noteworthy that the above results can be obtained by merely substituting $r_e + R_e$ for r_e where r_e is defined in Eq. (2.50).

2.4 LARGE-SIGNAL COMMON EMITTER BIPOLAR TRANSISTOR CHARACTERISTICS

2.4.1 Introduction

The parameters described up to now are valid only for small V_{be} signals up to about 10 mV. Some special cases of large signal characteristics will now be briefly treated. The treatment is based upon the work described by Holford in Ref. 2.1 and is further explained by Clarke and Hess in Ref. 2.3 and Frerking in Ref. 1.10. The treatment will only broadly present the derivation and the final results. Those interested in more detail may consult the cited references.

2.4.2 Basis and Results of the Derivation

The derivation is based upon the intrinsic model first shown in Fig. 2.5 and now repeated in Fig. 2.11 with slight changes in nomenclature. It is important to note that the derivation is valid only when $r_{bb'}$ and r'_e can be neglected.

It will be noted that some of the elements in Fig. 2.6 are absent in Fig. 2.11. This has been done to make the mathematics less complicated. Therefore, use of the results should include provisions for the missing elements or the circuits should operate at conditions where the missing elements are unimportant.

The derivation is as follows:

1. The results are assumed independent of V_{CE}. This assumption requires that $V_{CE} > 1.4(V_L + V_b)$.
2. β is assumed to be large, therefore

$$i_E \approx i_C$$

3. $r_{bb'}$ and $r'_e = 0$, by assumption.
4. $i_E = I_E + \sum_{M=1}^{M} I_{e_M} \cos(M\omega t)$ (2.66)
   Using standard notation, I_E is the mean emitter current, I_{e_M} is the rms component at frequency Mf; for example, I_{e_1} is the fundamental current.

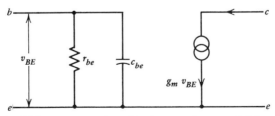

Figure 2.11 Hybrid-PI common emitter bipolar transistor model for determination of large-signal characteristics.

5. $$v_{BE} = V_{BE} + \sqrt{2}\, V_{be}(\cos \omega t) \qquad (2.67)$$
It is very important that the ac component V_{be} be highly sinusoidal, otherwise the results are not valid.

6. $$i_E = K_1 e^{K_2 v_{BE}} \qquad (2.68)$$
where K_1 and K_2 are constants of the transistor but functions of the temperature.

7. I_E and I_{e_M} for $M = 1, 2,$ and 3 are calculated from steps 4 to 6 by Fourier analysis, as functions of V_{be}.

8. From step 7 is calculated the applicable
$$\gamma_M = \frac{I_{e_M}}{I_E} \quad \text{for } M = 1, 2, 3 \qquad (2.69)$$

9. The ratio
$$\alpha = \frac{g_m}{g_{m_0}} \qquad (2.70)$$
is calculated from the relationship
$$g_m = \frac{I_{e_1}}{V_{be}} \qquad (2.71)$$
and
$$g_{m_0} = \frac{I_E}{26} \qquad (2.72)$$
from Eqs. (2.44) and (2.45a).

10. The derivation and Eqs. (2.46) and (2.47) also prove that at $T = 300°K$ and neglecting the 1 Ω
$$r_{be} = \frac{r_{be_0}}{\alpha} = \frac{\beta_o}{g_m} \qquad (2.73)$$

$$r_{ce} = \frac{r_{ce_0}}{\alpha} \qquad (2.74)$$

$$C_{bed} = \alpha C_{bed_0} = \frac{g_m(10^6)}{2\pi f_T} \qquad (2.75)$$

where the subscript 0 denotes the small-signal value as shown in Fig. 2.6. Or, in general,
$$Y_{be} = \alpha Y_{be_0} \qquad (2.76)$$

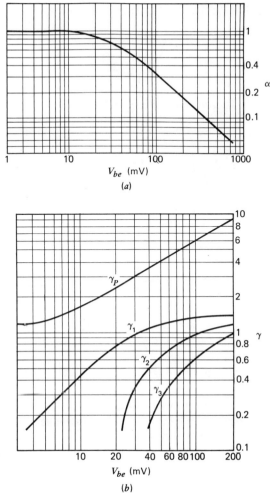

Figure 2.12 Common emitter bipolar transistor large-signal characteristics. (a) Variation of α with V_{be}. (b) Variation of γ with V_{be}.

Equations (2.69) and (2.70) are plotted in Fig. 2.12,[†] which also includes the plot

$$\gamma_P = \frac{i_{\text{peak}}}{I_E} \tag{2.69a}$$

[†]Holford, K., "Transistor *LC* Oscillator Circuits," *Mullard Tech. Commun.* 5, p. 19 (Dec. 1959), Figs. 2 and 5 (modified). Reprinted with permission.

From Eqs. (2.70), (2.71), and (2.72) one obtains the interesting relationship

$$\gamma_1 = \frac{\alpha V_{be}}{26} \tag{2.77}$$

From the curves, it is seen that for $\alpha \leq 0.3$,

$$\gamma_1 \approx 1.4 \tag{2.78}$$

therefore,

$$\alpha V_{be} \approx 36 \text{ mV} \tag{2.79}$$

Equation (2.79) is valid for $\alpha < 0.7$, where $\gamma \approx 1.4$ as shown in Figs. 2.12a and 2.12b.

It should be noted that all of the above relationships apply at $T = 300°K$ and slight revisions must be made for the other temperatures found in practice (-65 to $85°C$).

The analysis also shows an accompanying bias shift which tends to increase I_E as V_{be} is increased, but the circuits in this book are so designed that the effect of the bias shift is negligible, so it will not be considered.

Holford in Ref. 2.2 has also proposed similar techniques for transistors with emitter degeneration. Again Holford's work is further explained by Clarke and Hess in Ref. 2.3 and by Frerking in Ref. 1.10. However, this technique is not considered particularly useful except for low-frequency relatively high-power oscillators. It is also much more difficult to apply. For these reasons, it is not discussed further, but the results are summarized in Fig. 2.13.[†]

The exponential nature of Eq. (2.68) causes the emitter current, i_E, to be composed of pulses the width of which decreases sharply as the magnitude of V_{be} in Eq. (2.67) increases. The effect is due to the cut off of the emitter current during part of the cycle because of the swinging of the base emitter junction below the contact potential. This results in highly efficient conversion of dc power into ac power. This effect is identical to that which occurs in class C amplifiers and multipliers.

Figure 2.12b shows that γ_1 is 1.4 for V_{be} greater than 113 mV. Most engineers, when confronted with this relationship, find it very surprising and hard to believe that the fundamental component is greater than the dc current, but many times it has been experimentally demonstrated to be so.

Figure 2.12b can also be used in the design of frequency multipliers, and there will be occasion to use this information in Chapter 10.

[†]Holford, K., "Transistor *LC* Oscillator Circuits," *Mullard Tech. Commun.* 5, p. 64 (Feb. 1960), Fig. 6 (modified). Reprinted with permission.

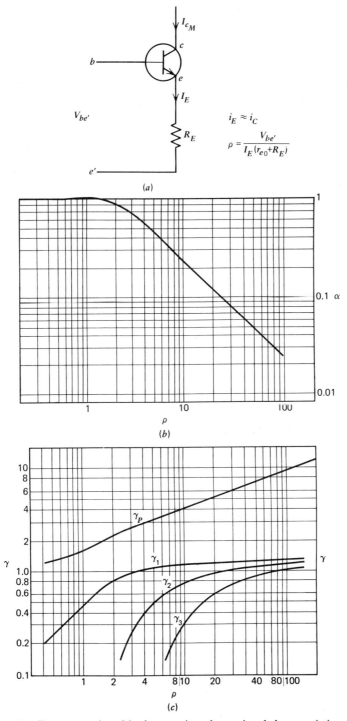

Figure 2.13 Common emitter bipolar transistor large-signal characteristics with emitter degeneration (a) Model of circuit and definition of ρ. (b) Variation of α with ρ. (c) Variation of γ with ρ.

2.5 THE COMMON BASE BIPOLAR TRANSISTOR

There are certain oscillators which find it more convenient to use the transistor in the common base configuration, particularly at high frequencies. The more important characteristics are briefly presented in this section.

2.5.1 Small-Signal Characteristics

2.5.1.1 y Parameters

$$\alpha = h_{21b} \tag{2.80}$$

and from Eq. (2.2)

$$\alpha = \frac{y_{21b}}{y_{11b}} \tag{2.81}$$

Figures 2.3p and 2.3r show that economical transistors may have

$$\alpha = -1 \tag{2.82}$$

for all frequencies to 200 MHz for $I_C \leq 2$ mA.

Figure 2.3p shows that at $f = 45$ MHz, y_{11b} has a phase angle of about $-5°$ and increases to about $-15°$ at 200 MHz. The minus sign signifies that y_{11b} is inductive.

Equation (2.17) states that

$$y_{11b} = y_{11e} + y_{12e} + y_{21e} + y_{22e} \tag{2.83}$$

$$\approx y_{21e} \tag{2.83a}$$

for the lower frequencies, where $y_{21e} \gg$ the sum of the remaining terms, which is seen from Figs. 2.3g, 2.3j, and 2.3k to be true up to about 100 MHz; so that for all frequencies < 100 MHz

$$y_{11b} = g_{m_0} \approx \left(\frac{26}{I_E} + 1 + \frac{r_{bb'}}{\beta_0}\right)^{-1} \tag{2.84}$$

from Eq. (2.54) at $T = 300°$K.

Equation (1.33) states that the input Y_{INb},

$$Y_{INb} = y_{11b} - \frac{y_{12b} y_{21b}}{y_{22b} + Y_L} \tag{2.85}$$

and Y_{INb} can be very small depending upon the value of the Y_b parameters and Y_L. This means that under certain conditions the input impedance can be quite

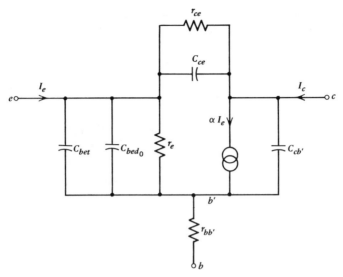

Figure 2.14 Hybrid-PI model of common base bipolar transistor.

large as contrasted to the low input impedance seen at most frequencies and loads. However, this effect will not be strong below 200 MHz in high-performance transistors such as that described in the curves of Fig. 2.3.

2.5.1.2 Hybrid-PI Model for the Common Base Transistor

Figure 2.14 shows a hybrid-PI model of the common base transistor. All symbols in this model, which are identical to those in Fig. 2.6a, have the same significance.

The major differences in the two models are shown in Table 2.2.

2.5.2 Large-Signal Characteristics

In general, Eq. (2.83) is still valid, that is,

$$\alpha = -1$$

The other characteristics depend upon whether v_{EB} is sinusoidal or i_E is sinusoidal. Both v_{EB} and i_E can be sinusoidal only in the small-signal case. In the large-signal case v_{EB} is sinusoidal when the emitter base circuit is effectively in parallel with the resonator. Similarly, i_E is sinusoidal in the dual case, that is, the emitter is effectively in series with the resonator.

2.5.2.1 Large-Signal Characteristics for Sinusoidal v_{BE}

The theory developed in Section 2.3 for the common emitter connection is also applicable to the common base connection. The only difference is that g_{m_0}

Table 2.2 Comparison of Common Emitter and Common Base Parameters

Parameter	Common Emitter	Common Base
Useful current generator representation	Characterized by g_m and V_{be}	Characterized by α and I_e
C_{bed_0}	Very important	Not so important as it is shunted by r_e which has a low value. This fact explains the superior high-frequency performance of the common base configuration.
$r_{bb'}$	Limits the high-frequency performance in conjunction with C_{bed_0}	More important in determining the value of y_{11b}
Important feedback capacitance	C_{cb}	C_{ce}

becomes the small-signal input conductance which has the same numerical value as the small-signal transconductance of the common emitter connection.

2.5.2.2 Large-Signal Characteristics for Sinusoidal i_E

These are very important for self-limiting oscillators, wherein the resonator circuit, usually a crystal network, is in series with the emitter, such as in the Butler oscillator described in Chapter 11.

In this case, I_e is known and it is desired to compute V_{eb} and the resultant R_{IN}. The method of performing these calculations will now be derived, in accordance with a modified form of the procedure beginning on p. 246 of Ref. 2.3,[†] as follows:

1. The results are assumed independent of V_{CE}. This assumption requires that $V_{CE} > 1.4(V_L + V_e)$
2. α is assumed ≈ -1 so that

$$i_E \approx -i_C \qquad (2.86)$$

3. r'_e and $r_{bb'}$ are assumed negligible so that

$$g_{m_0} = \frac{I_E}{26} \qquad (2.87)$$

4. $v_E = V_E + \sum_{m=1}^{M} V_{e_M} \cos(M\omega t)$ \qquad (2.88)

[†]Clarke/Hess, *Communication Circuits: Analysis and Design*, © 1971, Addison-Wesley, Reading, Mass. Fig. 6.7-4 (modified). Reprinted with permission.

2.5 The Common Base Bipolar Transistor

5 Let

$$i_E = I_E\left(1 + \sqrt{2}\,\frac{I_e}{I_E}\cos\omega t\right) \tag{2.89}$$

$$= I_{ES}e^{K_2 v_{EB}} \tag{2.89a}$$

from which

$$I_{ES}e^{K_2 v_{EB}} = I_E\left(1 + \frac{\sqrt{2}}{I_E}I_e\cos\omega t\right) \tag{2.89b}$$

where I_{ES} is the hypothetical value of I_E at $V_{EB} = 0$ and

$$K_2 = \frac{q}{kT} \tag{2.90}$$

$$= \frac{1}{26\text{ mV}} \quad \text{at } T = 300°\text{K} \tag{2.91}$$

Then solving Eq. (2.89b) for v_{EB}

$$v_{EB} = \frac{1}{K_2}\left[\ln\frac{I_E}{I_{ES}} + \ln\left(1 + \frac{\sqrt{2}I_e}{I_E}\cos\omega t\right)\right] \tag{2.92}$$

If the last term in Eq. (2.92) is expanded into a Fourier series, V_{e_1} is determined to be of the form

$$V_{e_1} = \frac{1}{\sqrt{2}\,K_2}\left(\frac{2}{\gamma}\left(1 - \sqrt{1-\gamma^2}\right)\right) \tag{2.93}$$

where

$$\gamma = \sqrt{2}\,\frac{I_e}{I_E} \tag{2.94}$$

The fundamental transistor input resistance is

$$R_{IN}(\gamma) = \frac{V_{e_1}}{I_e} = \frac{I_E}{I_e}\frac{V_{e_1}}{I_E} \tag{2.95}$$

$$= \frac{2\left(1 - \sqrt{1-\gamma^2}\right)}{\gamma^2 K_2 I_E} = \frac{2\left(1 - \sqrt{1-\gamma^2}\right)}{\gamma^2 g_{m_0}} \tag{2.96}$$

from Eqs. (2.95), (2.93), (2.92), (2.90a), and (2.87).

56 Transistor Properties Applicable to Oscillator Design

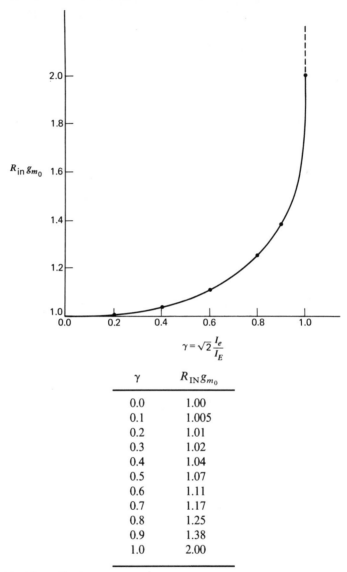

γ	$R_{IN}g_{m_0}$
0.0	1.00
0.1	1.005
0.2	1.01
0.3	1.02
0.4	1.04
0.5	1.07
0.6	1.11
0.7	1.17
0.8	1.25
0.9	1.38
1.0	2.00

Figure 2.15 Normalized input resistance for a sinusoidal emitter current in the common base bipolar transistor.

$R_{IN}(\gamma)g_{m_0}$ as given by Eq. (2.96) is plotted in Fig. 2.15.

It will be noted that the γ axis terminates at $\gamma = 1$, at which point the slope approaches infinity. Equation (2.96) states that, at $\gamma > 1$, $R_{IN}(\gamma)$ becomes complex. This, of course, is true in the ideal model. However, in the real case, this is impossible and, when γ tends to exceed 1, $R_{IN}(\gamma)$ progressively increases at a very rapid but finite rate.

Since i_E cannot be negative, it follows from Eqs. (2.89) and (2.94) that γ cannot exceed 1. Also,

$$I_E \geq 1.4 I_e \quad (2.97)$$

The latter provides a convenient means of fixing $I_{e_{max}}$.

2.6 BIASING OF BIPOLAR TRANSISTORS

In this section is considered the manner of biasing transistors to yield a given dc emitter current, I_E, from a power source, V_{BB}. The biasing circuit is to produce negligible change in emitter current upon variation of

1. Transistor characteristics.
2. Temperature.
3. Bias shifts due to the ac signals.

Figure 2.16 shows the dc biasing part of an oscillator circuit for a desired I_E. The procedure is as follows:

1. Specify $V_E \geq 2$ V. $\quad (2.98)$
2. $r_2 = \dfrac{V_E}{I_E} \quad (2.99)$
3. Determine the minimum value of β_o.
4. $r_b \approx \dfrac{\beta_o r_2}{5} \quad (2.100)$

where r_b is the Thévenin source resistance of the dc biasing source and

$$\frac{1}{r_b} = \frac{1}{r_{b_1}} + \frac{1}{r_{b_2}} \quad (2.101)$$

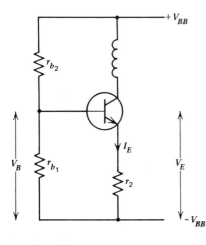

Figure 2.16 dc biasing circuit for the bipolar transistor. *Note:* r_2 is the sum of the bypassed and unbypassed emitter resistors.

58 Transistor Properties Applicable to Oscillator Design

5. $V_B \approx V_E + 700 \text{ mV}$ \hfill (2.102)

where 700 mV is the average voltage difference between base and emitter for a silicon bipolar transistor.

6. $r_{b_2} = 0.83 \dfrac{V_{BB}}{V_B} r_b$ \hfill (2.103)

7. $r_{b_1} = \left(\dfrac{1}{r_b} - \dfrac{1}{r_{b_2}} \right)^{-1}$ \hfill (2.104)

Equations (2.103) and (2.104) are valid only for $V_B < 0.83 V_{BB}$
The following example is investigated:

$V_{BB} = 10 \text{ V}$
$I_E = 2 \text{ mA}$
$V_E = 2 \text{ V}$
$\beta_o \min = 20$

then

$r_2 = 1000 \text{ }\Omega$
$r_b = 4000 \text{ }\Omega$
$V_B = 2.7 \text{ V}$
$r_{b_2} = 12{,}300 \text{ }\Omega$, make it 12 k$\Omega$
$r_{b_1} = 5.900 \text{ }\Omega$, make it 5.6 k$\Omega$

For the values chosen

$r_b = 3.8 \text{ K}$
$V_{B_{oc}} = 10 \left[\dfrac{5.6}{12 + 5.6} \right] = 3.18 \text{ V}$
$V_B = V_{B_{oc}} \dfrac{20}{20 + 3.8} = 2.67 \text{ V}$
$V_E = 1.97 \text{ V}$
$I_E = 1.97 \text{ mA}$

Let β_o change to 40. Then

$V_B = 3.18 \left[\dfrac{40}{40 + 3.8} \right] = 2.9 \text{ V}$
$V_e = 2.2 \text{ V}$
$I_E = 2.2 \text{ mA}$

so that a 100% change in β_o produces a 10% change in I_E.
Let V_{BE} change to 0.600 V due to temperature and/or ac signal change. For

the original transistor ($\beta_o = 20$)

$V_B = 2.67$ V
$V_E = 2.07$ V
$I_E = 2.07$ mA

so that a 100-mV change in V_{BE}, which is quite large, produces a 3% change in I_E.

2.7 THE JUNCTION FIELD EFFECT TRANSISTOR

2.7.1 Introduction

A brief presentation of the properties of the junction field effect transistor (JFET) is now made, as some of the important properties are considerably different from those of the bipolar transistor. The major differences are summarized in Table 2.3 for the common emitter/source connection at low and medium frequencies for high-performance transistors.

In addition, it should be noted that the bipolar transistor requires only knowledge of the β_o, f_T, and the power dissipation rating to be almost completely specified, while the JFET requires knowledge of many more parameters.

As shown in Table 2.3, the g_m of the JFET is much smaller than that of the bipolar transistor. Therefore, the JFET will be used primarily in those applications where the very high input impedance of the JFET is advantageous. This is true for crystals having very high resistances such as those at low frequencies.

Another important property of the JFET is its relatively low noise, as discussed in Chapter 14.

2.7.2 Ideal Small-Signal Relationships in JFETs at Low and Medium Frequencies

1 Variation of drain current with gate bias

$$I_D = I_{DSS}\left(1 - \frac{V_{GS}}{V_{GS(\text{off})}}\right)^2 \quad (2.105)$$

where I_{DSS} is the current at $V_{GS} = 0$ \quad (2.106)

$V_{GS(\text{off})}$ is the voltage at $I_D = 1\ nA$ \quad (2.107)

Table 2.3 Comparison of Bipolar and JFET Transistors

Parameter or Characteristic	Bipolar at Large V_{CE}	JFET in Pinch-Off Region	Typical value at $I_{C/D}$ = 5 mA	
			Bipolar	JFET
Ideal relationship $i_{C/D}$ versus $v_{BE/GS}$	Exponential	Square law		
Small-signal g_i	Large, proportional to I_C	Very small, weak function of I_D	2 m℧	< 0.001 m℧
Small-signal $g_f = g_{m_0}$	Large, proportional to I_C	Small, proportional to I_D	200 m℧	5 m℧
Small-signal C_i	Proportional to I_C	Weak function of I_D	35 pF	4 pF

2.7 The Junction Field Effect Transistor

2 Variation of g_{fs} with V_{GS}

$$(g_{m_0} \equiv g_{fs}) = g_{fs0}\left(1 - \frac{V_{GS}}{V_{GS(off)}}\right) \quad (2.108)$$

where

$$g_{fs0} \text{ is } g_{fs} \text{ at } V_{GS} = 0 \quad (2.108a)$$

3 Variation of g_{fs} with I_D

$$g_{m_0} \equiv g_{fs} = g_{fs0}\sqrt{\frac{I_D}{I_{DSS}}} \quad (2.109)$$

4 Relationship for g_{fs0}

$$g_{fs0} = \frac{2I_{DSS}}{V_{GS(off)}} \quad (2.110)$$

5 Relationship between I_{DSS} and $V_{GS(off)}$ for a given transistor type:

$$I_{DSS} \propto V_{GS(off)} \quad (2.111)$$

Thus, when $V_{GS(off)}$ increases, I_{DSS} also increases.

The above relationships demonstrate the great importance of I_{DSS} and $V_{GS(off)}$, which vary widely. It is therefore, at present, impractical to develop useful universal models similar to those for the bipolar transistor.

2.7.3 Small-Signal High-Frequency Characteristics

Figure 2.17 shows the y parameters versus frequency characteristics for the 2N4416A which is a very popular high-performance JFET. All these parameters are at $I_D = I_{DSS}$ which is an unusually large current.

Study of this figure discloses the following:

1. $g_{is} \approx 2.5(10)^{-5}f^3 \; \mu\mho$
   f in MHz
2. $b_{is} \approx 10f \; \mu\mho$ which is equivalent to a capacitance of 1.5 pF.
3. g_{os} is flat to 100 MHz at about 0.05 m$\mho$ and decreases with I_{DSS}.
4. $b_{os} \approx 5f \; \mu\mho$ which is equivalent to about 0.9 pF.
5. $g_{fs} \approx 5 \; \mho$ until about 400 MHz.
6. $b_{fs} \approx 10f \; \mu\mho$

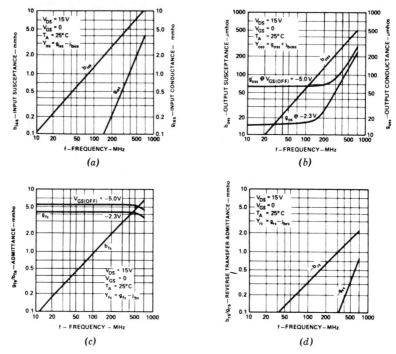

Figure 2.17 y parameters versus frequency for the 2N4416A common source junction FET. (a) y_{is}. (b) y_{os}. (c) y_{fs}. (d) y_{rs}. (Courtesy of Teledyne Semiconductor.)

7 Items 5 and 6 imply that

$$\theta_{y_{fs}} \approx (0.12f)°$$

8 $b_{rs} \approx 3f\,\mu\mho$ which is equivalent to about 0.45 pF.
9 Items 4 and 8 imply that

$$C_{dg} \approx C_{ds} \approx 0.45 \text{ pF}$$

It cannot be too highly stressed that the above values are subject to extremely wide variations due to the manufacturing process, but they give the orders of magnitude to be expected.

2.7.4 JFET Common Source Large-Signal Characteristics

As indicated in Table 2.3, the relationship between I_D and V_{GS} is square law. As a result, the large-signal characteristics, described in Section 2.4, which form the basis for one desirable type of limiting, are not applicable for the JFET. However, the very high input impedance of the JFET makes possible

2.7 The Junction Field Effect Transistor 63

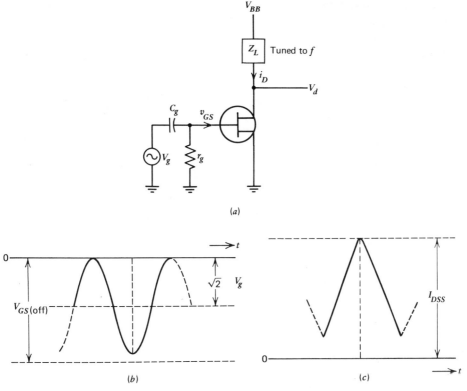

Figure 2.18 The JFET clamp biased amplifier. (*a*) Circuit diagram. (*b*) Wave shape of v_{GS}. (*c*) Wave shape of i_D.

the useful large-signal circuit, known as the clamp biased amplifier and described below. The treatment is based upon the procedure beginning on page 131 of Ref. 2.3,[†] and will present essentially only the final results. Those interested in more detail may consult the reference.

The Clamp Biased Amplifier Circuit

Consider the circuit shown in Fig. 2.18*a*. Clearly, in this circuit, if the $r_g C_g$ time constant is much greater than $T = 1/f$, the capacitor voltage charges to the peak value of v_G and remains constant at that value; hence

$$v_{GS} = \sqrt{2}\, V_g[\cos(2\pi f t) - 1] \tag{2.112}$$

The input voltage, v_{GS}, to the transistor is thus clamped to zero and has the

---

[†]Clarke/Hess, *Communication Circuits: Analysis and Design*, © 1971, Addison Wesley, Reading, Mass., Fig. 4.9-4 (modified). Reprinted with permission.

wave shape shown in Fig. 2.18b. The drain current i_D has the wave shape shown in Fig. 2.18c due to the square-law characteristics as given by Eq. (2.105). When $\sqrt{2}\,V_g = V_{GS(\text{off})}$, the width of the i_D pulse is 360°. However, when $\sqrt{2}\,V_g > V_{GS(\text{off})}$, the width decreases as V_g increases. The load, since it is tuned to f, will respond only to the fundamental component I_d of i_D.

Let

$$g_m = \frac{I_d}{V_g} \tag{2.113}$$

and

$$\alpha = \frac{g_m}{g_{m_0}} \tag{2.114}$$

where g_{m_0} is given by Eq. (2.110).

If a Fourier analysis is made of the i_D characteristics to obtain I_d as a function of V_g and α computed as a function of $V_g/-V_{GS(\text{off})}$ and plotted, Fig. 2.19 results.

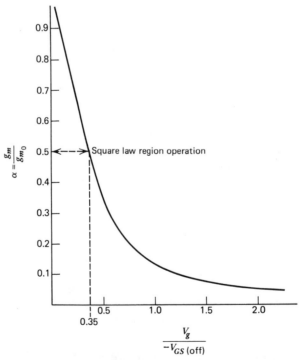

Figure 2.19 Plot of α versus $V_g/-V_{GS(\text{off})}$.

Table 2.4 Values of I_d/I_{DSS} and I_D/I_{DSS}

$\dfrac{V_g}{-V_{GS(\text{off})}}$	$\dfrac{I_d}{I_{DSS}}$	$\dfrac{I_D}{I_{DSS}}$
0	0	1
0.35	0.35	0.37
0.5	0.33	0.30
1.0	0.27	0.20

Additional Fourier analysis will produce curves for I_d/I_{DSS} and I_D/I_{DSS}. These curves are not shown, but some useful points are given in Table 2.4.

Another useful relationship is

$$r_{g_{ac}} = \frac{r_g}{3} \tag{2.115}$$

It should be noted that all the material in this section is valid only when

$$v_{D\,\min} < v_{G\,\max} \tag{2.116}$$

3

Piezoelectric Resonators

Contributed by **ARTHUR BALLATO**

3.1 INTRODUCTION

This chapter provides a brief applications-oriented treatment of crystal resonators. Its objective is to make it possible for the user to specify the resonators best suited to the application from the standpoints of performance and cost. It aims also at helping the user to avoid the pitfalls of costly overspecification.

Piezoelectric resonators are circuit components that provide combinations of electrical parameters and other features such as temperature stability, not obtainable using conventional capacitors, inductors, and resistors. Physically, they consist of carefully oriented and dimensioned pieces of quartz crystal or other suitable piezoelectric material to which adherent electrodes have been applied. The crystals are held within sealed enclosures by mounting supports that also serve as connections between the electrodes and the external leads.

The electroded crystal forms a special kind of capacitor. Because it is piezoelectric, the crystal changes shape when a signal is applied to the electrodes. The amount of motion varies over wide extremes depending upon how closely the applied signal frequency approaches a natural mechanical resonance of the crystal. In a properly designed resonator these regions of high-amplitude mechanical vibration are very narrow in frequency and are ideally suited for oscillator stabilization. The piezoelectric effect is responsible for converting the electrical signal to mechanical motion and it reconverts the vibratory motion of the crystal back into an electrical signal at the resonator terminals. Looking at the resonator simply as a circuit component consisting of an enclosure and leads, it is not necessary to know what the enclosure contains. Its electrical behavior can be represented completely in network terms. This equivalent circuit is discussed in the next section.

The piezoelectric resonator is unique not only because of the achievable combinations of circuit parameter values but also because of other important features such as cost, size, and stability with time, temperature, and other environmental changes. To see how these factors affect performance we will occasionally look within the crystal enclosure in subsequent sections. In the present section we discuss general features of crystal resonators.

3.1.1 Frequency Range

Crystal resonators are available to cover frequencies from below 1 kHz to over 200 MHz. At the low-frequency end much current development is directed toward wristwatch applications at 32.768 kHz and powers of two times this frequency. This work has led to new families of miniature resonators (see Section 3.5). More conventional resonators span the range 80 kHz to 200 MHz; these utilize bulk acoustic waves (BAWs) that propagate within the crystal (see Sections 3.2 and 3.3). Recent work with BAW resonators has produced units at UHF frequencies (see Section 3.5). Surface acoustic waves (SAWs) travel along the crystal surface. Devices based on SAWs are becoming increasingly more available for the range above 50 MHz into the low GHz region (see Section 3.4).

3.1.2 Frequency Accuracy[3.20]

The absolute frequency accuracy of crystal-stabilized commercial oscillators has been improving at the rate of a factor of 10 every 20 years, starting around 1940 when the accuracy ranged between 10^{-3} and 10^{-4}. By 1980 the figure ranged from 10^{-5} to 10^{-6}, with 10^{-6} to 10^{-7} projected by 2000. These figures include total fractional absolute frequency variations over all environmental ranges such as temperature, mechanical shock, and aging.

3.1.3 Frequency Stability[3.20]

Precision quartz oscillators, held at constant temperature and protected from environmental disturbances, have shown improvement at the rate of a factor of 10 every 10 years. In 1960, fractional stabilities of 10^{-10}, 10^{-8}, and 10^{-6} had been achieved with laboratory versions, limited commercial production units, and in large-scale production, respectively. By 1980 these figures had improved to 10^{-12}, 10^{-10}, and 10^{-8}, with values of 10^{-14}, 10^{-12}, and 10^{-10} projected for the year 2000. These values are for observation times in the neighborhood of 0.1 to 10 s.

3.1.4 Enclosures

The great majority of crystal enclosures are designated by HC- numbers (Holder, Crystal). Two of the most popular sizes are shown in Figs. 3.1 and 3.2 to establish the general size range of resonators.

Holders are distinguished as follows:

1. Material: metal, glass, ceramic.
2. Terminals: pins, leads, tabs.
3. Method of sealing: solder, resistance weld, cold weld, thermocompression, induction heating.
4. Shapes and sizes: conventional metal cans, miniature tubes, glass envelopes, microcircuit (DIP) packages, transistor cans, ceramic flatpacks.

3.1.5 Crystal Vibrator (Bulk Wave Types)

The vibrator sealed within the enclosure is distinguished as follows:

1. Material: quartz, piezoelectric ceramic, high coupling single crystals such as lithium niobate or lithium tantalate.
2. Type of growth: natural or cultured (synthetic) quartz, other synthetic crystals hydrothermal or melt grown.
3. Treatment: as-grown (unswept), swept (electrolysis), etched.

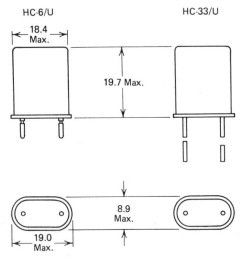

Figure 3.1 Standard crystal enclosures. Pin version HC-6/U, and wire lead version HC-33/U. Dimensions are in millimeters. Normally used for units in the frequency range 0.8 to 20 MHz.

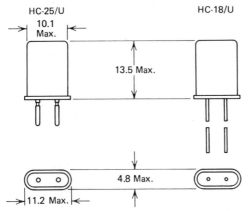

Figure 3.2 Standard crystal enclosures. Pin version HC-25/U and wire lead version HC-18/U. Dimensions are in millimeters. Normally used for units in the frequency range 10 to 200 MHz.

4. Crystal cut (orientation): unrotated, singly rotated, doubly rotated. Popular cuts are designated by names such as AT, BT, SC, and so on.
5. Vibrator geometry (plates, bars, forks, and so on) and dimensions.
6. Type of vibration: shear, extension, flexure, and so on. These may take place along one or more dimensions and may occur in combination.
7. Mode of vibration: what we have called the *type* of vibration is often called the *mode* of vibration. To avoid confusion we shall reserve the word *mode* to indicate one of the three polarizations of the vibrations that depend on the thickness dimension of a crystal plate. These are denoted the a, b, and c thickness modes, and are encountered in Section 3.3.
8. Overtone (harmonic) of operation.
9. Mounting supports: wire, strips, clamps, clips, and so on.
10. Crystal-to-mount bond: conductive bonding cement, epoxy, polymide; braze.
11. Electrode material: gold, aluminum, silver, copper; thin adhesion layer (Cr, Ti, Ni) plus gold.
12. Electrode deposition method: evaporation, sputtering, electroplating.
13. Ambient within enclosure: dry gas, for example, nitrogen, soft vacuum, high and ultrahigh vacuum.

These are discussed in greater detail in subsequent sections.

3.2 EQUIVALENT CIRCUITS

The symbol for a crystal resonator is shown in Fig. 3.3a. In precision applications, the metallic enclosure, leads and supports, and so on, require that

70   Piezoelectric Resonators

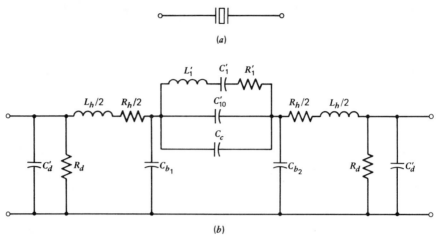

Figure 3.3 Crystal resonator. (a) Circuit symbol. (b) Complete equivalent circuit including the effects of holder and mounting supports. A single crystal resonance is represented by the L_1', C_1', R_1' arm. This circuit is valid in the vicinity of a single resonance up to UHF frequencies.

the resonator be treated as a two-port device. The complete circuit equivalent,[3.9] valid into the VHF range, is given in Fig. 3.3b and includes elements associated with the supports (R_h, L_h), enclosure and lead proximity (C_d', R_d), and electrode–enclosure capacitance (C_{b_1}, C_{b_2}, C_c). The crystal vibrator portion is represented in the vicinity of a single resonance by the C_{10}', C_1', R_1', and L_1' elements. It is virtually always the case that the influences of R_d and R_h can be neglected. Then, by suitable network manipulations, the complete circuit may be reduced to the circuit shown in Fig. 3.4, with the element values modified by the transformation.[3.9] The capacitors C_{d_1} and C_{d_2} range from a fraction of a picofarad to several picofarads; often they will be equal by manufacturing symmetry. Depending upon the method of utilizing the crystal, the presence of these capacitances will influence the resonator behavior. If the enclosure is metallic and is grounded to one lead, C_{10} is increased accordingly; if not, the shunt capacitors are lumped with the external circuit elements and have to be treated there. We are thus led to the four-element circuit of Fig. 3.5.[3.9,3.10]

It is often the case that treatments of crystal resonator parameters end with a discussion of the R_1, L_1, C_1, and C_0 circuit seen at the enclosure terminals and an enumeration of these values. This is considered to be as far as the resonator designer has to go. Since these numbers completely characterize the resonator, it is reasoned, then any other needed quantities can be calculated. From the point of view of the designer of the circuit into which the resonator is to be placed, these same four quantities (or an equivalent set with L_1 replaced by the nominal resonance frequency) are likewise adequate to specify the crystal. A simple relation for the approximate frequency shift of the resonator when operated with series load capacitor C_L completes the crystal's portion of

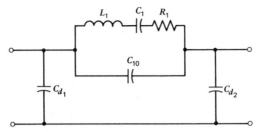

Figure 3.4 Equivalent electrical circuit of crystal resonator and enclosure. Below VHF frequencies the element values are constant; at higher frequencies the values are very weakly dependent on frequency, following from the constant values of Fig. 3.3b.

the design. Departures from the desired operating frequency and resonator power level are then corrected by mysterious time-consuming and costly "tweaking" procedures that result in nonoptimal designs.

In the systematic and straightforward procedures used in this book, somewhat more is asked of the resonator equivalent circuit, not the least of which is a reorientation of what is taken to be given in the resonator specification. Also required are more accurate expressions relating what is seen at the crystal terminals at the actual operating frequency; these are derived in the following sections.

Returning to Fig. 3.5, C_0 is called the *static* capacitance; it is the capacitance associated with the crystal and its adherent electrodes plus the stray capacitances internal to the crystal enclosure described above. It is a measured quantity that is specified. The value of C_0 does not include stray or wiring capacitance external to the enclosure. When these additional capacitances are added to C_0, the result is denoted as C_0'; this quantity is a specified value determined by the measured value of the latter effective stray capacitances added to C_0. The series R_1, L_1, C_1 portion is referred to as the *motional* arm of the circuit and arises from the mechanical crystal vibrations. C_1 and the resonator resistance R_L when operated with C_L in series, assuming $C_0' = C_0$ are both specified quantities. Also specified is df, a small frequency offset between the desired operating frequency, denoted simply as f, and the load frequency f_L obtained with C_L.

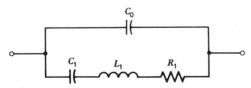

Figure 3.5 Simplified equivalent network of a crystal resonator. The element values are measured effective quantities that include various stray and parasitic effects, and are constant in the frequency regions centered about the resonance under consideration.

72   Piezoelectric Resonators

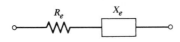

Figure 3.6   Resistive (R_e) and reactive (X_e) parts of the impedance represented by the network of Fig. 3.5. The quantities R_e and X_e are sensitive functions of frequency in the region of resonance.

3.2.1 R_e and X_e as Functions of Δf for the Case where $C_0' = C_0$

We assume that the presence of all stray capacitance internal to the enclosure has been resolved into the capacitor C_0, and that the resonator circuit plus internal strays remains of the form in Fig. 3.5. Expressions for the equivalent series resistance and reactance of the resonator are now derived using the $Z = R + \underline{X}$ representation covered in Section 1.2.1.5. Figure 3.6 is the network for the equivalent series resistance R_e and reactance X_e, which are functions of frequency.

The admittance of the network in Fig. 3.5 is

$$Y = \frac{1}{\underline{X}_0} + \frac{1}{R_1 + \underline{X}_1} = \frac{R_1 + \underline{X}_1 + \underline{X}_0}{\underline{X}_0(R_1 + \underline{X}_1)}. \tag{3.1}$$

The impedance $Z = Y^{-1}$ is then

$$Z = \frac{\underline{X}_0(R_1 + \underline{X}_1)}{R_1 + (\underline{X}_0 + \underline{X}_1)}. \tag{3.2}$$

Multiplication by $R_1 - (\underline{X}_0 + \underline{X}_1)$ gives

$$Z = R_e + \underline{X}_e = \frac{R_1^2 \underline{X}_0 + R_1 \underline{X}_0 \underline{X}_1 - R_1 \underline{X}_0(\underline{X}_1 + \underline{X}_0) - \underline{X}_1 \underline{X}_0(\underline{X}_1 + \underline{X}_0)}{R_1^2 - (\underline{X}_1 + \underline{X}_0)^2}. \tag{3.3}$$

The real and imaginary parts of Eq. (3.3) are the series resistance R_e and series reactance X_e:

$$R_e = \frac{-R_1 \underline{X}_0^2}{R_1^2 - (\underline{X}_1 + \underline{X}_0)^2} = \frac{R_1}{(R_1/X_0)^2 + ((X_0 + X_1)/X_0)^2}, \tag{3.4}$$

$$X_e = \frac{\underline{X}_0[R_1^2 - \underline{X}_1(\underline{X}_1 + \underline{X}_0)]}{R_1^2 - (\underline{X}_1 + \underline{X}_0)^2} = X_0 \left\{ \frac{R_1^2 + X_1(X_0 + X_1)}{R_1^2 + (X_0 + X_1)^2} \right\}. \tag{3.5}$$

These are sketched as functions of frequency over an extended range in Fig. 3.7. It is seen that X_e is zero at two values of frequency. The lower frequency is

3.2 Equivalent Circuits

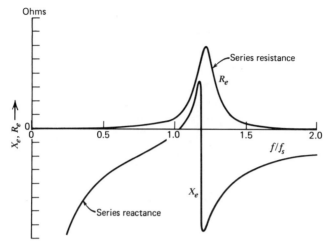

Figure 3.7 Resistive (R_e) and reactive (X_e) parts of the crystal resonator network of Fig. 3.5 as functions of frequency. X_e is zero at two frequencies; the lower one, denoted f_R, is the resonance frequency. At f_R, R_e is approximately R_1. Between the two zeros of X_e the resonator reactance appears inductive; the operating point of the resonator is normally somewhat above f_R.

just slightly above the series resonance frequency

$$f_s - \frac{1}{\left(2\pi\sqrt{L_1 C_1}\right)} \tag{3.6}$$

and is denoted the resonance frequency f_R. At f_R the equivalent resistance R_e is approximately R_1. The upper frequency at which X_e equals zero is usually denoted the antiresonance frequency f_A. At this point R_e is nearly a maximum. The high-impedance antiresonance frequency will not concern us further for oscillator applications.

With the assumption $|(X_0 + X_1)| \gg R_1$, Eqs. (3.4) and (3.5) become

$$R_e \approx \frac{R_1}{[(X_1 + X_0)/X_0]^2} \tag{3.7}$$

$$X_e \approx \frac{X_1}{[(X_1 + X_0)/X_0]} . \tag{3.8}$$

These expressions are simplified further as follows:

Let $\omega = \omega_s + \Delta\omega$. Then $X_1 = \omega L_1 - 1/\omega C_1$ becomes

$$X_1 = \frac{1}{\omega C_1}\left(\frac{2\Delta\omega}{\omega_s} + \frac{\Delta\omega^2}{\omega_s^2}\right) \approx \frac{1}{\omega_s C_1}\left(\frac{2\Delta\omega}{\omega_s}\right) \tag{3.9}$$

or

$$X_1 \approx \frac{2}{\omega_s C_1} \frac{\Delta f}{f_s} \quad \text{assuming } \omega_s \gg \Delta\omega \tag{3.10}$$

Consistent with the approximations to follow and with the assumption that $\omega_s \gg \Delta\omega$, we will drop the subscript s and use $\Delta f/f$, where f is the actual operating frequency. Since $X_0 = -1/\omega C_0$, one obtains

$$\frac{X_1 + X_0}{X_0} \approx 1 - \frac{2C_0 \Delta f}{C_1 f}. \tag{3.11}$$

Inserting Eq. (3.11) into Eqs. (3.7) and (3.8) gives

$$R_e \approx \frac{R_1}{(1 - 2C_0 \Delta f/C_1 f)^2} \tag{3.12}$$

and

$$X_e \approx \frac{(2/\omega C_1)(\Delta f/f)}{(1 - 2C_0 \Delta f/C_1 f)}. \tag{3.13}$$

Equation (3.13) may be further approximated by

$$X_e \approx \left(\frac{2}{\omega C_1}\right) \frac{\Delta f}{f}. \tag{3.13a}$$

The main assumption made in arriving at Eqs. (3.7) to (3.13) is that $|(X_0 + X_1)| \gg R_1$. This assumption is not universally valid. When $\Delta f = 0$, it is equivalent to the criterion

$$\frac{1}{\omega_s C_0 R_1} = M \gg 1. \tag{3.14}$$

The quantity M is one type of figure of merit and often is not large for overtone units (see Section 3.2.3.4). On the other hand, for any value of M, if $\Delta f \approx C_1 f/2C_0$, then $|X_0 + X_1| \approx 0$, and certainly is not $\gg R_1$. This frequency offset corresponds to the region of the antiresonance frequency, and by assumption is excluded from use.

3.2.2 R_e and X_e with Rated Load Capacitor C_L

The load capacitor C_L is a nonvariable specified quantity. Its presence in series with the resonator modifies R_e and X_e as follows: Assuming, as before, that $\Delta f/f \ll 1$, Eq. (3.10) gives an expression for X_1 in terms of Δf. This frequency shift is now taken such that the positive reactance of X_1 equals the reactance magnitude of C_0 in parallel with C_L, so that the total reactance is zero:

$$\frac{2}{\omega_s C_1} \frac{\Delta f_{Ls}}{f} = \frac{1}{\omega(C_0 + C_L)} \tag{3.15}$$

or

$$\frac{(f_L - f_s)}{f} = \frac{\Delta f_{Ls}}{f} = \frac{C_1}{2(C_0 + C_L)} \tag{3.16}$$

Sometimes Δf_{Ls} and C_L are specified in place of C_1; Eq. (3.16) then determines C_1. Substitution of this value of $\Delta f/f$ in Eqs. (3.12) and (3.13) gives the values of $R_L = R_e$ and $X_L = X_e$ at the load frequency f_L:

$$R_L \approx R_1 \left(\frac{C_0 + C_L}{C_L} \right)^2 \tag{3.17}$$

$$X_L \approx \frac{1}{\omega C_L} \tag{3.18}$$

As the load capacitance increases ($C_L^{-1} \to 0$), the load frequency f_L approaches the resonance frequency f_R, R_e approaches R_1, and X_L approaches zero.

In Fig. 3.8 the solid curves depict the variations of X_e and R_e in the vicinity of the operating point, wherein it is assumed that the static capacitance is the specified value C_0. The difference between f_s and f_R is exaggerated for clarity. At f_L the resistance is defined to be the load resistance R_L; its value is an input specification. In terms of the load resistance the quantity R_1 may be determined using Eq. (3.17):

$$R_1 = R_L \left(\frac{C_L}{C_0 + C_L} \right)^2 \tag{3.19}$$

An important derived quantity is the reactance change per change in fractional frequency. We denote this quantity as $\Delta X (\Delta f/f)^{-1}$ and calculate its value about the frequency f_s by using Eq. (3.13a). X_e equals zero at f_s, so the difference ΔX in reactance as the fractional frequency is changed by $\Delta f/f$ from this point is

$$\frac{\Delta X}{\Delta f/f} \approx \frac{2}{(\omega_s C_1)} = \frac{1}{\pi f C_1} \tag{3.20}$$

For f expressed in MHz and C_1 in pF, this becomes

$$\frac{\Delta X}{\Delta f/f} \approx \frac{10^6}{\pi f C_1}. \tag{3.20a}$$

76   Piezoelectric Resonators

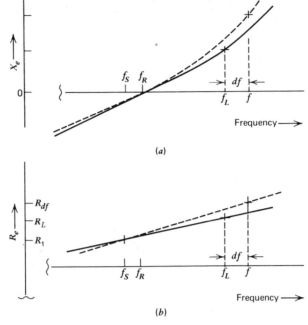

Figure 3.8 Impedance versus frequency. (a) X_e in the vicinity of resonance. The solid line is drawn for the case where the static capacitance is C_0; the dashed line represents the case where the static capacitance is increased to C_0'. At the actual operating point the frequency is f, and the value of X_e is X_{df}, corresponding to a static capacitance of C_0'. (b) R_e versus frequency in the vicinity of resonance. The solid and dashed lines have the same meaning as in (a). At the actual operating frequency (f), the value of R_e is R_{df}.

This expression is, strictly speaking, only accurate in the region close to f_s; we use it as a measure of the slope of the curve at the operating frequency f, although we recognize that it may underestimate the slope by a factor of 2 or 3. The slope of X_e versus frequency is a measure of crystal resonator *stiffness*. Operation of the resonator with C_L is chosen because the slope is increased, and changes in the llator do not affect f as much near f_L as near f_s.

If the crystal resonator, represented by the circuit of Fig. 3.5, is operated with series C_L, then the frequency will be f_L. Unfortunately, there is always additional stray capacitance shunting the crystal enclosure, for example, wiring capacitance. These strays are external to the enclosure and increase the value of C_0 in Fig. 3.5 to a larger value C_0'. Just as C_0 is a specification value, so too is C_0' a specification, but it depends upon the llator as well as C_0. The effects upon X_e and R_e of the change from C_0 to C_0' may be seen in Fig. 3.8, where the dashed lines represent the C_0' influence.

In order to produce the frequency f at which it is desired to operate, given that f may differ by a small amount df from f_L, as shown in Fig. 3.8, it is necessary to alter the value of the load capacitor from the rated C_L value. This is done with a trimmer capacitor. If, in addition to the shift in frequency by df, the fact that C'_0 and not C_0 is the true effective value of static capacitance to be taken into account, then the trimmer will have to accomplish both adjustments. The value of load capacitance required will be denoted $C_{L_{df}}$ and is determined by using Eq. (3.10) to compute the motional arm reactance $X_{1_{df}}$ ($= X_1$ at Δf taken to be $(f_L - f_s) + df$) and then using Eq. (3.16) to substitute for $(f_L - f_s)/f$.

$$X_{1_{df}} \approx \left(\frac{2}{\omega C_1}\right) \cdot \left(\frac{(f_L - f_s)}{f} + \frac{df}{f}\right) \qquad (3.21)$$

$$X_{1_{df}} \approx \left(\frac{2}{\omega C_1}\right) \cdot \left(\frac{C_1}{2(C_0 + C_L)} + \frac{df}{f}\right) \qquad (3.22)$$

The negative capacitance equivalent to $X_{1_{df}}$ is $(\omega X_{1_{df}})^{-1}$, or

$$\left(\frac{1}{C_0 + C_L} + \frac{2df}{C_1 f}\right)^{-1}, \qquad (3.23)$$

and $C_{L_{df}}$ is just this value minus C'_0:

$$C_{L_{df}} = \left[(C_0 + C_L)^{-1} + \frac{2df}{C_1 f}\right]^{-1} - C'_0 \qquad (3.24)$$

With f in MHz, and df in Hz, this is

$$C_{L_{df}} = \left[(C_0 + C_L)^{-1} + 2(10^{-6})\frac{df}{C_1 f}\right]^{-1} - C'_0 \qquad (3.24a)$$

The value of R_L is given as a specification. From it, the value of R_1 is found from Eq. (3.19). The actual resistance presented by the resonator at the frequency $f_L + df$, taking into account C'_0, will be denoted R_{df}. It is determined by using Eq. (3.12) with C_0 replaced by C'_0, and with Δf equal to $(f_L - f_s) + df$:

$$R_{df} = R_1 \left(1 - \frac{2C'_0[(f_L - f_s) + df]}{C_1 f}\right)^{-2} \qquad (3.25)$$

Equation (3.16) is used to eliminate $(f_L - f_s)/f$; after simplification one obtains:

$$R_{df} = R_1 \left(1 - \frac{C'_0}{C_0 + C_L} - \frac{2C'_0\,df}{C_1 f}\right)^{-2} \qquad (3.26)$$

With f expressed in MHz and df in Hz, Eq. (3.26) becomes

$$R_{df} = R_1\left(1 - \frac{C_0'}{C_0 + C_L} - \frac{2C_0' \, df}{C_1 f \times 10^6}\right)^{-2} \qquad (3.26a)$$

Power dissipation in the crystal is a specified quantity P_x. From P_x and the operating resistance R_{df}, the crystal current I_x follows:

$$I_x = \left(\frac{P_x}{R_{df}}\right)^{1/2} \qquad (3.27)$$

When P_x is given in mW and I_x in mA, the relation is

$$I_x = \left(\frac{1000 P_x}{R_{df}}\right)^{1/2} \qquad (3.27a)$$

A useful general expression, independent of the value of C_0', for the change in reactance with frequency is obtained from differentiating Eq. (3.22) and making use of Eq. (3.24) in the result to obtain[3.12]

$$\frac{\partial X_{df}}{\partial f} = \frac{2 Q_x R_{df}}{f} \qquad (3.28)$$

The quality factor Q_x is defined in Eq. (3.29); it is a constant for a given unit (see Section 3.2.3.1). R_{df} is the equivalent resistance of the osci at the frequency of measurement, at any value of C_0', and varies from approximately R_1 at f_R up to what is normally many times R_1 at f_A [see Eq. (3.39)]. With the measurement of R_{df} at the operating frequency, the reactance slope is simply and accurately determined by Eq. (3.28); it applies at any frequency in the resonance range.

3.2.3 Figures of Merit[3.11]

This subsection considers a number of figures of merit derived from the equivalent circuit parameters of Fig. 3.5, their importance, and application.

3.2.3.1 The Q Concept

The resonator *quality factor* Q_x is defined by the relation

$$Q_x = \frac{1}{\omega C_1 R_1} = \frac{1}{2\pi f C_1 R_1} \qquad (3.29)$$

where $\omega = \omega_s$, or, with f expressed in MHz and C_1 in pF,

$$Q_x = \frac{10^6}{2\pi f C_1 R_1} \qquad (3.29a)$$

Using Eq. (3.6), Q_x may be expressed in other equivalent forms, but involves only the parameters of the motional arm; it is the ratio of motional capacitive reactance to resistance at f_s. There is some confusion regarding the meaning and use of Q_x, and this can lead to costly overspecification. The Q concept for resonators has been carried over from its traditional use in connection with resonance in RLC circuits. In this latter case it is shown that the width of the resonance curve is inversely proportional to Q, so that the quality factor measures the *sharpness* of the resonance. With regard to the crystal resonator, the situation is complicated by the presence of C_0 (or C_0'). The frequency interval between resonance and antiresonance is the usual measure of *sharpness* in resonators. This is the region in Fig. 3.7 where X_e is positive. It depends upon the ratio of C_0 to C_1 almost exclusively; this is discussed in Section 3.2.3.3. The presence of loss ($Q_x^{-1} \neq 0$) blunts the extremes of X_e and causes X_e to be zero at the antiresonance point, but these regions are excluded from use in this book.

For filter applications the resonator Q shows up directly in the passband insertion loss. For oscillator applications, which are our concern here, it would appear from what has been said above that Q_x is not involved directly in determining the *sharpness* of the resonance, and hence the oscillator stability. This seems to be at variance with the usual statements regarding the importance of the high Q of quartz vibrators for frequency control. The confusion arises from two sources. First, it is certainly true that if a resonator is going to be operated in its inductive region, as is usually the case, then its frequency must lie within the limits given by the *sharpness* of the X_e curve. However, as seen in Fig. 3.8a, if the actual operating frequency is f and the resonator reactance is X_{df}, then any changes in X_e about the X_{df} point brought about by llator changes will produce frequency departures from f by amounts that depend not upon the resonance–antiresonance distance, but upon the slope of X_e at the operating point. Thus the stability is measured in part by Eq. (3.20), and this does not depend upon Q_x, but is rather inversely proportional to C_1.

The second source of confusion regarding Q stems from associating the crystal Q_x with the *operating Q* of the oscillator. The Q_x of the resonator is not the true measure of oscillator stability. The loaded or operating Q determines frequency stability, and this value depends upon both the osci and llator properties. The operating Q is proportional to the reactance slope [see Eq. (3.20)] and hence is determined by C_1; it is always less than Q_x. Because it is sometimes considerably less than Q_x, it will not pay in such cases to put a premium on the resonator Q_x. To do so will mean paying for an attribute that cannot be utilized. Considered from another viewpoint, the llator should be designed in such a manner that the loaded Q is as high as possible for greatest stability and most efficient use of the resonator Q_x.

The inverse of Q_x is a measure of loss in the vibrator. The loss is generally a combination of three contributions: (1) the internal friction of the crystal, (2) loss of vibratory energy to the mounting supports, and (3) energy loss to the gas within the enclosure. Most modern units are sufficiently evacuated that (3) plays no role. For quartz crystals, Q_x ranges from 10^4 to 10^7, with the lower values obtained at the lower frequencies, below about 1 MHz, because of mounting losses, and also at very high frequencies, above 200 MHz, because of internal friction. Realizable Q_x values are a maximum in the range 2.5 to 5 MHz as a tradeoff between external (mounting support) loss and internal (frictional) loss.

Above about 10 MHz the internal friction losses begin to dominate the Q_x for quartz. For AT-cut overtone units, Q_x is due almost exclusively to internal losses. A considerable portion of these losses arises from defects and impurities in the crystal lattice; these in turn are functions of the growth process and for cultured quartz can vary considerably in manufacture. The internal friction has been found to be correlated with infrared absorption, so a simple IR test during selection of material is used for quality control of Q_x. The Q_x figures are also found to be quite temperature sensitive and can lead to R_1 and R_L changes over operating temperature ranges. The R_L value specified should be the maximum permissible over the desired temperature range.

For AT-cut quartz at room temperature, the maximum intrinsic values of Q_x are found to follow the relation[3.13]

$$Q_x \cdot f \approx 16 \times 10^6 \qquad (3.30)$$

where f is in MHz. This relation is independent of overtone. It is given a simple explanation in the next section.

3.2.3.2 Motional Time Constant τ_1

Equation (3.29) is of the same form as Eq. (3.30) if the $R_1 C_1$ product is a constant. This is defined as the motional time constant τ_1 with a value corresponding to Eq. (3.30) of

$$\tau_1 = R_1 C_1 \approx 10^{-14} \text{ s} \qquad (3.31)$$

This quantity is simply related to the attenuation of an acoustic wave as it travels within the crystal to cause the vibration. It is a function of crystal cut, type, and mode of vibration, but does not vary by more than a factor of 2 or so. It is a function of temperature, but is independent of frequency and overtone as long as the losses arise only from the internal friction. As was stated earlier, this is usually only the case for quartz above 10 MHz where flat plates vibrating in thickness shear are used. These plates are mounted at the

Table 3.1 Frequency and Motional Time Constant, and Piezoelectric Coupling Factor of Popular High-Frequency Quartz Cuts

Cut	Frequency Constant, N_0 (MHz-mm)	Motional Time Constant, τ_1 (10^{-15} s)	Piezoelectric Coupling Factor, k (percent)
AT	1.661	11.8	8.80
SC	1.797	11.7	4.99
BT	2.536	4.9	5.62

edges where the vibratory motion is nearly zero provided the plate diameter-to-thickness ratio is greater than about 50 to 80. For such plates the fundamental frequency is given by

$$f = \frac{N_0}{t} \tag{3.32}$$

N_0, the frequency constant, is given for a number of popular cuts in Table 3.1; t is the plate thickness in mm and f is the nominal frequency in MHz. Values of τ_1 are also given in Table 3.1.[3.7]

As an exercise we will compute the lowest frequency AT-cut that will fit in an HC-6/U holder (Fig. 3.1) such that its Q_x will not be severely degraded by the mounting supports. From Fig. 3.1 we choose a plate diameter of 14 mm to permit room for the mounting posts. Taking $\frac{1}{60}$ of this for the thickness t of the plate, and using Eq. (3.32) and Table 3.1, we get approximately 7 MHz.

For diameter (or width)-to-thickness ratios less than about 60 normally, the mounting supports increase the loss and lead to an effective time constant τ that is described on the average by the relation

$$\tau = \tau_1 \left[1 + \left(\frac{60}{d} \right)^2 \right] \tag{3.33}$$

In Eq. (3.33), d is the diameter-to-thickness ratio. Continuing our example, if a 7-MHz AT-plate was contained in an HC-18/U holder, then from Fig. 3.2, a value of 7 mm for the diameter would be reasonable. This would give a value for d of nearly 30, and τ would be increased by a factor of 5; Q_x would be 20% of its maximum value at this frequency. Often Eq. (3.33) represents a lower bound of what is achieved in practice by contouring the plate to reduce edge motion (see Section 3.6).

From simple considerations of frequency and dimensions, reasonable estimates of Q_x may thus be found by the use of the motional time constant.

3.2.3.3 Capacitance Ratio r

The quantity r is defined as the ratio of the static to motional capacitances

$$r = \frac{C_0}{C_1} \tag{3.34}$$

It is a measure of the antiresonance–resonance (pole-zero) frequency separation that is important in filter applications

$$f_A - f_R \approx \frac{f_R}{2r} \tag{3.35}$$

and is related to a material constant called the piezoelectric coupling factor k which depends upon crystal cut, type, and mode of vibration. For the very popular thickness modes of plates,

$$2r = \left(\frac{\pi N}{2k}\right)^2 \tag{3.36}$$

where $N = 1, 3, 5, \ldots$, is the overtone number (see Section 3.2.4.1).

As was discussed in Section 3.2.2, the resonator is operated at a frequency f situated between f_R and f_A. The pole-zero distance, and hence r, is only important for oscillators to the extent that it qualitatively indicates how *stiff* the resonator is, that is, between what limits f may be *pulled* by a trimmer capacitor. Although it is often cited as one of the most important parameters for oscillator application, the really important parameter is C_1, and not r. One should also bear in mind that it is C_0' and not C_0 alone that must be considered in the oscillator design.

Equation (3.36) discloses that r increases with the square of the overtone order. From Eq. (3.35) we see that overtone units are more suitable for high-stability oscillator application; for temperature-compensated crystal oscillators (TCXOs) and variable crystal oscillators (VCOs) fundamental ($N = 1$) operation is preferred (see Section 3.2.5). Typical r values for quartz range from 100 to 20,000.

In applications where a capacitor in series with the crystal unit is varied for purposes of modulation or temperature compensation, the frequency–temperature behavior (see Section 3.7) will change somewhat with operating frequency. This effect is due to the temperature coefficient of k in Eq. (3.36) and must be accounted for in precision applications.

3.2.3.4 Figure of Merit M

The figure of merit M is defined as

$$M = \frac{Q_x}{r} \tag{3.37}$$

For quartz vibrators it ranges from less than unity to 10,000. Its chief use is to indicate when a vibrator ceases to possess an inductive region, that is, when f_R and f_A no longer exist as the zeros of X_e. Equation (3.35) is not valid if M becomes too small. In this case a more exact formula must be used that takes into account the blunting of the X_e curve by the finite Q_x around the maximum and minimum of X_e. It happens that when

$$M \leqslant 2 \tag{3.38}$$

the X_e curve no longer becomes positive (inductive), so f_R and f_A fail to be defined, and the analysis of Section 3.2.2 cannot be used. Because Q_x decreases with increasing frequency from Eq. (3.30), and r increases with the square of the harmonic number, from Eq. (3.36), there is a tradeoff to be considered for high-frequency resonators between use of high harmonics for increased stability, and the resulting rapid decrease in M. See Section 3.2.4.1. The maximum value of R_e occurs close to f_A and has the value

$$R_e(\text{maximum}) \approx R_1 M^2 \tag{3.39}$$

3.2.4 Typical Parameter Values

Ranges of parameter values for quartz resonators are given in this subsection. All generally used cuts and types of vibration are considered, including the increasingly important SC-cut, with the exception of SAW devices (see Section 3.4), and miniature resonators (see Section 3.5). Additional information on SC cuts is given in Section 3.3.

3.2.4.1 Overtones

Most types of vibration may be used on different overtones or harmonics. The most popular by far are the thickness vibrations of plates such as AT- and BT-cuts of quartz, where the thickness determines the fundamental frequency by Eq. (3.32). Odd overtones of this frequency may be excited in resonators as well, and for the overtone family of resonances of a given type the parameters of the equivalent circuit are simply related.

Figure 3.9 is the equivalent circuit of Fig. 3.5 extended to represent the family of overtones associated with the fundamental. C_0, C_1, L_1, and R_1 are unchanged in value from Fig. 3.5, but, as may be inferred from Eq. (3.36),

$$C_N = \frac{C_1}{N^2} \tag{3.40}$$

Since the frequency of the Nth overtone is N times the frequency of the fundamental, it follows from Eq. (3.6) that

$$L_N = L_1 \tag{3.41}$$

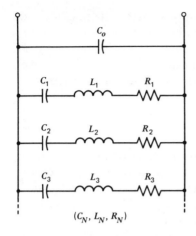

Figure 3.9 Equivalent circuit of Fig. 3.5 extended to represent overtones. Each series arm accounts for one resonance, in the vicinity of which the other arms may be neglected. In the region of the Nth overtone the circuit is accurately represented by C_0 plus C_N, L_N, and R_N.

Finally, from Eq. (3.31),

$$R_N = R_1 N^2 \tag{3.42}$$

As was mentioned in Section 3.2.3.3, overtones are used because the smaller C_N values lead to greater frequency stability because, by Eq. (3.20), the reactance slope increases with increasing N. The present discussion has considered only overtones belonging to a single family. In practice a number of families of types and modes of vibration are simultaneously present in the spectrum of the resonator, and the spectrum is usually more cluttered with unwanted responses as N increases. Special designs have been worked out to reduce the unwanted responses at a desired overtone. For this reason a crystal is usually designated for operation at a specific N and should be used only at this overtone for best results. This subject will be considered further in Sections 3.3 and 3.6.

3.2.4.2 Circuit Parameter Ranges for Quartz

Typical ranges of frequency and equivalent circuit parameters for traditional cuts of quartz and types of vibration are given in Table 3.2. Additional values for the newer miniature resonators are given in Section 3.5. Table 3.2 data of C_1 and R_1 are shown graphically in Figs. 3.10 and 3.11 as functions of frequency. Boundary lines are only approximate, and for special applications the boundaries may be stretched considerably. For example, recent developments of high-frequency plates have pushed the upper limit for fundamental AT-cut resonators well beyond 200 MHz.

From Figs. 3.10 and 3.11 we find two useful rules of thumb:

$$C_1 = 0.1 \text{ to } 0.001 \text{ pF} \tag{3.43}$$

Table 3.2 Quartz Vibrator Parameter Ranges

Cut	Type of Vibration	Frequency Range (MHz)	Approx. C_0/C_1	Range of C_1 (10^{-3} pF)	Range of R_1 (Ω)	Remarks
XY	Flexure	0.001–0.05	600	1–100	10^3–3×10^5	See Section 3.5 for frequency ranges and typical parameters of miniature resonators
NT	Flexure	0.005–0.14	900	1–30	10^3–3×10^5	
5°X	Extension	0.04–0.20	130	20–800	20–5000	
CT	Face shear	0.15–0.85	350	2–100	30–8000	
DT	Face shear	0.10–0.50	400	3–200	30–8000	
GT	Coupled extension	0.08–0.30	350	10–25	40–300	
SL	Face shear/ flexure	0.35–0.70	400	2–100	30–8000	
BT	Thickness shear	3–30	650	20–200	2–500	Fundamental
AT	Thickness shear	0.50–5	450–300	2–100	5–500	Fundamental
AT	Thickness shear	5–30	180–250	5–200	2–100	Fundamental
AT	Thickness shear	10–75	250×3^2	0.5–20.0	5–200	Third overtone
AT	Thickness shear	50–150	250×5^2	0.3–2.0	5–200	Fifth overtone
AT	Thickness shear	100–200	250×7^2	0.2–1	10–300	Seventh overtone

and

$$f \cdot R_1 = 10 \text{ to } 1000 \text{ Ω MHz} \tag{3.44}$$

with f expressed in MHz. These ranges hold from 1 kHz to over 50 MHz. In the absence of any additional information, good practical averages for R_1 and C_1 over these frequencies are

$$C_1 = 0.01 \text{ pF} \tag{3.45}$$

and

$$R_1 = \frac{100}{f} \tag{3.46}$$

One sees that these equations lead to time constant values that exceed the minimum [Eq. (3.31)] by factors of 1000 at 100 kHz, 100 at 1 MHz, and 10 at 10 MHz. Mounting losses are largely responsible for the high τ values; this is particularly true for low frequencies, but even for AT-plates the use of large electrodes to obtain a specified C_1 leads to vibratory motion at the plate edges and energy loss to the mounting supports. The data in Fig. 3.11 are for evacuated enclosures. Unevacuated units at frequencies below 0.5 MHz possess R_1 values that run about a factor of 2 higher than the upper limits in the figure, down to about a factor of 10 times higher than the lower limits in the figure. For example, at 20 kHz, R_1 ranges from about 1000 to 300,000 Ω in evacuated

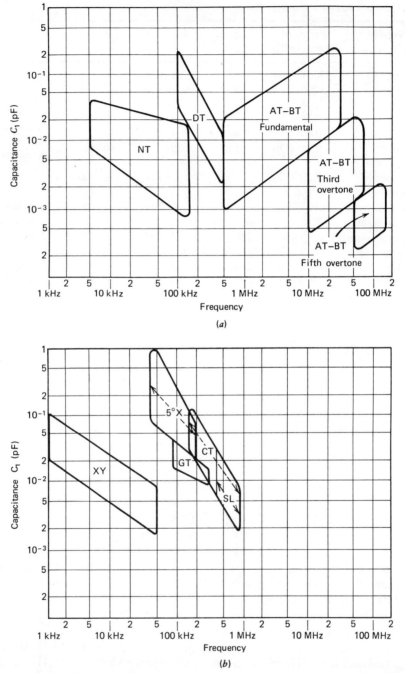

Figure 3.10 Typical ranges of motional capacitance C_1 in picofarads versus frequency. (a) For NT, DT, AT, and BT quartz cuts. (b) For XY, 5°X, GT, CT, and SL quartz cuts.

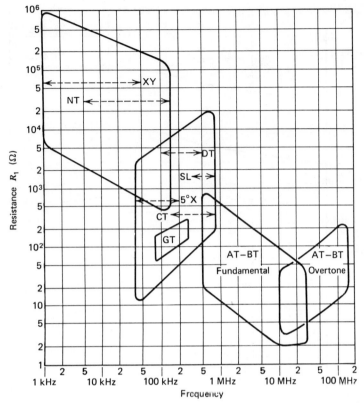

Figure 3.11 Typical motional resistance R_1 versus frequency ranges for NT, XY, 5°X, CT, DT, SL, GT, AT, and BT quartz cuts.

enclosures, and from about 8000 to 600,000 Ω in unevacuated enclosures. At frequencies above 0.5 MHz, the resistance change is about a factor of 2 at both the upper and lower resistance limits.

For the SC-cut one may arrive at parameter ranges using Table 3.2 and Figs. 3.10 and 3.11 for the AT-cut, by dividing the C_1 values by three and taking three times the R_1 values. These estimates follow from Table 3.1 and Eq. (3.36), and the fact that the dielectric constant of quartz determining C_0 is nearly independent of cut. Since the ratio of the squares of the coupling factors is about 3, the ratio of the r values is also; this is then the C_1 ratio since C_0 does not change for comparable resonator design geometry. The time constants are likewise approximately equal, so the resistance ratio follows. The above estimates neglect an 8% difference in plate thickness for similar AT and SC vibrators of the same frequency.

3.2.4.3 Preferred C_1 Values for AT- and SC-Cuts

Achievable motional capacitance values at different frequencies are given in Fig. 3.10 for standard crystal resonators. In different frequency ranges, size

88 Piezoelectric Resonators

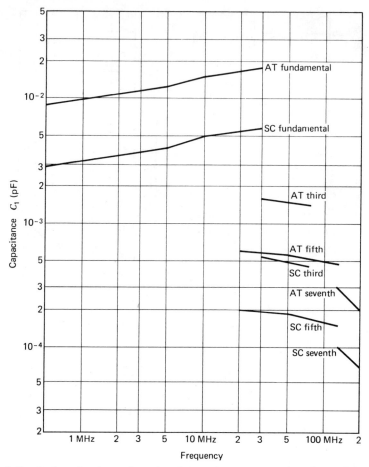

Figure 3.12 Preferred values of motional capacitance C_1, as functions of frequency and overtone, for AT and SC quartz cuts. Designs in a band about these values are most readily and economically produced and lead to the most cost-effective units.

and other considerations lead to the most economical designs if the C_1 values are limited to the preferred numbers indicated in Fig. 3.12. This figure is drawn for the predominantly used AT-cut and the newly introduced SC-cut that is becoming increasingly important for high precision applications. The figure may also be used to approximate the BT-cut by using the AT values; for other cuts and types of vibration, Fig. 3.10 is used to obtain approximate values by taking the range midpoints at each frequency.

Once a design C_1 value has been selected, the corresponding R_1 value can be estimated roughly from Fig. 3.11, or from Eq. (3.33) if the plate geometry is known.

3.2 Equivalent Circuits

An Illustrative Example

To summarize some of the developments of Sections 3.2.3 and 3.2.4 and to make additional illustrative points of a practical nature, we take as an example an AT-cut resonator with the following measured parameters:

$f = 5$ MHz
$Q_x = 750,000$
$R_1 = 4 \, \Omega$
$C_0 = 5.04$ pF.

The above equivalent circuit quantities are measured at the crystal enclosure terminals. From Eqs. (3.29) and (3.31) we obtain

$$C_1 = 10.6 \times 10^{-3} \text{ pF}$$

$$\tau_1 = 42.4 \times 10^{-15} \text{ s}$$

and Eq. (3.6) gives

$$L_1 = 95.5 \text{ mH}.$$

The C_0 value includes stray capacitance associated with the holder, but does not include strays that may be associated with the using circuitry. The capacitance ratio including resonator strays is

$$r = \frac{C_0}{C_1} = 475$$

This resonator is obviously a fundamental unit because if the k value from Table 3.1 is used in Eq. (3.36), it predicts a ratio, r_0, of 159 for the fundamental, and 1434 for the third overtone. Since 475 is less than the third overtone value, it must be a fundamental. The value of 475 is about three times higher than the theoretical value. This comes about for two reasons. First, presence of stray capacitance makes C_0 larger than the C_{10} that arises from the parallel-plate capacitor structure consisting of the crystal plus electrodes. Second, the value of k in Table 3.1 is a material constant for AT-cut quartz, whereas the k in Eq. (3.36) is an effective value that is reduced by a geometrical factor Ψ (less than 1) to account for the fact that the mechanical vibratory motion of the crystal plate is not of equal amplitude over the entire electrode area where the piezoelectric current is collected. When low r values are desired (see Section 3.2.5) a compromise is often necessary between having a relatively uniform distribution of motion over the electrodes, and having the motion taper off near the electrode edges so that the plate edges will be quiet. Then little energy is lost to the mounting supports, degrading τ_1, and hence R_1.

The rules governing the distribution of motional activity are called *energy trapping* laws and are touched upon in Section 3.6.

The figure of merit and maximum of R_e follow from Eqs. (3.37) and (3.39):

$$M \approx 1579$$

$$R_e(\text{maximum}) \approx 10 \times 10^6 \, \Omega$$

Equation (3.35) determines the pole-zero separation as

$$f_A - f_R \approx 5.26 \text{ kHz}$$

If the resonator is operated with load capacitance C_L of 32 pF, the frequency shift from f_R will be, from Eq. (3.16),

$$f_L - f_R \approx 716 \text{ Hz}$$

and the load resistance follows from Eq. (3.19) as

$$R_L \approx 5.3 \, \Omega$$

Normally this will be specified and R_1 will be derived. Using Eq. (3.32) the plate thickness is

$$t \approx 0.332 \text{ mm}$$

At this point little further of an electrical nature may be determined without looking within the resonator enclosure. We will assume the following physical attributes:

Enclosure: HC-6/U
Plate diameter: $D = 14$ mm
Electrode diameter: $D_e = 5.6$ mm

With this additional information we are able to determine

$$d = \frac{D}{t} \approx 42.2$$

which, from Eq. (3.33), yields the estimate

$$\tau \approx 35.7 \times 10^{-15} \text{ s}$$

this compares with 42.4 fs computed from the $R_1 C_1$ product and discloses that with greater attention to mounting the R_1 could be reduced somewhat. Such a specification could result in much greater cost and would be of doubtful utility since C_1, the more important parameter, is almost unaffected by the mounting loss.

The average dielectric permittivity of quartz is

$$\varepsilon \approx 40 \times 10^{-3} \text{ pF/mm}$$

and the static capacitance C_{10} of the electrodes is therefore

$$C_{10} = \frac{\varepsilon \pi D_e^2}{4t} \approx 3 \text{ pF}$$

The stray capacitance due to the mount and enclosure is approximately

$$C_{\text{stray}} = C_0 - C_{10} \approx 2 \text{ pF}$$

and the true capacitance ratio, excluding the strays, is

$$r = \frac{C_{10}}{C_1} \approx 280$$

With the effect of strays excluded, we may estimate the geometrical factor Ψ that measures the distribution of motion under the electrodes compared to the ideal case of uniform amplitude where Ψ is unity:

$$\Psi = \frac{r_0}{r} = \frac{159}{280} \approx 57\%$$

The Ψ value varies with D/t and D/D_e ratios, as well as electrode thickness and plate contour (see Section 3.6). In theoretical calculations of resonator equivalent circuit parameters, ideal values for $\Psi = 1$ are obtained. Then a realistic value of Ψ is obtained from the particular design at hand. This leads to effective values by multiplying the theoretical C_1 values by Ψ and dividing both L_1 and R_1 quantities by this factor. C_0 is unaffected by the motional distribution.

3.2.4.4 Some Standard Units Designed for Operation with Load Capacitor C_L

Table 3.3 lists typical values associated with standard crystal units designed to operate with load capacitance C_L as tabulated. These are described by CR number (see Ref. 3.62) and are referred to in the popular literature as *antiresonant* units. From the discussion in prior sections we see that this designation is not correct; the units operate near f_L and not f_A and should more properly be called *load-resonant* units. In Table 3.3 the units selected are all fundamentals, with the exception of the third overtone CR-33/U entries, and the fifth overtone CR-71/U unit. Standard crystal units have not yet been

Table 3.3 Typical Values of Standard Units Designed to Operate with Load Capacitor C_L

CR Number	Cut	f (MHz)	C_0 (pF)	C_L (pF)	R_L (ohms)	R_1 (ohms)	r	C_1 (10^{-3} pF)	Q (10^3)	M
18	AT	1	7	32	575	388	250	28	14.7	59
27	AT	5	7	32	60	40	250	28	28.4	114
36	AT	10	7	32	24	16	250	28	35.5	142
62	AT	20	7	32	20	13	250	28	21.9	88
29	DT	0.200	11	32	6000	3,324	400	27.5	8.7	22
33	AT	10	12	32	65	34	2,500	4.8	97.5	39
33	AT	25	12	32	17	9	2,500	4.8	147.	59
36	AT	0.800	7	32	625	421	250	28	16.9	68
37	5°X	0.090	6.2	20	5000	2,910	120	51.7	11.8	98
37	5°X	0.250	2.5	20	5500	4,345	120	20.8	7.0	59
38	NT	0.016	7.6	20	110,000	57,860	900	8.4	20.4	23
38	NT	0.100	4	20	90,000	62,460	900	4.4	5.7	6
42	5°X	0.090	6.2	32	4500	3,159	120	51.7	10.8	90
42	5°X	0.250	2.5	32	5000	4,305	120	20.8	7.1	59
57	CT	0.500	7	32	3000	2,022	350	20	7.9	23
68	AT	3	7	32	40	27	250	28	70	281
71	AT	5	4	32	150	119	35,000	0.11	2341	67

established incorporating SC-cuts (see Section 3.3), SAW resonators (see Section 3.4), or miniature resonators (see Section 3.5), but will in the future.

3.2.4.5 Some Standard Units Designed for Operation near f_R

Table 3.4 lists typical values associated with standard crystal units designed to operate near f_R. These are referred to in the popular literature as *series resonant cuts*, indicating that the llator series capacitance is very large ($C_L^{-1} \to 0$). In addition to the circuit values, the operating temperature range is given; the resistance figure is a maximum value over the temperature range. Where a single temperature is listed, the unit is designed to be operated in an oven at that temperature. The frequency deviation column ($\Delta f/f$) lists the rated frequency tolerance, in parts per million, for the unit due to the sum of all environmental disturbances, for example, temperature, vibration, and so on.

In both Tables 3.3 and 3.4, a C_0 value of 7 pF is often given for AT-cut units. This value occurs as a maximum limit of C_0 for most standard AT units and was established at one time along with minimum R_1 values in order to guarantee sufficient Q_x and reactance slope.

3.2.5 Circuit Modifications

This subsection deals with frequently used resonator circuit alterations. The changes are made to the circuit of Fig. 3.5 either by the deliberate addition of

Table 3.4 Typical Values of Standard Units Designed to Operate near f_R

CR Number	Cut	f (MHz)	C_0 (pF)	R_1 (Ω)	r	C_1 (10^{-3} pF)	Q (10^3)	M	T^a (°C)	$\Delta f/f^b$ (10^{-6})
50	NT	0.100	4	60,000	900	4.4	6.0	7	−40 to +70	±20
26	DT	0.300	9	4,000	400	22.5	5.9	15	+75	±20
45	DT	0.455	5	3,300	385	13	8.2	21	−40 to +70	±200
19	AT	0.800	7	520	300	23.3	16.4	55	−55 to +105	±50
157	AT	5	7	37	180	38.9	22.1	123	−55 to +105	±50
79	AT	10	7	30	200	35	15.2	76	−55 to +105	±50
79	AT	20	7	20	250	28	14.2	57	−55 to +105	±50
65	AT	20	7	40	2150	3.3	60.3	28	+75	±10
84	AT	60	7	40	2300	3.0	22.1	10	+85	±20
67	AT	30	7	40	2200	3.2	41.4	19	−55 to +105	±25
81	AT	65	7	40	2100	3.3	18.5	9	−55 to +105	±50
74	AT	125	4.5	25	5800	0.78	65.3	11	+80 to +90	±10
80	AT	50	7	50	6250	1.1	57.9	9	−40 to +90	±20
80	AT	95	7	60	5700	1.2	23.3	4	−40 to +90	±20
82	AT	80	7	50	6000	1.2	33.2	6	−55 to +105	±50

[a] Operating temperature range.
[b] Frequency tolerance over temperature range.

other circuit elements, or because of the intrinsic properties of the crystal unit under consideration.

3.2.5.1 Addition of a Series Element and/or a Parallel Element

Figure 3.7 shows the crystal series reactance plotted over a wide frequency range. The reactance slope is relatively small and constant from somewhat below to the immediate vicinity of f_R, and increases steadily in the inductive region above f_R. As discussed earlier, operation at f_L in the inductive region leads to greater frequency stability because of the higher reactance slope; incidental changes in llator reactance produce smaller frequency changes about f_L than about f_R.

Variable crystal oscillators (VCOs) are an example where operation is desirable in the region of lower reactance slope, that is, at, or somewhat below, f_R. Changes in the osci total series reactance in this region will produce the largest frequency shifts for a given reactance change (greatest "pullability"). Figure 3.13a shows schematically the crystal unit alone, and Fig. 3.13b shows the resonator with series inductor. Looking at the total series reactance of inductor plus crystal as a function of frequency in a narrower range centered about f_R gives the result sketched in Fig. 3.14 for "series inductor." It is the solid "crystal alone" curve translated upward by the inductor reactance, which may be considered constant over this narrow frequency range. If, instead, a

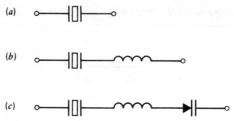

Figure 3.13 (*a*) Crystal resonator. (*b*) Resonator with series inductor. (*c*) Resonator with inductor and varactor in series to adjust the frequency of operation for modulation and compensation applications.

series capacitor is used, the osci total reactance will be reduced at each frequency. This is shown in Fig. 3.14 by the "series capacitor" curve, drawn here for the case of capacitor reactance magnitude equal to that of the inductor. When both inductor and capacitor are placed in series with the resonator, the resulting reactance curve is determined by the sum of the reactances; in particular, if the capacitor is a variable reactance (varactor), as in Fig. 3.13c, operation may be made to take place in the region near f_R where the relatively small reactance slope produces large frequency shifts for given varactor changes made for purposes of frequency modulation, temperature compensation, and so on.

By definition, the osci consists in this case of the crystal network—the resonator plus inductor plus varactor, although for practical measurement purposes it might be advisable to include the varactor with the llator.

Inclusion of an inductor in series with the resonator has, in general, two broad functions. The first, discussed above, is to permit the crystal to operate

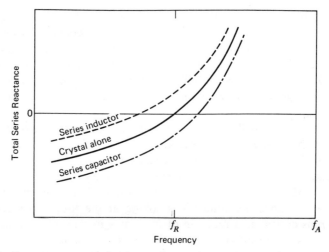

Figure 3.14 Reactance versus frequency for oscis consisting of crystal plus series inductor (short-dashed curve) and crystal plus series capacitor (long-dashed curve).

in the region around f_R where the reactance slope is small, so that large "pullability" is achieved. The second function served by a series inductor is to force the crystal network to look inductive in those applications where it must look inductive, for example, in the Pierce family of oscillators. The resonator may not become inductive at high overtones (see Sections 3.2.3.4 and 5.3.5, item 2).

The heart of the oscillator is the crystal; it should be operated in the frequency region at, or near, f_L. Deficiencies in crystal manufacture, such as poor lapping and plating tolerances resulting in f_L values that deviate from specifications, can be corrected only to a very limited extent by circuit changes, and only at the expense of performance, such as decreased pullability. This situation becomes more critical for units with high r values, especially overtone units. Crystal resonator manufacturing tolerances exert a pervasive influence on the performance of the using oscillator.[3.69]

An inductance is sometimes placed in parallel with the crystal to *antiresonate* C_0. When this is done f_R is unchanged, but the pole is moved upward in frequency, increasing the separation and reducing the apparent capacitance ratio.

3.2.5.2 High-Loss Quartz and Ceramic Resonators

In lower-frequency applications where moderate stability can be used and where cost is an overriding consideration, ceramic resonators are often the choice. They are made of polycrystalline ferroelectric materials in the barium titanate and zirconate families. The ceramic mixture is formed to its desired shape and converted to an artificial piezoelectric by slow cooling from a temperature above the ferroelectric transition point in the presence of a strong electric field. In a manner analogous to the formation of a magnet by aligning molecular dipoles in a strong field, the molecular electric dipoles are oriented. When the high voltage is disconnected from the cooled sample, the preferential orientations of the electric dipoles of the polycrystalline aggregate yield a permanent electric moment which is equivalent to the presence of piezoelectricity.

The equivalent electrical circuit of a ceramic resonator is shown in Fig. 3.15. It is the same as that in Fig. 3.5 except for the addition of the shunt resistance R_0. It comes about from the presence of dc conduction paths around the polycrystalline domains and is intrinsic to the material. The circuit of Fig. 3.15 is sometimes encountered with quartz resonators when there is a leakage path between terminals resulting from faulty construction or damage to the enclosure.

Table 3.5 lists a set of typical ceramic resonator parameters.[3.8] For ordinary applications the frequency range of ceramics ranges from tens of kHz to about 10 MHz. Q_x values vary between 80 and 1000, while r values between 1 and 80 may be achieved. Figure of merit values range from 1 to 12. Because of the high coupling values that may be obtained (low r values), ceramics are suitable

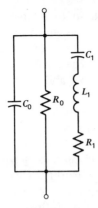

Figure 3.15 Equivalent circuit of a ceramic resonator. In the limit of high dc resistance it becomes Fig. 3.5 for highly insulating dielectric crystals such as quartz.

for medium and wideband low-loss filters. One drawback of these elements is their temperature coefficient of frequency, which is large, being in the range -40 to -80 parts per million per °C. For wideband applications this is not a severe limitation, but for narrowband oscillator application temperature regulation or compensation would more than cancel any cost savings of the element over a corresponding quartz unit.

3.2.5.3 Multimode Resonators

The resonator circuit of Fig. 3.5 is modified by the intrinsic presence of additional modes of motion. These will be treated in Sections 3.3 and 3.6.

3.3 DOUBLY ROTATED CUTS

In the range 0.5 to 200 MHz the resonator of choice is the thickness shear plate. At present the AT-cut accounts for more than 90% of the usage, but for

Table 3.5 Typical Ceramic Resonator Parameters

Quantity	Value	Unit and Remarks
f	452	kHz
C_1	15	pF
r	12.3	C_0/C_1 (often 2 to 30)
R_1	25	Ω
R_0	190	kΩ
Q_x	940	Generally 80 to 1000
$\Delta f/f$	0.003	Temperature stability (20–60°C)
$\Delta f/f$	0.0013	Time stability
$\Delta k/k$	-0.016	per
$\Delta Q_x/Q_x$	0.015	decade
ac depoling	> 1	kV/mm

high- and moderate-precision applications the SC-cut promises to become increasingly popular. The AT-cut is a member of the family of singly rotated cuts; it is formed by aligning the plane of the saw blade with the X-Z crystal axes, and then rotating the blade about the X axis until the angle is reached where the resulting quartz slice has good frequency–temperature properties. Two such cuts exist, the AT-cut with angle $\theta \approx +35°$ and the BT-cut with angle $\theta \approx -49°$. These single rotations are the same as the carpenter's *miter cuts*. The more general case of doubly rotated cuts is obtained by starting, as before, with the blade of the saw aligned with the X-Z crystal axes; the blade is then rotated about its Z axis by an angle ϕ. This first rotation leaves the blade in the X'-Z plane. The second rotation, by angle θ, takes place about the new X' axis. Doubly rotated cuts correspond to the carpenter's *compound cuts*.

Singly rotated quartz cuts with electrodes on the major faces vibrate in thickness shear, with motion along the X axis and acoustic waves travelling in the thickness direction at resonance. In doubly rotated plates three families of motions are in general produced piezoelectrically. These are called the a, b, and c modes, and correspond respectively, to thickness extension, fast thickness shear, and slow thickness shear waves. When $\phi = 0°$ (the singly rotated cuts), the piezoelectric coupling coefficients driving the a and b modes are zero; only the c (slow shear) mode is driven. This mode, with its overtones, was shown earlier to be represented by the circuit of Fig. 3.9. Apart from the static C_0, any mode may be represented by the network of Fig. 3.16a. When three mode families are piezoelectrically excited, the complete circuit is that given in Fig. 3.16b.

In Fig. 3.16b, each mode has the parallel string of series arms as seen in Fig. 3.16a. The three families are distinguished in resonance frequencies, mode strengths, and pole-zero distances as described below.

Consider the fundamental of each mode first. Corresponding to each mode will be a separate piezoelectric coupling factor. By Eqs. (3.34) and (3.36) each mode will have its own particular value of motional capacitance. Because the waves travel at different speeds, each mode will have a different frequency constant (the frequency constant is equal to one-half the wave velocity). From Eq. (3.32), the modal frequencies will differ, and, by Eq. (3.6), the inductances will also be different. The loss is usually unequal for each mode as well; this is reflected in three unequal time constants, which, by Eq. (3.31), leads to three differing resistance values.

When the overtones of each mode are considered, Eqs. (3.40), (3.41), and (3.42) determine the remaining circuit elements in Fig. 3.16.

The SC-Cut

As the angle ϕ is increased from zero (AT-cut), if θ is kept approximately constant, it is found that the resulting cuts have relatively good frequency–temperature behavior, but several other things also take place. First of all, the strengths of the a and b modes increase from zero, while the strength of the c mode decreases. The separation of the three modes change, as well as

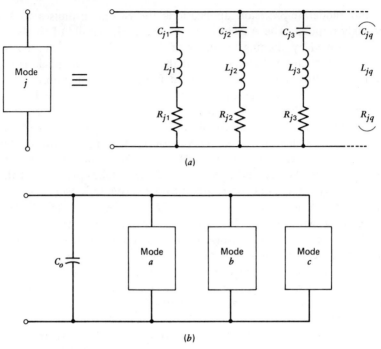

Figure 3.16 (a) Motional arms representing the fundamental and overtones of one mode. The overtone elements are related to those of the fundamental by Eqs. (3.40), (3.41), and (3.42). (b) Complete equivalent circuit for the thickness modes of a doubly rotated plate such as the SC-cut. Each mode has the circuit representation of the (a) part of the figure. The fundamental motional arms differ for each mode; the overtone relations given in (a) hold for each mode separately.

the absolute frequencies. The most important changes that take place, from the standpoint of precision resonators, as ϕ is varied, are the changes in sensitivity to electrode stresses and to thermal transients (see Section 3.9). At $\phi \approx 21.9°$ the c-mode stress sensitivity becomes zero, and this is the orientation of the SC (stress compensated) cut.[3.15]

Figure 3.17 is a theoretically calculated mode spectrograph of an SC-cut.[3.7] For various practical purposes, such as controlling the temperature coefficient, the ϕ and θ angles vary somewhat in design, so the figure is to be regarded as somewhat variable in mode separations. The frequency of each resonance is indicated above the peak, normalized to the fundamental of the c mode. The modal attenuations are much more variable than the frequencies; these can be modified in manufacture, for example, by bevelling or contouring, whereas the calculations are for a flat plate with uniform motion ($\Psi = 1$); these are indicated adjacent to each resonance and are in dB.

The separation between the b- and c-mode fundamental resonances normally ranges from 8.5 to 10%; the mode strengths will likewise vary by ±5 dB

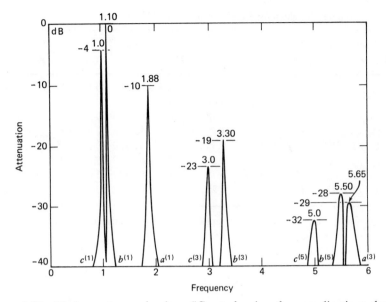

Figure 3.17 Mode spectrograph of an SC-cut showing the complications that arise from the multimode situation. The Nth overtone of mode m is labeled $m^{(N)}$. At the top of each resonance peak is the frequency of the mode, with $c^{(1)}$ taken as unity. The theoretically calculated attenuations are noted at the sides of the peaks; these are in dB. In practice the amplitudes will vary with design, sometimes considerably. The frequency separations will also vary with change in angle ϕ.

with respect to each other. The c-mode SC is about 10 dB weaker than the AT-cut having the same frequency and electrode area.

When designing with SC-cuts, it is of great practical importance to realize that a unit designed for one overtone is not necessarily useful for another overtone. The mode spectrum of resonances and their strengths have to be considered carefully. It may be that a simple mode selection circuit in the llator will suffice for operation on overtones of the stress-compensated c mode. This will be considered further in the circuitry sections.

At present the SC-cut has not been subjected to standardization, but a number of designs at standard frequencies are commercially available. Table 3.6 compares precision AT and SC units.[3.14] Both types are operated in an oven, the AT at its higher-temperature zero temperature coefficient point and the SC at its lower point (see Section 3.7). These units begin to approach the limiting τ_1 values and show attention to this aspect of design; all units are in holders of the HC-6/U size.

SC cuts are more difficult to x-ray than ATs, but this cost factor will diminish considerably with the advent of microprocessor-controlled *smart* x-ray machines currently being developed. From a material point of view, the piezocoupling produces resonances in the SC c mode that have about one-third

Table 3.6 Parameter Values for Precision AT and SC Units

f (MHz)	Cut/ N	C_0 (pF)	R_1 (ohms)	r	C_1 (10^{-3} pF)	Q (10^3)	M	T^a (°C)	$\Delta f/f^b$ (10^{-6})	T_i^c (°C)
5	SC/1	6.8	7	1545	4.4	1033	669	+30 to +75	±10	+100
5	AT/3	3.1	30	5167	0.60	1768	342	+60 to +90	±2	+26
5	SC/3	3.6	85	20,000	0.18	2080	104	+30 to +75	±1.5	+98
10	AT/3	5.0	15	4167	1.2	884	212	+60 to +90	±4	+26
10	SC/3	5.2	55	14,857	0.35	827	55.7	+30 to +75	±1.5	+95
10	AT/3	5.0	12	2273	2.2	603	265	+60 to +90	±4	+26
10	SC/3	5.2	35	6933	0.75	606	87.4	+30 to +75	±3	+95
10	AT/5	3.7	75	18,500	0.20	1061	57.4	+60 to +90	±1.5	+26

[a] Lower turning point range (SC-cuts); upper turning point range (AT-cuts).
[b] Calibration at turning point.
[c] See Section 3.7 for definition.

the pole-zero spacing of the corresponding AT; this increased *stiffness* means greater stability for a given overtone. The multiplicity of modes makes the circuit design more difficult for SCs, but the fact that these resonators are compensated against frequency changes caused by electrode stresses (a large component of long-term aging in ATs) and compensated against thermal transient effects (a contributor to short-term instabilities in ATs) more than makes up for these inconveniences for high-stability applications.

3.4 SURFACE ACOUSTIC WAVE (SAW) DEVICES[3.16, 3.19]

The waves considered in prior sections are called bulk acoustic waves (BAWs) because they travel within the bulk of the resonator and are reflected at the surfaces to form standing waves at resonance. Another species of wave motion exists and is called a surface acoustic wave (SAW). As its name implies, the motion is concentrated at the surface of the crystal; the wave motion decays exponentially with distance from the surface, so that typically 90 to 95% of the energy is contained within one acoustic wavelength of the surface. Corresponding to the BAW AT quartz cut is the SAW ST-cut. This is a singly rotated cut with $\theta \approx +42°$ to 43°, and has favorable frequency–temperature behavior (see Section 3.7).

Excitation of SAWs piezoelectrically is accomplished by means of interdigital transducers (IDTs). These are a series of electrode stripes of alternating polarity, shown schematically[3.19] in the center of Fig. 3.18a. In practice the stripe widths and the gaps between stripes are equal to one-quarter of an acoustic wavelength at resonance. In AT and SC BAW resonators the thickness determines the frequency; with IDT patterns the frequency is determined by

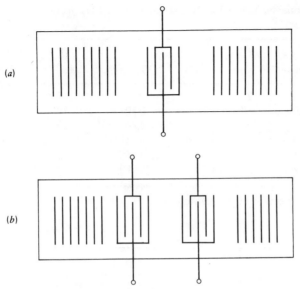

Figure 3.18 (*a*) Schematic representation of a one-port surface acoustic wave resonator (SAWR). Electrode finger spacings, rather than crystal dimensions, determine the resonance frequency. Frequencies well into the GHz region are possible using microelectronic fabrication procedures. (*b*) Schematic representation of a two-port SAWR.

the finger and gap widths. Since these can be made very narrow by photolithographic means, SAW frequencies into the GHz region can be realized on robust substrates.

The earliest SAW devices consisted of two IDT structures separated by hundreds of wavelengths, thus forming an acoustic delay line. These two-port devices were then used as filters by adjusting the IDT finger lengths and separations. Delay-line oscillators were also used, but the circuit element of greatest interest in oscillator design is the SAW resonator (SAWR). Figure 3.18 shows both one- and two-port varieties. The unconnected fingers that appear at the ends of the SAWRs are reflectors that may be either deposited metallic fingers or etched grooves. Either structure presents a slight acoustic discontinuity to the waves, which when summed over many hundreds of reflectors, returns almost all of the energy to the central *cavity* and produces a high Q_x value. Aluminum is the electrode material preferred for SAW application with quartz because its acoustic impedance is within a few percent of that of quartz. At a high power level it has been observed that the electric fields produced near the finger edges tend to make the aluminum migrate, to the detriment of the aging and device properties; doping the aluminum with copper greatly reduces the effect.[3.21]

The device in Fig. 3.18*a* has exactly the same equivalent circuit as that of a BAW resonator, namely, that of Fig. 3.5. The circuit values are functions of the number and spacing of the IDT and reflector fingers and of the material

Table 3.7 Parameter Values for ST-Cut Quartz One-Port Surface Wave Resonators[3.19]

f (MHz)	C_0 (pF)	R_1 (Ω)	r (10^3)	C_1 (10^{-3} pF)	Q (10^3)	M	$\Delta f/f^a$ (10^{-6})
50	10	25	11.8	0.85	150	12.7	± 15
50	20	50	25.1	0.80	80	3.2	± 15
500	2	20	2.5	0.80	20	7.9	± 15
500	4	50	7.5	0.53	12	1.6	± 15
1000	1	100	6.3	0.16	10	1.6	± 15
1000	2	250	25.1	0.080	8	0.32	± 15

^{a}Frequency tolerance over temperature range $0°-+55°C$.

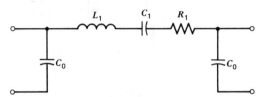

Figure 3.19 Equivalent circuit of the two-port SAWR. Because C_0 does not shunt the motional arm, this device is preferred in many applications over the one-port SAWR, whose equivalent circuit is that shown in Fig. 3.5.

constants of the substrate crystal. Representative values[3.19] are given in Table 3.7.

A two-port SAWR[3.19] is shown in Fig. 3.18b and represented by the circuit of Fig. 3.19. Because the C_0 does not shunt the motional arm as it does with Fig. 3.5, compensation of C_0 by a shunt inductor is not necessary with this device. Element values from Table 3.7 are representative for the motional arm. In both types of SAWRs, the R_1 value is very temperature dependent and must be carefully regarded in practical designs. The frequency tolerance given is for the temperature range 0 to 55°C.

3.5 MINIATURE RESONATORS

In the design of most kinds of traditional resonators, the emphasis is placed on obtaining as uncomplicated a type of vibratory motion as possible, for example, pure thickness shear, or flexure. In some cases coupled motions are employed, as in GT- and SL-cuts, but even here the pattern of the motion is not too complicated. Most recently, driven by the need to obtain resonators of small size for wristwatches, much effort has been placed on the analysis and design of units that employ more complicated coupled motions, for example, coupled thickness shear and flexure together. The coupling affords the possibil-

Table 3.8 Representative Parameters for Miniature Quartz Resonators

Shape	Vibration Type	f Range (kHz)	C_0 (pF)	R_1 (Ω)	C_1 (10^{-3} pF)	Q (10^3)	XYZ Dimensions (mm)	Temperature Coefficient	Ref.
Fork	Overtone flexure	10–1000		10k–40k	0.4–2.5	80–250	$1.0 \times 2.6 \times 0.13$	$a_0 = 0; b_0 = -30$	3.22
Rod	Length extension	1000		2.5k/600	0.3/1.3	200	$0.9 \times 3.4 \times 0.10$	$a_0 = 0, b_0 = -40$	3.23
Bevelled rectangle	DT Contour shear	500–1000			14–4.5		$40 \times 0.8 \times 3.0$		3.24
Rectangular plate	Edge	712.6/1180	0.13/0.080	40k	0.13/0.086	40	$15.2 \times 3.2 \times 0.8$ $9.5 \times 2.0 \times 0.5$	$a_0 = 0, b_0 = -39$ to -64	3.25
Cylindrical strip	AT shear	4194	1.35	53.9	3.45	223	$6.0 \times 0.4 \times 1.6$	$a_0 < \pm 0.06$	3.26
Rectangular plate	GT coupled extension	2300		110	2.5	200–300	$1.41 \times 1.47 \times 0.07$	$a_0 = b_0 = 0; c_0 = 4.5$	3.27
Fork	Flexure	200	1.0	3k	1.0	300	$0.9 \times 2.9 \times 0.16$	$a_0 = 0; b_0 = -15, -30$	3.28
Fork	Flexure	100 (50–200)	1.0	10k	1.0	200	$0.9 \times 4.7 \times 0.10$	$a_0 = 0; b_0 = -10, -20$	3.29
Rectangular plate	Width extension	1049			1.2	200–300	$2.7 \times 4.0 \times 0.20$	$a_0 = b_0 = 0; c_0 = +50$	3.30
Fork	NT flexure	32.768	1.05	28k	1.76	>60	$1.1 \times 4.7 \times 0.10$		3.31

Table 3.9 Typical Parameter Values for Thin-Membrane Plate Resonators of Quartz and Lithium Tantalate

Cut	N	f (MHz)	C_0 (pF)	R_1 (ohms)	r (10^3)	C_1 $(10^{-3}$ pF)	Q (10^3)	M	C_{d_1} (pF)
AT	3	450	0.6	250	10.7	0.056	24	2.24	0.4
BT	3	465	0.7	116	18.4	0.038	76	4.13	0.8
AT	1	100	1.34	30	1.8	0.74	67	37.2	
AT	3	450	1.83	250	30.	0.061	24	0.80	
AT	5	500	1.27	550	60.	0.021	25	0.42	
BT	1	575	1.25	20	3.0	0.42	30	10	
BT	3	470	1.69	90	35.	0.048	80	2.29	
BT	5	475	1.65	420	70.	0.024	36	0.51	
LiTaO$_3$	1	240	1.58	150	0.10	15.8	0.30	3.00	
LiTaO$_3$	3	720	1.90	150	0.75	2.53	0.75	1.00	

ity of obtaining good temperature behavior along with useful motional parameters.

Many of the miniature resonators employ novel support structures that are part of the resonator design. Most are produced by mass-production techniques employed by the microelectronics industry, such as photolithography. The units are sized to fit into tubular watch crystal enclosures (2 mm diameter, 6 mm length) or tiny ceramic packages (1.5 × 2.4 × 6.7 mm).

Table 3.8 contains parameter values on a representative set of miniature resonators. The next-to-last column lists the temperature coefficients (see Section 3.7). The units are a_0 in $10^{-6}/°C$, b_0 in $10^{-9}/°C^2$, and c_0 in $10^{-12}/°C^3$.

In the category of miniature resonators are the thin-membrane plate vibrators.[3.32] These are ordinary crystal plates that have had the material in the central part of the plate removed by ion or chemical milling to form a thin membrane that is supported along its periphery. The membrane vibrates in thickness shear like a conventional plate vibrator. Table 3.9 lists parameter values for AT- and BT-cut quartz and for lithium tantalate membranes.[3.33] The equivalent circuit for the first two entries is that of Fig. 3.4 with $C_{d_1} = C_{d_2}$.

3.6 UNDESIRED RESPONSES

Finite crystal plates in general have a complicated mode spectrum. One often strives to have a simple distribution of motion and as little coupling as possible and so obtain a clean spectrum. Recipes for the design of AT and BT plate resonators have been worked out that minimize the presence of undesired responses; these criteria are known as *energy trapping* rules.[3.34] These rules require a relationship between the electrode and plate geometry to be satisfied. If a plate resonator is designed without trapping, then the resulting spectrum

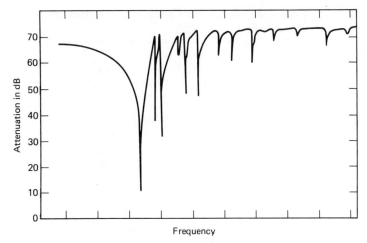

Figure 3.20 Mode spectrograph of a plate resonator. The unwanted modes placed above the main resonance make this unit unfit for filter application, but the 20-dB suppression of the strongest unwanted mode is usually sufficient to guarantee proper operation in an oscillator. Specifying units suitable for filters in oscillator applications would result in greatly increased cost.

will appear like that in Fig. 3.20; often the responses above the main, desired mode will be as strong, or stronger, than the main mode. If the unwanted responses are 10 dB, or more, weaker than the main response, then for oscillator applications the unit is generally acceptable. Certain obscure effects are dealt with in Chapter 18 where the unwanted modes pose a problem. Otherwise, such responses need to be minimized only for filter application. For use in oscillators the required amount of unwanted mode suppression may be simply obtained by contouring[3.35] or bevelling the plate edges and by keeping the ratio $D/D_e > 2.5$, as is the case in the illustrative example in Section 3.2.4.3. The equivalent circuit for the unwanted modes is that of Fig. 3.16a, but with much more complicated relations between the parameters of the motional arms than Eqs. (3.40), (3.41), and (3.42).

3.7 STATIC TEMPERATURE EFFECTS

When temperature is slowly varied it is found that the frequency shift due to the variation can be expressed in the form:

$$\frac{\Delta f}{f} = a_0(T - T_0) + b_0(T - T_0)^2 + c_0(T - T_0)^3 \qquad (3.47)$$

T is the temperature variable and T_0 is a reference temperature, often taken as 25°C. The coefficients a_0, b_0, and c_0 are the first-, second-, and third-order

Table 3.10 Higher-Order Temperature Coefficients of Quartz Cuts

Cut	b_0 $(10^{-9}/°C^2)$	c_0 $(10^{-12}/°C^3)$
AT	0.4	110
BT	−40	−128
SC[a]	−12.3	58.2
ST	−34	
CT	−57	−161
DT	−19	75
5°X	−25	
GT	0.2	−26

[a] c mode; for mode b, $a_0 \approx -26 \times 10^{-6}/°C$.

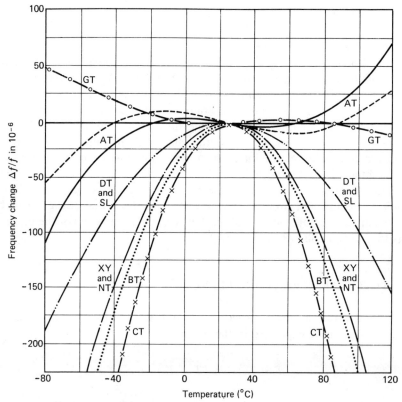

Figure 3.21 Frequency–temperature characteristics of quartz cuts when temperature is changed slowly. The curves may be shifted by small changes in orientation angles.

3.7 Static Temperature Effects

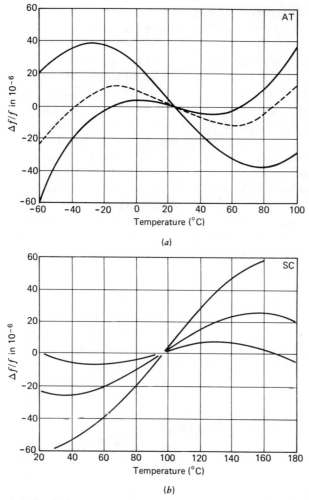

Figure 3.22 (a) Static frequency–temperature curves for members of the AT-cut family. Changes in angle θ produce the changes in maximum and minimum points permitting an orientation to be found that minimizes frequency change over a specified temperature range. The dotted curve is the *optimum* AT-cut, with variation of only ±12 ppm from −50° to +100°C. (b) Static frequency–temperature curves for members of the SC-cut family. The inflection temperature is approximately 96°C.

temperature coefficients of frequency, and are usually expressed in units of $10^{-6}/°C$, $10^{-9}/°C^2$, and $10^{-12}/°C^3$, respectively. The temperature coefficients a_0, b_0, and c_0 and the thickness modes a, b, and c share a somewhat common notation that arose for historical reasons, but no confusion should arise. The cuts of interest for oscillator application are those having either zero or small a_0 values, and reasonably small b_0 and c_0 values. Some examples occur in the next-to-last column of Table 3.8.

If T_0 is taken where the frequency–temperature (f–T) slope is zero, then $a_0 = 0$, and the curves are determined by the b_0 and c_0 values. Table 3.10 lists values, for a number of quartz cuts, of the quantities b_0 and c_0.[3.37] Frequency–temperature graphs are given in Figs. 3.21 and 3.22.

When T_0 is taken such that b_0 is zero, the temperature is referred to as the inflection temperature T_i; an example appears in the last column of Table 3.6. The inflection temperature has been used for T_0 in Fig. 3.22. In both parts of this figure is seen the effect on the cubic curve of changes in the angle θ. By adjustment of θ during manufacture the f–T curve may be controlled to provide the minimum frequency excursion over a specified operating temperature range. The normal low and high limits on the f–T curves are shown in the figure for the AT and SC families of cuts. Changes in orientation angle provide means to adjust the cuts having parabolic curves in Fig. 3.21; the turn-over temperature can be moved 20 or 30°C for special applications. The dashed curve in Fig. 3.22a shows the *optimum* AT-cut, which has the least frequency change (about ± 12 parts per million) over the range -50 to $+100$°C.

Figure 3.23 relates the frequency tolerance, for AT-cuts, to be expected in manufacture when the angle tolerance $\Delta\theta$ is specified. The curve labeled A covers the temperature range -55 to $+105$°C. If an angle tolerance $\Delta\theta$ due to x-raying, lapping, polishing, and so on, is known for a particular product line, say ± 3 min of arc, and if the desired operating range is that specified by A, then the frequency tolerance of the finished units will be within ± 30 ppm over that temperature range. Tightening the angular tolerance to ± 1 min of arc will

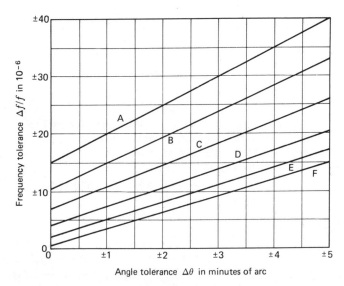

Figure 3.23 Frequency tolerance obtained with units manufactured with a specified angle tolerance $\Delta\theta$. The curves indicate the operating temperature range in °C as follows: A = -55 to $+105$; B = -45 to $+95$; C = -35 to $+85$; D = -25 to $+75$; E = -15 to $+65$; F = -5 to $+55$.

reduce the $\Delta f/f$ tolerance to ± 15 ppm. The practical consequences of this graph are obvious. Specifying a certain frequency tolerance that is too tight and/or specifying an operating temperature range that is too broad will require units that are costly to produce.

By means of Eqs. (3.16) and (3.36) one may show that the difference in the slopes of the f–T curves of a resonator operated at f_R and at f_L depends upon the temperature coefficient of the piezoelectric coupling factor k. This effect is non-negligible for quartz and must be taken into account in the design of TCXOs.[3.36]

3.8 AGING EFFECTS

Slow changes in frequency with time are referred to as resonator aging. The principal causes of aging are contamination within the enclosure that is redistributed with time, slow leaks in the enclosure, mounting and electrode stresses that are relieved with time, and *oilcanning*, where the enclosure is stressed by changes in atmospheric pressure outside the envelope. Changes in the quartz are usually negligible for most applications.

A typical aging curve is given in Fig. 3.24.[3.38] The curve marked A ages upward in frequency in approximately logarithmic fashion. The positive aging

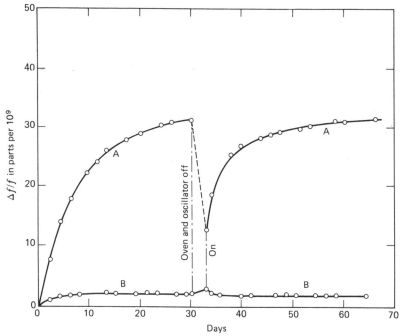

Figure 3.24 Frequency changes in precision quartz crystal units due to interruptions in oscillation and oven control. A: solder bonding and glass encapsulation. B: thermocompression bonding and high-temperature processing.

shown is the most common form; it might arise from contamination on the vibrator surfaces during inoperation being redistributed to the other portions of the enclosure when the crystal is in motion.[3.39] This is seen to be probable from the response to turn-off and subsequent restart. The cleaner crystal of curve B has a much reduced aging. Negative aging is usually caused by enclosure leakage.

Aging is also a function of drive level of the crystal unless contamination and stresses have been minimized by use of high-temperature high-vacuum bakeout and sealing and use of SC-cuts.

In high-volume production aging rates less than 1 part in 10^8 per day can be achieved; for high-precision units the figure can be improved to 5 parts in 10^{11} per day on a low-yield selective basis.

3.9 ENVIRONMENTAL EFFECTS (EXCLUDING STATIC TEMPERATURE EFFECTS)

Figure 3.25 shows an idealized curve of frequency versus time for a crystal resonator subjected to a variety of environmental disturbances.[3.40] A thermal transient occurs at t_1; mechanical accelerations in the forms of vibration, shock, and turning the crystal over in a gravitational field (tip-over) occur between t_2 and t_7. At t_5 the oscillator is turned off, causing another thermal transient, and at t_6 it is turned back on again. Superimposed on these disturbances is the long-term aging slope discussed briefly in Section 3.8. If the curve is sufficiently magnified, it will be found to contain noisy fluctuations (see Section 3.13) about the mean frequency.

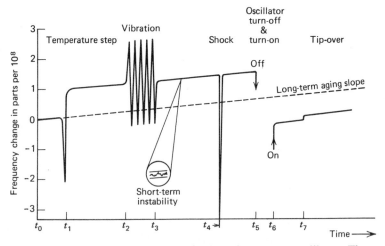

Figure 3.25 Idealized frequency–time behavior of a quartz oscillator. The greatest contributors to frequency change are thermal transients, accelerations, cessation, and reestablishment of vibration, and long-term aging.

3.9.1 Thermal Transients[3.43–3.45]

The f–T curves in Section 3.7 assumed slow temperature variations. If $dT/dt = \dot{T}$ is sufficiently large to produce thermal gradients across the resonator, then it is found that the curve of Eq. (3.47) must be modified to read

$$\frac{\Delta f}{f} = a_0 \Delta T + b_0 \Delta T^2 + c_0 \Delta T^3 + (\tilde{a}\,\Delta T + \hat{a})\dot{T} \qquad (3.48)$$

where $\Delta T = \Delta T(t) = T(t) - T_0$. In Eq. (3.48), $\tilde{a}$ and $\hat{a}$ depend upon material constants of quartz as well as the resonator design. For tests involving a ramp function of temperature, the $\tilde{a}$ coefficient has the effect of changing the apparent angle of orientation of the resonators; that is, it rotates the S-shaped cubic curve. The $\hat{a}$ coefficient translates the cubic curve up or down in frequency. When an abrupt temperature change is encountered, as with oven turn-on, AT-cuts will behave as seen in Fig. 3.26, producing an overshoot and slow approach to equilibrium.[3.41] SC-cuts, on the other hand, experience no such transients and simply follow the static f–T curve irrespective of $\dot{T}$.

The consequences of this difference in behavior are striking when one has to adjust the oven of a high-precision unit to the turnover (zero-slope) point of the f–T curve. Even minute changes in temperature will send the AT-cut oscillator's frequency far from where the static curve would predict, and many oven time constants have to be waited for the frequency to settle finally at the

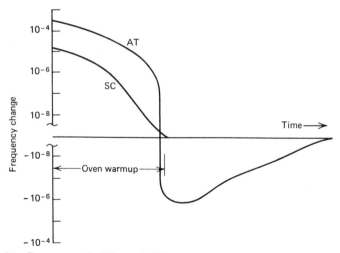

Figure 3.26 Response of AT- and SC-cuts to a thermal transient, due to oven warmup. Thermal gradients in the AT produce a dramatic overshoot and a long approach to steady state; the thermally compensated SC approaches final frequency in a time dictated solely by the oven parameters.

static point. Creeping up on the turn-over temperature in this manner is exceedingly time-consuming and costly.

Moreover, when the oven has been set at its reference point, it can only maintain the temperature within a finite range. If this range is used in Eq. (3.47), the $\Delta f/f$ so predicted will be found to be many times smaller than that observed. The $\tilde{a}$ and $\hat{a}$ terms can cause substantial frequency departures in high-precision AT oscillators.

In SC-cut resonators the effect is nearly absent, and the setting the oven temperature to the turn-over point is a simple rapid procedure.

3.9.2 Thermal–Frequency Hysteresis[3.42]

Figure 3.27 gives an example of thermal hysteresis encountered when cycling a quartz resonator over an extended temperature range. The effect is a function of the temperature range; its causes are poorly understood. It should be borne in mind that the effect varies from unit to unit and is very difficult to model for the purpose of compensation by microprocessor programming.

3.9.3 Acceleration Effects

Shock, vibration, and change in attitude (tip-over) are acceleration effects. For a given design and direction of applied acceleration, the frequency shift is

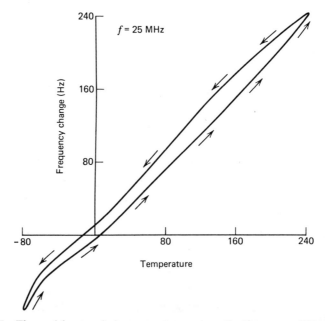

Figure 3.27 Thermal hysteresis in quartz resonators. Cycling over wide temperature extremes produces hysteretic behavior that approaches a semireproducible loop after a number of repetitions; changing the temperature excursions modifies the loop.

3.9 Environmental Effects (Excluding Static Temperature Effects)

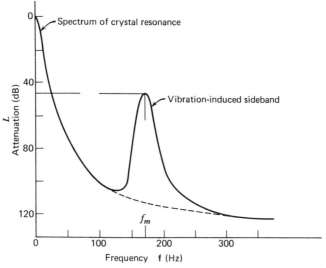

Figure 3.28 Crystal resonance spectrum showing vibration-induced sideband at modulation frequency f_m. Sideband level, referred to the crystal resonance, is a measure of the crystal's acceleration sensitivity.

proportional to the acceleration $\bar{a}$:

$$\frac{\Delta f}{f} = \bar{a}\gamma \qquad (3.49)$$

$\bar{a}$ is expressed in g units and γ is the acceleration sensitivity coefficient. Values for γ normally run a few parts in 10^9 per g for the AT-cut. Recent work has led to units with γ values in the 10^{-10} range, but with relatively labor-intense methods.

If a resonator operating at its resonance frequency is subjected to a sinusoidal vibration at frequency f_m, where $f_m \ll f_R$, then sidebands will be produced at the carrier frequency $\pm f_m$. An example is shown in Fig. 3.28. A measurement of the sideband level L, in dB, normalized to the carrier level, determines γ from

$$\gamma = \left(\frac{2 f_m}{\bar{a} f}\right) \times 10^{L/20} \qquad (3.50)$$

3.9.4 The Polarization Effect

A dc voltage impressed on the electrodes of an AT-cut will produce a very small frequency shift; for an SC-cut the effect is considerably larger and may be utilized for compensation purposes. Expressed in terms of a constant $\bar{e}$, we obtain

$$\frac{\Delta f}{f} = \bar{e}E \qquad (3.51)$$

E is the vacuum electric field strength. For AT-cuts, $\bar{e} = 0.04$ and for SC-cuts, $\bar{e} = 2.3$; the units of $\bar{e}$ are 10^{-9} mm/V, and of E are V/mm. If 10 V dc is applied to a typical 5-MHz fundamental SC-cut, the normalized frequency shift is approximately 6×10^{-8}. In some circuits static electricity can accumulate across the crystal terminals, leading to a frequency shift caused by the polarization effect. If this charge fluctuates with time, an added contribution to noise will then be present (see Section 3.13). By shunting the crystal by a high resistance, this influence can be overcome.

3.10 DRIVE LEVEL EFFECTS

The apparent resistance of a resonator is often found to be a function of the crystal current I_x. The effect is sketched in Fig. 3.29. The motional capacitance C_1 may also vary with drive, but its variation is much less than that of R_1.

3.10.1 Second Level of Drive[3.43, 3.48]

At low drive levels, the resistance R_1 of many crystals increases appreciably, and very often abruptly. Examples are given in Figs. 3.30a and 3.30b. In Fig. 3.30a, as the voltage across the crystal is increased from 0 to B, the current

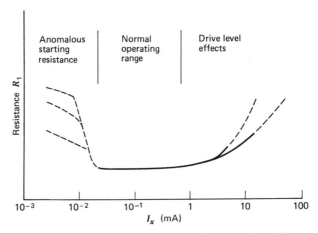

Figure 3.29 Schematic representation of the variation of resonator motional resistance with crystal current. At very low values anomalously high resistances are sometimes encountered; at high drive levels nonlinear effects increase the resistance.

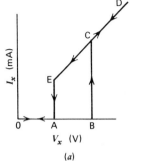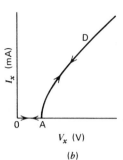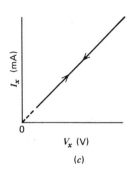

Figure 3.30 (*a*) Anomalous starting resistance occurring as abrupt transitions leading to a hysteresis loop. (*b*) Nonlinear starting resistance occurring as a single-valued function of crystal voltage V_x. (*c*) Normal linear crystal resistance extending down to very low values of V_x.

remains zero. It rises abruptly to point C, where its value is consistent with the rated R_1. Further increases in V_x produce no change in R_1, nor do decreases until point E is reached when I_x again falls to zero. The hysteresis-like effect is called *second level of drive*.[3.46] In some resonator units the behavior shown in Fig. 3.30*b* is encountered where the curve shows a continuous increase in R_1 with decrease in V_x from point D to point A, and no crystal activity for V_x values below this.

This effect is almost exclusively due to the surface preparation of the resonator. It becomes apparent at current densities below about 10 $\mu A/mm^2$, which, for units in HC-6 and HC-18 enclosures, and for frequencies in the 1–30-MHz range, works out to mA currents and μW power levels.

It is obvious that the presence of this effect can produce serious starting and operating effects in the oscillator. Furthermore, the effect can be insidious and not show up in tests on a newly made resonator, only to appear after a period of storage.

Production of units without this defect requires attention to surface cleanliness. Maintaining the resonator surface free from contaminating particles of lapping and polishing compound and use of chemical etching to remove microcracks effectively removes the phenomenon and results in units with the characteristic shown in Fig. 3.30*c*. The effect is thought to be much less severe with SC-cuts than with AT. It may be dealt with in the specification of the unit.

3.10.2 Amplitude–Frequency Effect[3.49, 3.50]

At higher levels of drive the resonance curve ceases to be symmetric and bends to one side as shown in Fig. 3.31.[3.47] The AT-cut behaves as a *hard spring* and bends toward higher frequencies; the BT-cut acts like a *soft spring* and bends toward lower frequencies. The locus of the maxima of the resonances is given

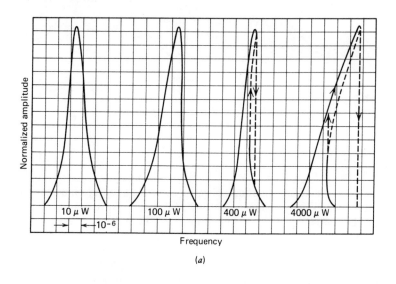

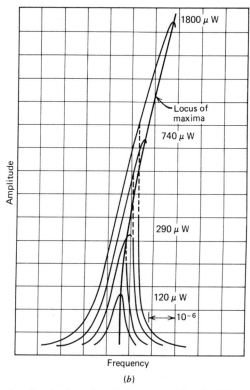

Figure 3.31 (a) Amplitude–frequency effect. As drive level is increased the crystal resonance curve becomes a multiple-valued function of frequency. Curve heights have been scaled to the same value. (b) Amplitude–frequency curves superimposed, showing the locus of maxima. The resonator is an AT-cut at 1 MHz.

by

$$\Delta f/f = aI_x^2. \quad (3.52)$$

For conventional AT designs at 5 and 100 MHz, a is approximately 0.20 to 0.25/A^2. Values of a between 0.02/A^2 and 0.05/A^2 have been reported,[3.49, 3.50] for doubly rotated cuts, with ϕ angles in the range ±2° of the SC-cut, at both third and fifth overtones. With planoconvex designs, a decreases with decreasing curvature and is smallest in flat plates. The effect depends upon geometrical design, material, overtone, and mode of motion. The lowest a value measured to date on a BAW resonator is 0.004/A^2 on a third overtone, doubly rotated plate having $\phi = 20°$. ST-cut SAW devices have even lower values; a value of $a = 0.0011/A^2$ at 110 MHz has been measured.

The amplitude–frequency effect is more correctly characterized by use of current density rather than I_x in Eq. (3.52); when put on this basis the effect becomes perceptible at densities of roughly 50 $\mu A/mm^2$ for AT-cuts. Since in most designs electrode area decreases with increasing frequency, keeping I_x constant for units of increasing frequency would mean increasing the current density. In order to keep the maximum frequency change $\Delta f/f$ constant, I_x should therefore decrease in higher-frequency units.

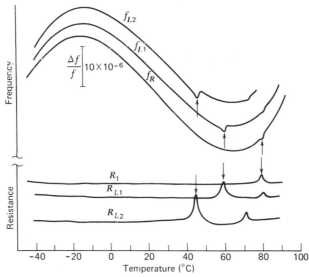

Figure 3.32 Activity dips in the frequency–temperature plots of an AT-cut when operated without and with load capacitors. Crystal parameters: $f = 20$ MHz; $C_0 = 2.1$ pF; $R_1 = 3\ \Omega$; $C_{L1} = 35$ pF; $C_{L2} = 22$ pF; $I_x = 2$ mA. All curves have been vertically displaced for clarity. The activity dip temperature is a function of C_L indicating that the dip is caused by coupling to a flexure mode with a large negative temperature coefficient.

3.11 ACTIVITY DIPS[3.51,3.70]

An example of an activity dip in the f–T curve of an AT resonator is shown in Fig. 3.32. It appears in both the frequency and in the apparent crystal resistance. It is a potential source of difficulty for oscillator applications and is dealt with in the specification of the unit. No activity dips have yet been observed in SC-cuts.

Slow changes in R_1 with temperature are often encountered; these are not activity dips. In the resonator specification the maximum value of resistance over the temperature range should be specified, not simply the resistance at a reference temperature. In oscillators that are operated over a range of temperatures, changes in crystal resistance with temperature will produce drive level output power changes.

3.12 RADIATION EFFECTS[3.52–3.55]

Two types of resonator radiation effects are usually distinguished: (1) collision processes due to neutrons and α-particles, and (2) ionizing radiation effects due to x-rays, gamma rays, and electrons.

3.12.1 Neutrons

Displacement damage produces permanent frequency shifts in quartz resonators because it modifies the effective elastic constants. Since the effect is very structure sensitive, the type and quality of the quartz material is extremely important. Permanent displacement damage due to neutrons shifts the resonator frequency at the approximate rate $\Delta f/f \approx 10^{-21}/\text{cm}^2/\text{neutron}$.

3.12.2 Ionizing Radiation

Most resonator radiation sensitivity work has been done in this area. The ionizing radiation effect depends upon (1) the amount of impurities present in the quartz and (2) upon the temperature profile created. Aluminum often substitutes for silicon in quartz, where a proton, or ion of lithium, sodium, or potassium compensates for the additionally needed charge. The radiation makes the monovalent ions migrate along the relatively open Z-axis channels within the quartz, and produces frequency changes.

Pulsed ionizing radiation produces frequency changes in quartz resonators, even in the absence of impurities, because thermal gradients are set up in the quartz, leading to changes in the nonlinear elastic constants. The effect depends on the crystal cut, being negative for the AT-cut, but is almost insensitive to the type (natural or cultured) of quartz used, and to the quality, although Q degradation and even cessation of the oscillation has been observed when impure quartz is used. Temperature gradient effects on resonator frequency, arising from pulsed ionizing radiation, can be largely compensated by utilizing SC-cut resonators. These will, however, still be subject to frequency shifts due to impurities, if present.

Steady-state radiation effects depend strongly on defects in the material. Just as annealing has been used to increase the Q of quartz, a combination of high temperatures (from 350 to 550°C) and strong electric fields (referred to as *sweeping*) has been found to produce material with superior purity and thus superior radiation hardness. Swept Z-growth cultured quartz is the most radiation tolerant; this is a consequence of the removal of sodium, lithium, and potassium ions. The sweeping process (done in a vacuum) has also been shown to produce material having fewer etch channels than nonswept cultured quartz. Etch channels degrade the strength (shock resistance) and serve as repositories of etchant that can produce long-term aging.

Typical values for the frequency change due to ionizing radiation are

$$\frac{\Delta f}{f} \approx \begin{cases} \text{natural quartz: } 10^{-11}/\text{rad} \\ \text{swept cultured quartz: } < 10^{-12}/\text{rad} \end{cases}$$

3.13 NOISE IN QUARTZ CRYSTALS[3.43, 3.44, 3.57, 3.59]

This topic is just beginning to receive the attentions of researchers to the extent it deserves. For some years it has been known that one source of noise arising from the crystal itself is the thermal noise associated with its temperature of operation. The equivalent noise resistance of a quartz crystal is the same as its effective series resistance R_1 at the using temperature. This leads to a noise voltage of [3.57]

$$v(\text{noise}) = \sqrt{4k_B T B} \tag{3.53}$$

where $k_B T$ is 5×10^{-21} W-s at room temperature, and B is the bandwidth in which the voltage is measured. The crystal can be considered at resonance to be a narrowband filter for the additive wideband noise voltage of Eq. (3.53).

In addition to this noise source, it is suspected that stress relaxation in the electrodes, as well as correlations with vibrations, via the coefficient γ [see Eq. (3.49)], and correlations with temperature fluctuations via the coefficients $\tilde{a}$ and $\hat{a}$ [see Eq. (3.48)] are operative to produce noise effects. An oven improvement can lead to considerably better noise performance when using resonators with non-negligible $\tilde{a}$ values.[3.60]

The short-term stability is mostly limited by the thermal noise and can be improved by increasing the crystal drive; however, the circuit noise is usually very much greater than the crystal noise unless one deliberately overdrives the resonator. If the crystal drive level is too high, however, the flicker (f^{-1}) noise and the higher-order noise components ($f^{-2}, f^{-3}, f^{-4}, f^{-5}, \ldots$) all become much higher; this includes the amplitude-to-frequency noise conversion via the coefficient a [see Eq. (3.52)].

Static charges may build up across the resonator in certain circuit configurations, and if these fluctuate, an additional contribution to noise will be produced; a high shunting resistance will prevent this (see Section 3.9.4).

3.14 GENERAL SPECIFICATIONS

A specification is a set of basic ordering information. The simplest such set for crystal resonators might consist of only three numbers: holder type, nominal frequency, and C_L. Such a specification, or *spec*, would be appropriate only for the least-accurate applications. For the resonators required for the designs contained in this book, it is necessary to specify explicitly the crystal parameters mentioned as inputs in the design algorithms. These are the cut, overtone, R_L, C_L, C_1, and C_0.

In addition to the above, several other important characteristics are in general called for; a representative list might be as follows:

Description

1. Type of crystal enclosure.
2. Nominal frequency.
3. Overtone order, N.
4. Marking.

Operating Circuit Conditions

1. Crystal enclosure grounded or not.
2. Load capacitance C_L.
3. Level of drive.
4. Reference temperature.
5. Operating temperature range.

Frequency Tolerance

1. Overall tolerance.
2. Adjustment tolerance and tolerance over the temperature range.

Electrical Characteristics

1. Parallel capacitance C_0.
2. Motional capacitance C_1.

3 Rated resistance at C_L, R_L.
4 Insulation resistance.
5 Effect of drive level.
6 Unwanted responses.
7 Aging.
8 Permissible deviation of frequency from a smooth function over the operating temperature range.
9 Permissible deviation of resistance from a smooth function over the operating temperature range.
10 Absence of high starting resistance effects.

Mechanical and Climatic Characteristics

1 Robustness and tensile strength of terminations.
2 Flexibility of wire terminations.
3 Bond test of pin terminations.
4 Soldering ⎫ Alternatively, an aging specification
5 Sealing ⎭ can be given.
6 Vibration.
7 Acceleration.
8 Shock.
9 Operable temperature range.
10 Storage temperature range.
11 Temperature cycling.

There is a general need within the crystal industry today to increase the absolute accuracy of parameter measurements. Because of this, the engineer should fully specify those characteristics that are important in his application. This may include a description of the measurement procedure and possibly specification of an apparatus for the determination of these parameters.

3.15 MILITARY SPECIFICATIONS

Resonators meeting military specifications are required to undergo rigorous inspection and are designed for ruggedness and quality. The user begins with a military standard, or MIL-STD. This is a document for the equipment designer to pick out standard items with desired parameters, but is not used for procurement. It contains only those items recommended for new designs. For crystal units the MIL-STD is given in Ref. 3.61.

Then the user progresses to the military specification, or MIL-SPEC. This is a procurement document for use in purchasing items. It lists parts numbers, parameters, and tolerances. The crystal specification is given in Ref. 3.62; crystal holders are given in Ref. 3.63.

Finally, one requires a qualified products list, or QPL. This lists those manufacturers qualified to supply the specified items meeting the military requirements.

Crystal resonators are grouped in Federal Supply Classification (FSC) Class 5955: Piezoelectric Crystals.

The agency responsible for military specifications is: Defense Logistics Agency, DESC, Dayton, Ohio 45444. The documents in Refs. 3.61, 3.62, and 3.63 may be obtained through this agency or from Naval Publication and Form Center, 5801 Tabor Ave., Philadelphia, Pa. 19120.

To date, there are no military specifications governing doubly rotated cut BAW resonators (specifically, the SC-cut), SAW resonators, or miniature resonators.

3.16 NATIONAL AND INTERNATIONAL SPECIFICATIONS

The principal nonmilitary sources of crystal resonator specifications and standards in the United States are (1) The IEEE, (2) EIA, and (3) ANSI.

Pertinent IEEE standards are given in Refs. 3.64 and 3.65. They may be obtained from IEEE Service Center, 445 Hoes Lane, Piscataway, N.J. 08854.

The EIA has specifications dealing with holders and sockets, production tests, application information, and cultured quartz. They may be obtained from Electronic Industries Association, 2001 Eye St., NW, Washington, D.C. 20006.

ANSI interfaces with international standards groups such as the International Electrotechnical Commission (IEC). Typical IEC publications are given in Refs. 3.66, 3.67, and 3.68. These IEC documents may be obtained in the United States from American National Standards Institute, 1430 Broadway, New York, N.Y. 10018, or directly from International Electrotechnical Commission, 1-3 rue de Varembé, Geneva, Switzerland.

To date, no standards exist to cover doubly rotated cut BAW resonators (specifically the SC-cut), SAW resonators, or miniature resonators.

4

Other Resonators

Chapter 3 treated piezoelectric resonators in detail. This chapter will briefly describe other types of resonators.

4.1 INDUCTANCE CAPACITANCE CIRCUITS (*LC* RESONATORS)

This resonator is used in low- and medium-performance oscillators and is probably the most popular. It is useful from very low to extremely high frequencies. The application of the LC resonator is covered in detail in Chapters 5 to 10.

The physical form of the resonator is principally determined by the oscillator's frequency and function. As the name indicates, it consists of a capacitor and inductor. The inductor and capacitor may be combinations of fixed and/or variable elements depending upon whether the oscillator is fixed or variable frequency.

A tremendous amount of information on the construction and performance characteristics of inductors and capacitors suitable for use as oscillator resonators has been published and is readily available. This information will not be repeated here.

The most important performance characteristics are:

1. Stability as a function of time and environmental conditions.
2. The Q of the resonator which limits the operating Q of the oscillator. The Q of the resonator approximates the Q of the inductor since the Q of the capacitor normally greatly exceeds that of the inductor. In order to increase the resonator Q, considerable study has been made of their use at cryogenic temperatures where the Q is increased by many orders of magnitude.

In designing *LC* oscillators, it is important to choose values of *L* and *C* and configure the circuit so that the contributions of the transistor to the operating frequency is very small compared to that of the resonator.

4.2 ELECTROMECHANICAL RESONATORS

The performance of this class of resonators is intermediate between that of the *LC* and piezoelectric resonators. It is mostly of historical interest since it has been almost completely superceded by the crystal oscillator, with frequency dividers if necessary, and *RC* type of oscillators where the superior performance crystal oscillator is not required.

Typical examples of this class of resonators are the vibrating reed and the tuning fork equipped with driving and pickup coils. The frequency is determined by the dimensions and material of the device. In order to simplify the driving and pickup processes, the material usually has magnetic properties.

The oscillator circuit block diagram is that of Fig. 1.16a where the resonator is the β network.

4.3 THE MAGNETOSTRICTION RESONATOR

The magnetostriction effect is the expansion or contraction of magnetic materials as the result of magnetization, and the converse change of magnetization as the result of strain. This effect may be used as the basis for a resonator. The performance approaches that of low-performance crystal resonators. The practical range of oscillation frequency is within the audio and supersonic frequency bands. The oscillator block diagram is similar to that of the electromechanical resonator described above. The magnetostriction resonator is only of historical importance as it has been supplanted by the crystal oscillator, combined with frequency dividers if necessary.

5

The Pierce, Colpitts, Clapp Oscillator Family

5.1 INTRODUCTION

The types of oscillators mentioned in the chapter title are members of a large family of oscillators which have essentially the identical mathematical model, but which differ in practical realization.

Figure 5.1 shows the elementary schematic diagrams of these oscillators. It will be noted that they are quite similar, but differ in two respects:

1. The method of feeding the dc power and biasing the transistor.
2. The method of delivering the RF power output to R_L.

These differences appear minor, but they seriously influence the performance and the selection of the circuit to be used in a particular application. The effects of items 1 and 2 will be considered in detail in later chapters. However, it is worthwhile mentioning at this point that, except in unusual circumstances, the Pierce is preferred over the Colpitts, and the Clapp is rarely used for crystal oscillators.

Figure 5.2 is the composite schematic showing the ac portion of the three oscillators of Fig. 5.1, including the load R_L. The table indicates the oscillator node which when connected to the datum converts the oscillator of Fig. 5.2 into that of Fig. 5.1. The datum is normally the familiar *ground* point or plane. The connection of the appropriate node to ground has important effects upon the following:

1. The manner in which the dc power is fed to the oscillator and the consequent loading effects on the ac circuitry.

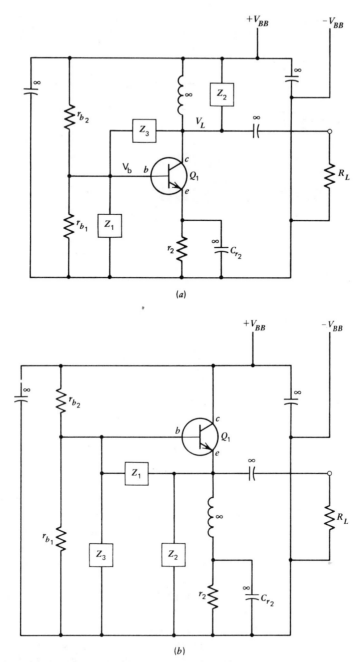

Figure 5.1 Schematics of the Pierce family of oscillators. (*a*) Pierce. (*b*) Colpitts. (*c*) Clapp.

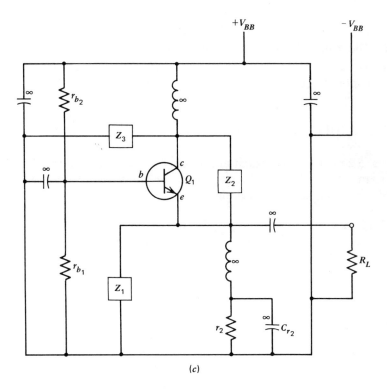

Figure 5.1 (Continued).

Oscillator Type	Datum Connected to
P (Pierce)	e
CO (Colpitts)	c
CL (Clapp)	b

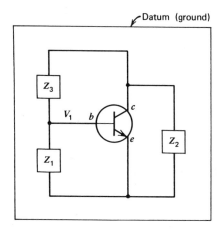

Figure 5.2 ac representation of the Pierce family of oscillators.

128 The Pierce, Colpitts, Clapp Oscillator Family

2. The manner in which the output power is fed to the external load.
3. The distribution of the stray elements to the ground plane which has important effects, particularly at high frequency.

5.2 THE IDEALIZED OSCILLATOR

In Fig. 5.2 the transistor immittances have been incorporated into Z_1, Z_2, and Z_3, so that $Y_{be} \to 0$. Following the procedure outlined in Section 1.3.2 and Fig. 1.16b, and assuming g_m is real, then

$$V_1 = -g_{m_0} V_1 \cdot \frac{Z_2(Z_3 + Z_1)}{Z_1 + Z_2 + Z_3} \cdot \frac{Z_1}{Z_3 + Z_1}$$

$$= -g_{m_0} V_1 \frac{Z_1 Z_2}{Z_1 + Z_2 + Z_3}$$

and

$$A_{L_0} = -g_{m_0} \frac{Z_1 Z_2}{Z_1 + Z_2 + Z_3} \tag{5.1}$$

writing

$$Z_1 + Z_2 + Z_3 = Z_s \tag{5.2}$$

Eq. (5.1) becomes

$$A_{L_0} = -g_{m_0} \frac{Z_1 Z_2}{Z_s} \tag{5.1a}$$

To start oscillation,

$$\left[|A_{L_0}| = g_{m_0} \frac{|Z_1||Z_2|}{Z_s}\right] > 1 \tag{5.3}$$

and

$$\theta_{Z_1} + \theta_{Z_2} - \theta_{Z_s} \approx \pi \tag{5.4}$$

or

$$\theta_{Z_s} \approx \theta_{Z_1} + \theta_{Z_2} - \pi \tag{5.4a}$$

At steady-state oscillation

$$A_L = 1$$

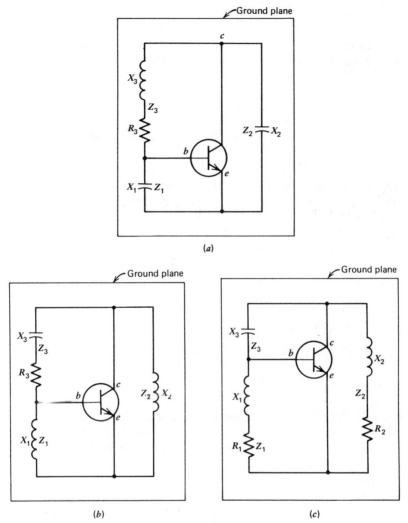

Figure 5.3 The idealized oscillator (a) X_1 and X_2, capacitive. (b) X_1 and X_2, inductive. (c) Practical equivalent of (b).

so that Eq. (5.3) becomes

$$|Z_s| = g_m|Z_1||Z_2| \tag{5.5}$$

If it is assumed that Z_1 and Z_2 are purely reactive and if Z_1 and Z_2 are both inductive or capacitive, then from Eq. (5.4a)

$$\theta_{Z_s} = 0 \quad \text{or} \quad 2n\pi \quad \text{which is equivalent to 0} \tag{5.6}$$

where n is an integer.

If it is assumed that Z_1 is inductive and Z_2 is capacitive or vice versa, then $\theta_{Z_s} = -\pi$, which is impossible since Z_1, Z_2, and Z_3 are real passive elements.

It is therefore seen that Z_1 and Z_2 must *both* be either inductive or capacitive and Z_s must be resistive, and the frequency of oscillation will be that at which

$$X_3 + (X_1 + X_2) = 0 \qquad (5.7)$$

Figure 5.2 therefore, for the idealized case described above, becomes Fig. 5.3a or 5.3b.

It is evident that Fig. 5.3b is somewhat impractical because it is impossible to make lossless inductors, but the basic reasoning will not be rendered invalid if X_1 and X_2 are somewhat lossy.

A more practical realization of Fig. 5.3b would be that shown in Fig. 5.3c.

Going back to Fig. 5.2 recalls that Figs. 5.3a and 5.3c each represents three possible oscillator configurations so that now there are a total of six configurations. Which one should be used depends upon the application, and a few examples are now considered.

5.3 EXAMPLES OF IDEALIZED OSCILLATOR APPLICATIONS

5.3.1 Wide Range Local Oscillator for Receiver

In this case the frequency is tunable over a wide range by a single variable capacitor. The most practical arrangement is that one side of that capacitor be grounded. Looking at Figs. 5.3b and 5.3c, it is immediately seen that the only circuits satisfying the requirements are Fig. 3c with collector or base grounded; that is, the Colpitts or Clapp oscillator.

Figure 5.4a is the oscillator in the Colpitts configuration. Figure 5.4b is the same oscillator in the Clapp configuration.

In both circuits X_1 and X_2 can be replaced by a single tapped coil.

The above example is useful for demonstrating the application of the theory developed in Section 5.2.

5.3.2 Example of Narrow Range Local Oscillator for Receiver (in the FM Band); Also Using a Grounded Variable Capacitor

Obviously this can be accomplished by the same circuits described in Section 5.3.1. However, it is desired to demonstrate some additional possible circuits using the Fig. 5.3a configuration. This is made possible by the fact that a fixed inductor in parallel with a variable capacitor is equivalent to a variable inductor as shown in Fig. 5.5a provided that $X_c > X_L$. By using this fact, the circuits of Fig. 5.5 can be obtained from the configurations of Fig. 5.3a.

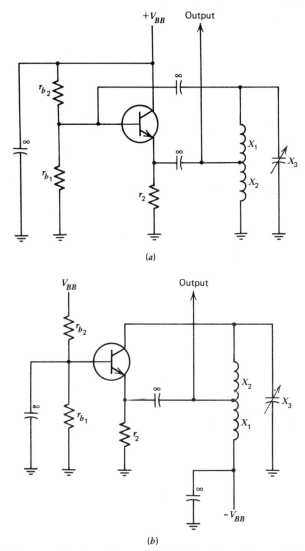

Figure 5.4 Wide-range local oscillators. (*a*) Colpitts (sometimes called Hartley). (*b*) Clapp.

5.3.3 Stable Fixed Frequency *LC* Oscillator

Because it is easier to obtain high Q capacitors than high Q inductors, the choice of oscillator configuration is the Pierce version of Fig. 5.3*a*. The resulting schematic is shown in Fig. 5.6.

For maximum stability, X_1 and X_2 should be small compared to the transistor capacitive reactances. The Q of X_3 should be as high as possible, and the circuit should be designed so that the loaded Q is almost equal to Q_{X_3}.

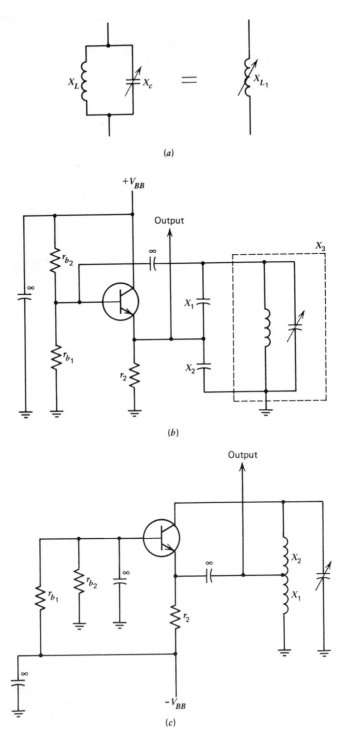

Figure 5.5 Narrow range local oscillators. (a) Synthesis of variable inductor. (b) Colpitts oscillator. (c) Clapp oscillator.

5.3 Examples of Idealized Oscillator Applications

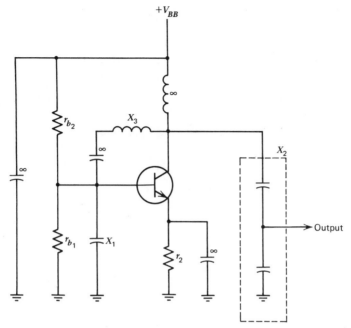

Figure 5.6 Stable Pierce oscillator.

The output power is shown as being extracted from X_2, but occasionally it is extracted from X_3 as in induction welding machinery.

In this connection, it is very important to note that the main factor in determining the stability of the oscillator is the *stability of the components*. In the past, and also at present, excessive attention was and is being paid to maximizing the Q of the oscillator circuit, in the mistaken notion that high Q always results in high stability. It can be experimentally demonstrated that operating Q's as low as 4 can exist in LC oscillators having stabilities better than 10^{-6}. This subject is discussed further in Chapter 17.

5.3.4 Crystal Oscillator with the Crystal Network Operating in the Inductive Region

In this case the crystal can be Z_3 in Fig. 5.3a or (Z_1 or Z_2) in Fig. 5.3c. In actual practice, the configuration in Fig. 5.3c is rarely, if ever, used because of the greater number of inductors and the greater difficulty of supplying the necessary biasing conditions. It is noteworthy, for historical reasons, that the circuit, wherein the crystal is Z_2 in Fig. 5.3c, is called the Miller oscillator.

Replacing Z_1, Z_2, and Z_3 in Figs. 5.1 and 5.3a with the appropriate components, the circuits of Fig. 5.7 result.

Examining Fig. 5.7, it will be noted that the Clapp circuit is more complicated and is therefore rarely used. The Pierce circuit is more stable since the stray elements to ground are across X_1 and Z_2, which may be relatively low

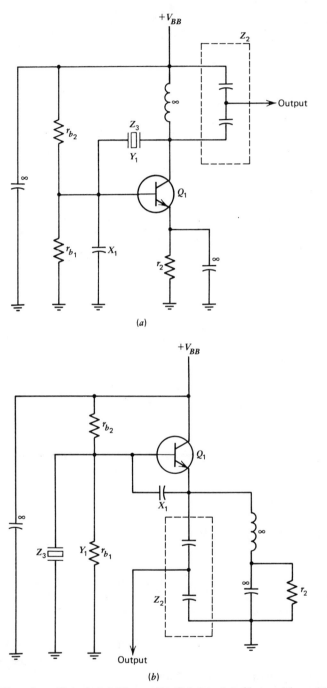

Figure 5.7 Crystal oscillators. (*a*) Pierce. (*b*) Colpitts. (*c*) Clapp with positive power supply. (*d*) Clapp with negative power supply.

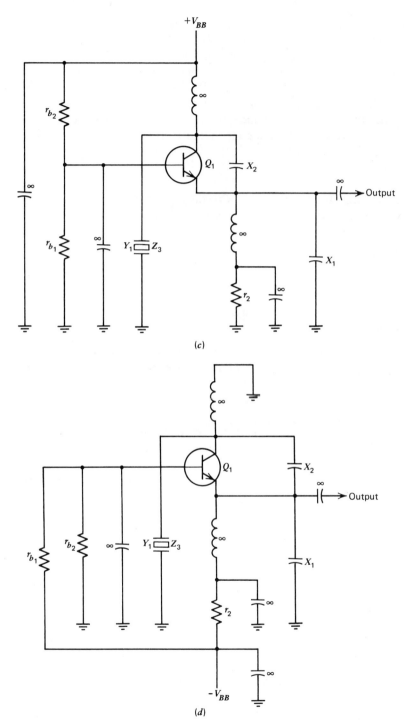

Figure 5.7 (Continued).

impedances, while in the Colpitts circuit the stray elements to ground are across the crystal, which is very sensitive. On the other hand, there are many cases where it is desired to ground one side of the crystal, and for that reason the Colpitts circuit is used. Also, the bias circuit composed of r_{b_1} and r_{b_2}, forming r_b, is across X_1 in the Pierce circuit and therefore usually has a negligible effect. In the Colpitts circuit, r_b is across the crystal which deteriorates the circuit performance. It will also be appreciated that the Colpitts circuit is more difficult to design because of these reasons. An important variation of the Colpitts circuit called the semi-isolated Colpitts oscillator is discussed in detail in Chapter 10.

5.3.5 Crystal Oscillators with the Crystal in the Zero or Negative Reactance Region

Because the $\partial X/\partial f$ of crystals in the inductive region is higher than that in the zero or negative reactance region, oscillators of this design are relatively rare. However, there are some such designs and among those are the following:

1 In VCOs where a low $\partial X/\partial f$ is desirable in order to facilitate the ease of *pushing* the frequency.

2 The standard antiresonant crystals are rated in government and industrial specifications up to about 30 MHz. Above 30 MHz only standard series resonant crystals are available. To use these crystals at their rated frequency, it is necessary for them to operate at zero reactance.

It is evident that when an inductor is connected in series with the crystal, the resulting combination can be considered a crystal operating in the inductive region and the circuits described in Section 5.3.4 can then be used. It is important that the inductor not be made too large, as spurious oscillations may result. In this application, the Pierce circuit usually works best.

5.4 EFFECT OF THE PHASE ANGLE OF g_m IN THE IDEALIZED OSCILLATOR

Repeating Eq. (5.1a)

$$A_{L_0} = \frac{-g_{m_0} Z_1 Z_2}{Z_s}$$

and at steady-state oscillation, making $Z_1 = -jZ_1$, $Z_2 = -jX_2$

$$Z_s = g_m X_1 X_2$$

from which

$$|Z_s| = |g_m|X_1 X_2 \qquad (5.8)$$

or

$$|g_m| = \frac{|Z_s|}{X_1 X_2} \qquad (5.8a)$$

and

$$\theta_{Z_s} = \theta_{g_m} \equiv \theta \qquad (5.9)$$

but

$$\theta_{Z_s} = \cos^{-1} \frac{R_s}{|Z_s|} \qquad (5.10)$$

or

$$|Z_s| = \frac{R_s}{\cos \theta} \qquad (5.10a)$$

so that Eq. (5.8a) becomes

$$|g_m| = \frac{R_s}{X_1 X_2 \cos \theta} \qquad (5.11)$$

Denote the value of g_m, for which $\theta_{g_m} = 0$, as $g_{m(\theta=0)}$ then from Eq. (5.7)

$$g_{m(\theta=0)} = \frac{R_s}{X_1 X_2} \qquad (5.12)$$

From Eqs. (5.11) and (5.12)

$$\frac{|g_m|}{g_{m(\theta=0)}} = \frac{1}{\cos \theta} \qquad (5.13)$$

which states that the amplitude of g_m must increase in accordance with Eq. (5.13).

For example, if $\theta = 20°$, then

$$|g_m| = 1.064 g_{m(\theta=0)}$$

which is not a very large increase.

The phase angle of g_m will also introduce a frequency shift which can be computed thus:
from Eq. (5.2)

$$Z_s = \{R_3 + j[X_3 - (X_1 + X_2)]\} \tag{5.2a}$$

when $\theta = 0°$,
$$X_3 = X_1 + X_2.$$

If Z_3 is a crystal, the impedance of which is

$$R_1\left(1 + 2jQ_x\frac{\Delta f}{f}\right) \tag{5.14}$$

then for $\theta = 0°$,

$$2R_1Q_x\left|\frac{\Delta f}{f}\right|_{\theta=0} = X_1 + X_2 \tag{5.15}$$

For $\theta \neq 0°$, from Eqs. (5.10a) and (5.2a),

$$\frac{R_1}{\cos\theta} = |Z_s| = \sqrt{R_1^2 + \left(\left(2R_1Q_x\frac{\Delta f}{f} - (X_1 + X_2)\right)\right)^2} \tag{5.16}$$

and, from Eq. (5.15)

$$\frac{R_1}{\cos\theta} = \sqrt{R_1^2 + \left(\frac{2R_1Q_x}{f}(\Delta f - \Delta f_{(\theta=0)})\right)^2} \tag{5.17}$$

Solving Eq. (5.17) for $(\Delta f - \Delta f_{(\theta=0)})/f$

$$\frac{\Delta f - \Delta f_{(\theta=0)}}{f} = \frac{\sqrt{(1/\cos^2\theta_{g_m} - 1)}}{2Q_x} = \frac{\tan\theta_{g_m}}{2Q_x} \tag{5.18}$$

Again for $\theta = 20°$

$$\frac{\Delta f - \Delta f_{(\theta=0)}}{f} = \frac{0.18}{Q_x} \tag{5.19}$$

For example, for $Q_x = 100,000$

$\theta_{g_m} = 20°$ will result in a fractional frequency shift of 1.8×10^{-6} \quad (5.20)

In the real oscillator, R_1 should be replaced by $R_1 + (R_T - R_1)$, where R_T is the total circuit loss resistance and Q_x should be replaced by Q_{op}. In that case, the frequency shift will be R_T/R_1 times the shift calculated above.

5.5 THE REAL OSCILLATOR

5.5.1 Introduction

In this section the theory of the real oscillator is developed, taking into account all the losses present in the oscillator circuit. As explained previously, all the circuit configurations and circuit types of this chapter have the same mathematical model. Therefore, it is sufficient to analyze one particular configuration and circuit type knowing, *a priori*, that the theory will apply equally well to all other configurations and types. It is advisable to choose the particular oscillator on the basis of its popularity and its ease of design. The circuit chosen is the Colpitts, shown in Figs. 5.1a and 5.3a, because of its considerable popularity and its greatest difficulty of design. The theory can be then applied to the other popular types with rather obvious simplifications and additions, as will be done in Chapters 7 to 10.

Figure 5.8 is a schematic diagram of the Colpitts oscillator of the type to be analyzed. Figure 5.9 is the schematic diagram of the ac portion of Fig. 5.8. The differences between Figs. 5.8 and 5.9 are:

1. Figure 5.8 shows the circuit as it is physically constructed. Each component shown on this figure is a physical component and the design problem consists of determining these physical components. Capacitors shown as ∞ are meant to be capacitors, the reactance of which are essentially 0 at the operating frequency f.

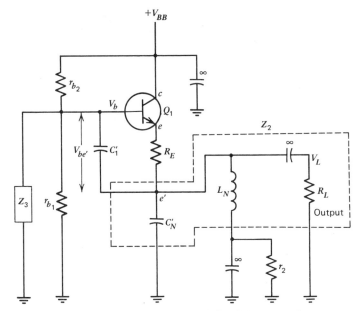

Figure 5.8 Detailed schematic of the Colpitts oscillator.

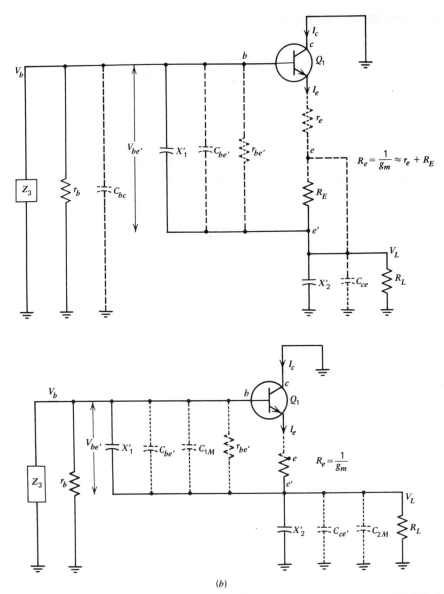

Figure 5.9 Colpitts oscillator. (*a*) Detailed schematic of the ac circuitry. (*b*) Modified schematic, Step 1. (*c*) Modified schematic, Step 2.

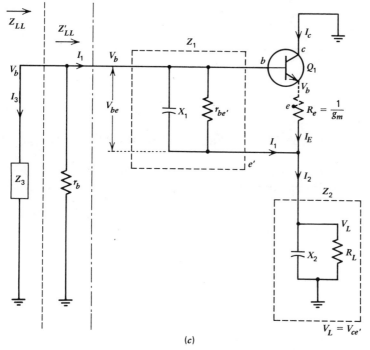

(c)

Figure 5.9 (Continued).

2   The power supply is not shown in Fig. 5.9.
3   Figure 5.9a includes the parameters of the transistor shown as dashed lines, using the Hybrid-PI model.
4   X_1' in Fig. 5.9 is the reactance of C_1' of Fig. 5.8 at the operating frequency f.
5   X_2' in Fig. 5.9 is the combined reactance of C_N' and L_N of Fig. 5.8 at f, or

$$X_2' = \frac{1}{2\pi f C_N' - 1/2\pi f L_N} \tag{5.21}$$

5.5.2   Assumptions and Conditions

1   The values of all immittances, voltages, and currents are those at the operating frequency f.
2   V_{be} is sinusoidal. (5.22)
3   g_m is real, so that $\theta_{g_m} = 0$. (5.23)
4   $g_m = \dfrac{I_e}{V_{be'}}$ (5.23a)
5   r_{ce} is either infinite or lumped into R_L.

142 The Pierce, Colpitts, Clapp Oscillator Family

6 β is fairly large, so that $i_C \approx i_E$. (5.24)
7 The effect of $r_{bb'}$ is negligible.
8 r_b is the combination of r_{b_1} and r_{b_2}

$$r_b = \frac{1}{1/r_{b_1} + 1/r_{b_2}} \qquad (5.25)$$

5.5.3 Calculation of the Oscillatory Equations of the Circuit of Fig. 5.9a

1 The first step is to allocate C_{bc} in Fig. 5.9a, to portions in parallel with X_1' and X_2 in the manner described in Section 5.5.4.2.5. This results in Fig. 5.9b.

2 The second step is to combine X_1' and X_2' with its parallel capacitors to form X_1 and X_2. This results in Fig. 5.9c. Obviously

$$X_1 = \frac{1}{2\pi f\left((2\pi f X_1')^{-1} + C_{be'} + C_{1M}\right)} \qquad (5.26)$$

$$X_2 = \frac{1}{2\pi f\left((2\pi f X_2')^{-1} + C_{ce'} + C_{2M}\right)} \qquad (5.27)$$

3 In Fig. 5.9c we obtain from Eq. (1.8a), assuming $R_L > 5X_2$, $r_{be'} > 5X_1$,

$$Z_1 = \underbrace{\frac{X_1^2}{r_{be'}}}_{R_{in}} - jX_1 \qquad (5.28)$$

$$= R_{in} + \underline{X}_1 \qquad (5.28a)$$

$$Z_2 = \underbrace{\frac{X_2^2}{R_L}}_{R_2} - jX_2 \qquad (5.29)$$

$$= R_2 + \underline{X}_2 \qquad (5.29a)$$

(See Section 1.2.1.5 for the $\underline{X}$ notation.)

4 Z_{LL}' is now calculated

$$Z_{LL}' = \frac{V_b}{I_1} \qquad (5.30)$$

By definition of R_e, R_e and Z_1 are effectively in parallel as shown in Fig. 5.9c,

5.5 The Real Oscillator

where the voltage at the base and intrinsic emitter are both V_b,

$$V_L = V_b \frac{Z_2}{Z_1 \| R_e + Z_2} \tag{5.31}$$

where $Z_1 \| R_e$ denotes the parallel combination of Z_1 and R_e

$$\frac{Z_1 R_e}{Z_1 + R_e} \tag{5.32}$$

By inspection

$$I_1 = \frac{V_b - V_L}{Z_1} \tag{5.33}$$

$$= \frac{V_b}{Z_1}\left[1 - \frac{Z_2}{Z_1 \| R_e + Z_2}\right] \tag{5.34}$$

from Eq. (5.31);

$$= \frac{V_b}{Z_1}\left(\frac{Z_1 \| R_e}{Z_1 \| R_e + Z_2}\right) \tag{5.34a}$$

Therefore, from Eq. (5.30)

$$Z'_{LL} = Z_1\left(1 + \frac{Z_2}{Z_1 \| R_e}\right) \tag{5.35}$$

$$= Z_1 + Z_2 + \frac{Z_1 Z_2}{R_e} \tag{5.35a}$$

from Eq. (5.32)

$$\left.\begin{array}{l}\text{and defining } Z_1 = R_{\text{in}} + jX_1 \\ \text{for other } Z_n, \ Z_n = R_n + jX_n\end{array}\right\} \tag{5.35b}$$

$$Z'_{LL} = (\underline{X}_1 + \underline{X}_2) + R_{\text{in}} + R_2 + g_m(R_{\text{in}} + \underline{X}_1)(R_2 + \underline{X}_2) \tag{5.36}$$

from Eqs. (5.28a) and (5.29a) or

$$Z'_{LL} = \underbrace{(1 + g_m R_{\text{in}})\underline{X}_2 + (1 + g_m R_2)\underline{X}_1}_{\underline{X}_a} + \underbrace{R_{\text{in}} + R_2 + g_m \underline{X}_1 \underline{X}_2 + g_m R_{\text{in}} R_2}_{R_a}$$

$$\tag{5.36a}$$

and $Z_{LL} = r_b \| Z'_{LL}$ is now calculated to be

$$Z_{LL} \approx \underline{X}_a + R_a + \frac{(\underline{X}_a)^2}{r_b} \qquad (5.37)$$

assuming

$$|\underline{X}_a| > 5R_a,$$

$$r_b > 5|\underline{X}_a|$$

Converting Eq. (5.37) to $Z = R + jX$ notation,

$$Z_{LL} = -j((1 + g_m R_{in})X_2 + (1 + g_m R_2)X_1) + R_{in} + R_2 + g_m R_{in} R_2$$

$$- g_m X_1 X_2 + \frac{[(1 + g_m R_{in})X_2 + (1 + g_m R_2)X_1]^2}{r_b} \qquad (5.37a)$$

Noting that

$$g_m R_{in} \ll 1$$

$$q_m R_2 \ll 1$$

$$X_1 X_2 \gg R_{in} R_2$$

and substituting for R_{in} and R_2 from Eqs. (5.28) and (5.29)

$$Z_{LL} = -j\left[(1 + g_m R_2)X_1 + (1 + g_m R_{in})X_2\right] + \frac{X_1^2}{r_{be'}}$$

$$+ \frac{X_2^2}{R_L} + \frac{(X_1 + X_2)^2}{r_b} - g_m X_1 X_2 \qquad (5.38)$$

It is interesting to note that $-g_m X_1 X_2$ may be interpreted as the negative resistance without which oscillations cannot exist.

5 For the circuit in Fig. 5.9c to oscillate, Eqs. (1.47) and (1.48) state that

$$Z_3 = R_3 + jX_3 = -Z_{LL} \qquad (5.39)$$

Substituting for Z_{LL} and separating the resulting equation into real and imaginary parts

$$R_T = R_3 + \underbrace{\frac{X_1^2}{r_{be'}}}_{R_{in}} + \underbrace{\frac{X_2^2}{R_L}}_{R_2} + \underbrace{\frac{(X_1 + X_2)^2}{r_b}}_{R_b} = g_m X_1 X_2 \qquad (5.40)$$

and
$$X_3 = -[(1 + g_m R_2)X_1 + (1 + g_m R_{in})X_2] \tag{5.41}$$

Solving for g_m in Eq. (5.40)
$$g_m = \frac{R_T}{X_1 X_2} \tag{5.42}$$

Equation (5.41) becomes, from Eq. (5.42),
$$X_3 = -\left[\left(1 + \frac{R_2 R_T}{X_1 X_2}\right)X_1 + \left(1 + \frac{R_{in} R_T}{X_1 X_2}\right)X_2\right] \tag{5.43}$$

$$= -(X_1 + X_2)\left(1 + \frac{R_T(R_2 X_1 + R_{in} X_2)}{X_1 X_2 (X_1 + X_2)}\right) \tag{5.43a}$$

$$= -(X_1 + X_2)(1 + \varepsilon) \tag{5.44}$$

where
$$\varepsilon = \frac{R_T(R_2 X_1 + R_{in} X_2)}{X_1 X_2 (X_1 + X_2)} \tag{5.45}$$

and is usually less than 0.01, so Eq. (5.43) becomes
$$X_3 \approx -(X_1 + X_2) \tag{5.46}$$

As has been shown previously, g_m is a function of amplitude. Therefore Eq. (5.40) can be used to determine the amplitude of oscillation. Equation (5.46) can be used to determine the frequency of operation and Eq. (5.45) can be used to determine the effect of small variations in the parameters upon the frequency.

5.5.4 Derivation of the Circuit Design Equations

5.5.4.1 Introduction

The derivation is made for the Colpitts circuit of Fig. 5.9c because, as previously explained, of its relative popularity and greater difficulty of design. As will be seen later, the derivation is equally valid for all the other circuit types. For experimental verification of the theory presented here, it is preferable to use the Pierce circuit because the e' point is then at ac ground potential and $V_{be'}$, the most important ac voltage in the circuit, can then be easily measured.

The derivation is based upon the concept that the most important quantities are the Z_3 current, I_3, and the output power, P_L. All other circuit parameters are derived from these quantities.

5.5.4.2 Derivation of the Basic Theory

5.5.4.2.1 Calculation of Output Power

The output power is given by

$$1000 P_L = \frac{V_L^2}{R_L} \tag{5.47}$$

5.5.4.2.2 Calculation of the Currents

Assuming $Z_3 \ll r_b$,

$$\left(1 + \frac{Z_3}{r_b}\right) I_3 = -I_1 \approx I_3 \tag{5.48}$$

$$V_b = I_3 Z_3 = I_3 (R_3 + jX_3) \tag{5.49}$$

and

$$|V_b| = |I_3|\sqrt{R_3^2 + (X_3)^2} \tag{5.50}$$

from Eq. (5.46)

$$= I_3 \sqrt{R_3^2 + (X_1 + X_2)^2} \tag{5.50a}$$

$$V_L = V_b - I_1 Z_1 \tag{5.51}$$

from Eq. (5.48)

$$= V_b + I_3 Z_1 \left(1 + \frac{Z_3}{r_b}\right) \tag{5.51a}$$

$$V_{be'} = I_1 Z_1 \approx I_3 Z_1 \tag{5.52}$$

and

assuming $|Z_1| \approx |X_1|$

$$|V_{be'}| \approx |I_3||X_1| \tag{5.52a}$$

$$V_L = I_2 Z_2 \tag{5.53}$$

From Eqs. (5.49), (5.51a), and (5.53) one obtains

$$I_2 = \frac{I_3}{Z_2}\left[Z_3 + Z_1\left(1 + \frac{Z_3}{r_b}\right)\right] \tag{5.54}$$

$$I_e = I_2 - I_1 \tag{5.55}$$

$$= \frac{I_3}{Z_2}\left[Z_3 + Z_1\left(1 + \frac{Z_3}{r_b}\right)\right] + I_3\left(1 + \frac{Z_3}{r_b}\right) \tag{5.56}$$

from Eqs. (5.48) and (5.54)

$$= \frac{I_3}{Z_2}\left[Z_1 + Z_2 + Z_3 + \frac{Z_3(Z_1 + Z_2)}{r_b}\right] \qquad (5.56a)$$

or from Eq. (5.46) and approximations

$$I_e \approx \frac{I_3}{Z_2}\left[R_{in} + R_2 + R_3 + \underbrace{\frac{(X_1 + X_2)^2}{r_b}}_{R_b}\right] \qquad (5.57)$$

$$\approx \frac{I_3}{Z_2} R_T \qquad (5.58)$$

from Eqs. (5.28), (5.29), and (5.40); and

$$\frac{|I_3|}{|I_e|} \approx \frac{|X_2|}{R_T} \equiv a_i \qquad (5.58a)$$

assuming $|X_2| \approx |Z_2|$. Also

$$I_2 = \frac{I_e}{R_T}[R_T - Z_2]$$

from Eqs. (5.48), (5.55), and (5.58)

$$= I_e \frac{R_{df} + R_{in} + R_b - jX_2}{R_T} \qquad (5.59)$$

and

$$|I_2| = |I_e|\frac{\sqrt{(R_{df} + R_{in} + R_b)^2 + X_2^2}}{R_T} \qquad (5.59a)$$

5.5.4.2.3 Calculation of X_2

Let

$$R_p = R_3 + R_{in} + \underbrace{\frac{X_1^2}{r_b}}_{R_{bp}} \qquad (5.60)$$

and let

$$\left(R_2 = \frac{X_2^2}{R_L}\right) \to 0 \tag{5.61}$$

Occasionally, Eq. (5.61) is true, r_b and R_p are known, and it is desired to calculate X_2. Then, from Eq. (5.58a)

$$a_i = \frac{X_2}{R_p + (2X_1X_2 + X_2^2)/r_b} \tag{5.62}$$

solving for X_2

$$X_2 \approx \frac{a_i R_p}{1 - 2a_i X_1/r_b} \approx a_i R_p \left[1 + \frac{2a_i X_1}{r_b}\right] \tag{5.63}$$

because $2a_i X_1/r_b \ll 1$.

5.5.4.2.4 Calculation of V_L

$$V_L = I_2 Z_2 = I_3 \left[Z_3 + Z_1\left(1 + \frac{Z_3}{r_b}\right)\right] \tag{5.64}$$

from Eq. (5.54) so that

$$V_L = I_3 \left[R_3 + R_{in} + j(X_3 - X_1) + \frac{Z_1 Z_3}{r_b}\right] \tag{5.64a}$$

$$= I_3 \left[R_3 + R_{in} + \frac{X_1(X_1 + X_2)}{r_b} + jX_2\right] \tag{5.65}$$

from Eq. (5.46) and approximations; and

$$V_L = I_3 \sqrt{\left(R_{in} + R_3 + \frac{X_1(X_1 + X_2)}{r_b}\right)^2 + X_2^2} \tag{5.66}$$

so that

$$\frac{|V_b|}{|V_L|} = \frac{\sqrt{R_3^2 + (X_1 + X_2)^2}}{\sqrt{\left(R_{in} + R_3 + \frac{X_1(X_1 + X_2)}{r_b}\right)^2 + X_2^2}} \tag{5.67}$$

from Eqs. (5.50a) and (5.66). If the resistance components are neglected

$$\frac{|V_b|}{|V_L|} \approx \frac{X_1 + X_2}{X_2} \tag{5.67a}$$

Similarly, from Eqs. (5.52) and (5.56)

$$\frac{|V_{be'}|}{|V_L|} = \sqrt{\frac{R_{in}^2 + X_1^2}{\left(R_{in} + R_3 + \frac{X_1(X_1 + X_2)}{r_b}\right)^2 + X_2^2}} \tag{5.68}$$

$$\approx \frac{X_1}{X_2} \tag{5.68a}$$

when the resistance components are neglected.

5.5.4.2.5 Calculation of the Approximate Contribution of C_{bc}

In Section 5.4.3, C_{cb} was replaced by C_{1M} and C_{2M} as shown in Fig. 5.9b. The manner in which C_{1M} and C_{2M} are calculated will be shown in this section.

The general procedure consists of partitioning C_{cb} into C_{1M} and C_{2M}, connected as shown in Fig. 5.10.

The potential at the junction of C_{1M} and C_{2M} is the same as that of the emitter e', V_L. This fact gives the clue as to how C_{1M} and C_{2M} are calculated.

From Fig. 5.10 it is seen that

$$X_{C_{1M}} + X_{C_{2M}} = X_{cb} \tag{5.69}$$

Let

$$\frac{V_{be'}}{V_L} = M_M \tag{5.70}$$

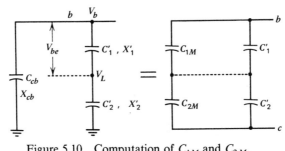

Figure 5.10 Computation of C_{1M} and C_{2M}.

then
$$X_{C_{1M}} = M_M X_{C_{2M}} \tag{5.71}$$

from which
$$X_{C_{2M}} = \frac{X_{C_{cb}}}{1 + M_M} \tag{5.72}$$

and
$$X_{C_{1M}} = \frac{M X_{C_{cb}}}{1 + M_M} = \frac{X_{C_{cb}}}{1 + 1/M_M} \tag{5.73}$$

From Eq. (5.72),
$$C_{2M} = C_{cb}(1 + M_M) \tag{5.74}$$

and from Eq. (5.73),
$$C_{1M} = C_{cb}\left(1 + \frac{1}{M_M}\right) \tag{5.75}$$

The above analysis assumes that $V_{be'}$ and V_L are in phase, which is not strictly true. The computation therefore is only an approximation. Also, the possible absence of phase coherence introduces positive and negative resistive components which may, at times, be significant. Furthermore, the analysis only accounts for the Miller effect contributions of C_{cb} to C_1 and C_2 and it does not consider all of the direct transfer of energy between the collector and the base by C_{cb}.

5.5.5 Calculation of $C_{ce'}$

It is obvious that when $R_E = 0$, $C_{ce'} = C_{ce}$. When R_E is not zero, the potential at point e almost equals the potential at point e', because of the limiting action; therefore, $C_{ce'} \approx C_{ce}$.

5.5.6 Calculation of Oscillator Operating Q, Q_{op}

The circuit Q_{op} due to the loss in Z_3 can be shown to be

$$Q_{Z_3} = \frac{f}{2R_3} \frac{\partial [(X_3 - (X_1 + X_2)]_{OL}}{\partial f} \tag{5.76}$$

provided that no additional losses exist.

The total circuit Q is deteriorated by the presence of the other circuit losses. The total equivalent circuit resistance is shown to be R_T in Eq. (5.40).

Therefore,

$$Q_{op} = \frac{f}{2R_T} \frac{\partial [X_3 - (X_1 + X_2)]_{OL}}{\partial f} \qquad (5.77)$$

For a crystal oscillator, from Eq. (3.28),

$$Q_{op} \approx Q_x \cdot \frac{R_{df}}{R_T} \qquad (5.77a)$$

since $\partial (X_1 + X_2)/\partial f \approx 0$ for the crystal operating frequency range.

Equations (5.76) to (5.77a) require clarification.

$[X_3 - (X_1 + X_2)]_{OL}$ signifies the open loop value of $[X_3 - (X_1 + X_2)]$ as measured in a manner similar to that described in Section 1.3.2. When the loop is closed, the circuit functions as the desired oscillator, $[X_3 - (X_1 + X_2)] \to 0$, and therefore its differential also approaches zero. However, the value of $[X_3 - (X_1 + X_2)]_{OL}$ as well as that of R_T must be calculated at the signal amplitudes that exist in the oscillating mode, as these values may be strong functions of the signal amplitudes.

In a well-designed high-performance crystal oscillator, Q_{op} in Eq. (5.77a) is a weak function of df since R_{df} is by far the major constituent of R_T.

See also Section 7.7.2.

5.6 THE THEORY APPLIED TO CRYSTAL OSCILLATORS

5.6.1 Introduction

All the theory developed in this chapter applies completely to the crystal oscillator. In this case, Z_3 is the crystal resonator network, I_3 is the crystal current I_x, and R_3 is the effective crystal resistance R_{df}. However, the theory requires several additions to include the circuit modifications necessary to ensure that the circuit operates at the crystal correct overtone and mode.

At this point, it cannot be stressed too highly that the circuit designer should have a full knowledge of the various crystal responses and their magnitude, gathered from measurements on a representative sampling of the entire production run of the crystals, before proceeding with the design of the circuit for that run of crystals. Too often circuits have been unnecessarily complicated by assuming responses that, in fact, do not exist. For example, normally one would expect crystals to have a stronger response at the fundamental frequency than at the third overtone; so the designer includes a third overtone selector circuit. However, it is possible to build crystals which have much stronger third overtone than fundamental responses. If the designer had investigated and discovered that this condition exists for the run of the crystals to be used, the third overtone mode selector could have been eliminated, with the consequent improvement in performance, reduction in cost, and greater reliability.

The approach adopted in this book is that the circuit is designed for frequency, frequency stability, power output, crystal resistance, and crystal power, all of which are prespecified.

5.6.2 Crystal Fundamental Wide Frequency Range Oscillator Circuits

These oscillators may be the Pierce and Colpitts circuits shown in Figs. 5.7a and 5.7b. However, oscillators using the circuits over wide frequency ranges will have wide variations in frequency stability, power output, and crystal power drive. If the designer considers these variations acceptable, he should design the circuits at several extreme conditions, for example, lowest and highest frequency, lowest and highest power output, lowest and highest crystal resistance, and come up with a compromise circuit. The theory in this book will help the designer predict the oscillator behavior under all these varying conditions.

In view of the fact that wide range fundamental oscillators are in substantial use, some observations will therefore be made about them.

In the circuits of Fig. 5.7 the loop gain, for a constant crystal resistance

$$A_L \propto \frac{1}{f^2} \tag{5.78}$$

in accordance with Eq. (5.40), since both

$$X_1, X_2 \propto \frac{1}{f} \tag{5.79}$$

Therefore, the circuit will oscillate at the lowest frequency which will provide an R_T sufficient to satisfy Eq. (5.40). It is recommended that a margin of safety of 3 be provided; for example,

$$\frac{R_{T_1}}{R_{T_N}} = \frac{3}{N^2} \tag{5.80}$$

where N is the overtone order.

For example, R_T at the fundamental should not exceed $(1/3)R_T$ at the third overtone, $3R_T/25$ at the fifth overtone, and so on. Also, L_1 in Fig. 5.7 should be large enough so that its reactance will be higher than the capacitive reactance it parallels at the fundamental frequency.

5.6.3 Fundamental Oscillator Circuits with Prespecified Parameters and Performance Objectives

The design of this oscillator is fully covered by the theory already presented.

5.6 The Theory Applied to Crystal Oscillators

5.6.4 Overtone Oscillator Circuits (See Chapter 3)

As described in Sections 5.6.1 and 5.6.2, the oscillator will tend to operate at the crystal fundamental frequency unless steps are taken to prevent it by providing an overtone selector circuit. The selector may take one of two forms.

1. If the third overtone is desired, the selector may be a series resonant circuit tuned to the fundamental and connected between the base and emitter or the collector and emitter. This type of selector is much less desirable than type 2 since, from a practical point of view, it will require more components and since it is useful only for rejecting the fundamental.

2. The second type of selector is a parallel circuit, tuned to a frequency below f_N, the desired overtone, and above the next lower overtone and connected between base and emitter or collector and emitter. Usually, it is connected between collector and emitter because, in practice, it requires fewer additional components. The following analysis will therefore be made on the basis of the connection of the parallel tuned circuit between collector and emitter, but it should be noted that exactly the same analysis applies to the other connection.

Section 5.2 and Fig. 5.3a show that if X_1 is capacitive, X_2 must also be capacitive. If X_2 is inductive, then oscillations cannot take place. Suppose X_2 is the parallel tuned circuit shown in Fig. 5.11a, the reactance characteristics of which are shown in Fig. 5.11b. In Fig. 5.11b, X is inductive from $f \to 0$ to $f = f_s$. Above f_s, X is capacitive and oscillations will then be possible.

If f_s is set below f_N, which is the desired overtone, and above f_{N-2}, which is the nearest lower overtone, then oscillations can take place at f_N and above, provided the loop gain is sufficient. It will also be noted from Fig. 5.11b that above f_s $|X|$ continually decreases as f increases, thus decreasing the loop gain.

Figures 5.7a and 5.7b show oscillators originally designed for fundamental operation. To convert them to overtone operation, all that is necessary is to adjust the values of L_2 and the capacitors so that it resonates at f_s and has the value X_2 at f_N. To aid in calculating the necessary values of L_N and C_N, the following relations have been formulated: let

$$s = \frac{f_s}{f_N} \qquad (5.81)$$

For $N = 1$, fundamental operation,

$$s^2 = 0.2 \qquad (5.82a)$$

154 The Pierce, Colpitts, Clapp Oscillator Family

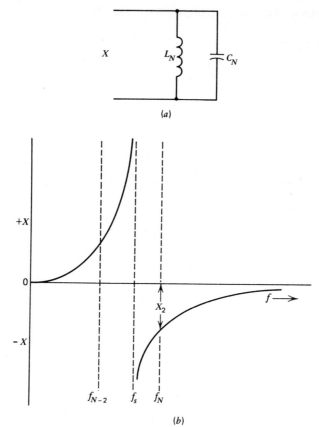

Figure 5.11 Reactance of parallel tuned circuit. (*a*) Schematic. (*b*) Reactance versus f.

For $N \neq 1$, overtone operation,

$$s = 1 - \frac{1.5}{N} \tag{5.82b}$$

$$X_{C_N} = (1 - s^2) X_2 \tag{5.83}$$

$$C_N = \frac{159{,}000}{X_{C_N} f}, \quad C \text{ in pF}, f \text{ in MHz} \tag{5.84}$$

$$X_{L_N} = \frac{X_{C_N}}{s^2} \tag{5.85}$$

$$L_N = \frac{X_{L_N}}{2\pi f}, \quad L_N \text{ in } \mu\text{H}, \tag{5.85a}$$

5.6 The Theory Applied to Crystal Oscillators

It should be noted that for fundamental operation L_N acts only as a dc path, and f_s has been chosen to make L_N a reasonably small value. Also, f_s has been so fixed by Eq. (5.82) that it is close to f_{N-2} than to f_N. This is done to reduce the value of $\partial X / \partial f$ at f_N in order to increase the stability of this circuit.

When the crystal cut is doubly rotated such as the SC-cut, additional modes are present, as described in Chapter 3. This requires that s of Eq. (5.82) be selected so that f_s is comfortably above the nearest lower strong mode. In Section 5.6.5.2, it will be shown that, for the third and fifth overtone SC, s should be 0.75. Therefore, Eq. (5.82b) should be modified to, for SC cuts,

$$S = 0.75 \quad \text{for } N = 3, 5 \quad (5.82c)$$

5.6.5 Crystal Mode Selection Circuits (See Chapter 3)

Assuming $f_b / f_c = 1.088$.

5.6.5.1 Introduction

As mentioned in Section 5.6.4, doubly rotated crystals have additional strong modes. When these crystals are used, the circuit must be modified so that oscillations at any of these additional undesired modes do not take place. The treatment now being developed applies to the SC-cut crystal at the first, third, and fifth overtones, but the necessary modifications to the treatment for cuts having different modes will be obvious to the reader.

5.6.5.2 Basic Requirements

It can be seen from Fig. 3.17 that, for third overtone operation, the desired mode is $c3$ at f_{c3}. Adjacent are modes $a1$ at $f_{a1} \approx 0.63 f_{c3}$ and $b3$ at $f_{b3} \approx 1.088 f_{c3}$. Similarly for fifth overtone operation, the desired mode is $c5$ at f_{c5}. Adjacent are modes $b3$ at $f_{b3} \approx 0.64 f_{c5}$ and $b5$ at $f_{b5} \approx 1.088 f_{c5}$.

Therefore, for both third overtone and fifth overtone operation, the circuit should be designed so that oscillations can take place only between $0.75 f_c$ and $1.088 f_c$. It is not difficult to suppress the oscillation below $0.75 f_c$ as discussed in Section 5.5.3, but it is much more difficult to suppress the oscillation above f_c because the f_b nearest strong response, which is very often stronger than the f_c response, is so close to f_c. To be safe, it might be desirable to make the upper limit of oscillation $1.06 f$ rather than $1.088 f_c$, but this will result in sharper slope of the reactance curve and greater instability.

In Section 5.6.3 it was shown that the circuitry to suppress the oscillations below f_c by choosing a proper value of s, which, as stated above, should be 0.75. Additional circuitry is necessary to suppress the oscillation above f_c. The additional circuitry may take several different forms. One form is shown in Fig. 5.12.

In Section 5.2 it was shown that X_1 and X_2 must both be inductive or capacitive in order to oscillate. If X_2 takes the form shown in Fig. 5.12a and X_1 takes the form of Fig. 5.12b, then Fig. 5.12c shows that X_1 and X_2 are never

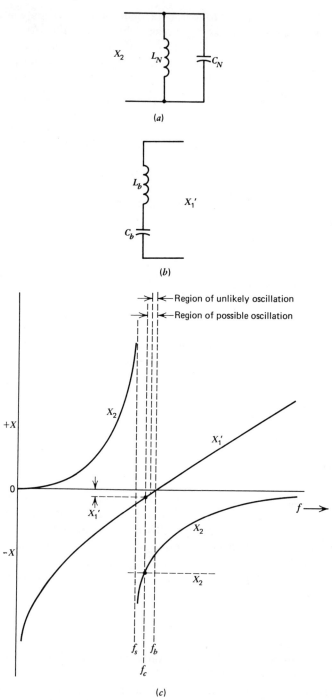

Figure 5.12 Two network type mode selection circuit. (a) Schematic of X_2. (b) Schematic of X_1'. (c) Reactance versus f of X_2 and X_1'.

both inductive, but are both capacitive within a narrow frequency region. Therefore, oscillations will take place within that region provided the loop gain is sufficiently high.

5.6.5.3 The Design of X_1

The design of X_2 was described in Section 5.6.4. The design of X_1' is now given for the SC-cut crystal, where $f_b = 1.088 f_c$.

$$C_b = 0.156 C_1' \tag{5.86}$$

where C_1' is the capacity of X_1' in Fig. 5.9a.

$$L_b = \frac{10^6}{C_b (6.84 f)^2} \quad \mu\text{H} \tag{5.87}$$

It will be noted that C_b is about one-sixth of C', and sometimes it is so small that it is not practical. In that case, if X_2' is much smaller than X_1', the networks are reversed. Often, this solution also is not practical, particularly for low crystal currents. In that case the high-reactance element remains a single capacitor and the low-reactance element is replaced by the networks shown in Fig. 5.13, which provide the combined rejection functions of X_1 and X_2 of Fig. 5.12.

5.6.5.4 The Design of the Single Overtone and Mode Selector Circuit

In the cases where this circuit is used, it usually takes the place of X_2 or the equivalent C_N and L_N in Fig. 5.8.

The reactance curves of Fig. 5.13c apply to both the networks of Fig. 5.13a and 5.13b. The choice of which network to use depends upon the practicability of the components making up the network.

Since X_1 is a capacitive reactance, Fig. 5.13c shows that oscillation is possible only in the frequency region between f_s and f_c because of the low value of X_2 which causes a low loop gain above f_c.

1 The design of network form 1,

$$X_{L_N} = \frac{(f/f_s)^2 - 1}{(f/f_b)^2 - 1} X_2 \tag{5.88}$$

or

$$X_{L_N} = -5.01 X_2 \tag{5.88a}$$

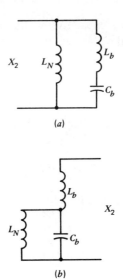

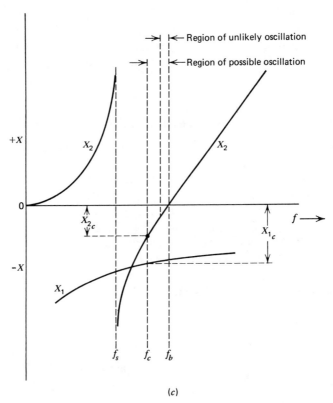

Figure 5.13 The single network overtone and mode selector (a) Schematic of form 1 network. (b) Schematic of form 2 network. (c) Reactance versus f.

5.6 The Theory Applied to Crystal Oscillators

for

$$\frac{f_s}{f} = 0.75, \quad \frac{f_b}{f_c} = 1.088, \quad f_c = f \quad (5.89)$$

and

$$L_N = \frac{X_{L_N}}{2\pi f} \quad (5.90)$$

$$L_b = \frac{L_N}{(f_b/f_s)^2 - 1} \quad (5.91)$$

$$= 0.905 L_N \quad (5.91a)$$

for the same conditions.

$$X_{C_b} = \left(\frac{f_b}{f_c}\right)^2 X_{L_b} \quad (5.92)$$

$$= 5.37 X_2 \quad (5.92a)$$

for the same conditions and

$$C_b = \frac{159{,}000}{X_{C_b} f} \quad (5.93)$$

2 The design of network form 2,

$$X_{L_N} = \frac{X_2}{\left(1 - (f_c/f_s)^2\right)^{-1} + \left((f_b/f_s)^2 - 1\right)^{-1}} \quad (5.94)$$

$$= -2.629 X_2 \quad (5.94a)$$

for the conditions of Eq. (5.89) and

$$L_N = \frac{X_{L_N}}{2\pi f} \quad (5.95)$$

$$L_b = \frac{L_N}{(f_b/f_s)^2 - 1} \quad (5.96)$$

$$= 0.905 L_N \quad (5.96a)$$

Table 5.1 Mode Selection Networks for $f_b/f_c = 1.088$

Network Form	$\dfrac{X_{L_N}}{X_2}$	$\dfrac{L_b}{L_N}$	$\dfrac{X_{C_b}}{X_2}$
1	−5.01	0.905	5.37
2	−2.62	0.905	1.478

for the same conditions

$$X_{C_b} = -X_{L_N}\left(\frac{f_s}{f_c}\right)^2 \tag{5.97}$$

$$= 1.478 X_2 \tag{5.97a}$$

for the same conditions, and

$$C_b = \frac{159{,}000}{X_{C_b} f} \tag{5.98}$$

3 Comparison of network forms 1 and 2.

From Table 5.1 it will be noted that components of network form 2 are of considerably lower impedance levels, especially the capacitor. This can be very useful in improving the stability of many circuits. As a rule, network form 1 is preferable when X_{C_2} and/or f are low, while network form 2 is preferable when X_{C_2} and/or f are high.

It is very important to note that the form 1 and form 2 networks are valid only when $C_2' \ggg C_{ce'} + C_{2M}$, in Fig. 5.9b. This is usually true for $C_{ce'}$. However, there are circuits where C_{2M}, which is the Miller effect capacitance, is comparable in value to C_2': see, for example, Sections 10.2.3 and 10.2.4, and Design Example 10.1 in Chapter 10. In those circuits, the form 1 network may still be used provided $C_{ce'} + C_{2M}$ is lumped into L_N at f_c. The form 2 network should not be used as it is too complicated to compensate for large values of C_{2M}. In any event, oscillator circuits having large values of C_{2M} should be avoided as they tend to be unstable and difficult to manufacture, as pointed out in Chapter 10.

5.7 LOAD IMPEDANCE TRANSFORMATION

Section 1.4.2(2) states that one important function of the oscillator is to transform the oscillator load impedance, R_L, into the actual load impedance,

5.7 Load Impedance Transformation

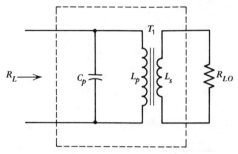

Figure 5.14 Transformer coupling network.

R_{LO}, that the oscillator will feed. In this section, some circuits for the transformation are briefly described.

5.7.1 The Transformer Coupling Network

Figure 5.14 shows the transformer coupling network. T_1 is the main component and usually consists of a toroid having a turns ratio

$$n \approx \sqrt{\frac{R_L}{R_{LO}}} \qquad (5.99)$$

Care must be taken that the windings are tightly coupled, otherwise Eq. (5.99) will be incorrect.

$$2\pi f L_p \text{ is usually made about } R_L/10 \qquad (5.100)$$

producing a working Q of about 10

$$C_p = \frac{25{,}300}{f^2 L_p} \qquad (5.101)$$

5.7.2 Capacitive Divider Coupling Networks

The network is shown in Fig. 5.15, where

$$\frac{R_L}{R_{LO}} \approx \left(\frac{C_1 + C_2}{C_1}\right)^2 \qquad (5.102)$$

$$L_p = \frac{25{,}300}{f^2 C_1 C_2 / (C_1 + C_2)} \qquad (5.103)$$

$$\frac{159{,}000}{fC_2} < \frac{R_{LO}}{5} \qquad (5.104)$$

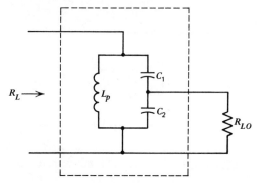

Figure 5.15 Capacitive divider coupling network.

5.7.3 Capacitive Coupling Network

This network is the simplest and is shown in Fig. 5.16.

$$\frac{159{,}000}{fC_1} = \sqrt{R_L R_{LO}} \tag{5.105}$$

$$\frac{159{,}000}{fC_1} > 5R_{LO} \tag{5.106}$$

$$L \approx \frac{25{,}300}{f^2 C_1} \tag{5.107}$$

It has the disadvantage that it increases the harmonic content of the output.

A similar circuit is where C_1 and L are interchanged. This circuit has the advantage that the output harmonic content is reduced.

5.7.4 PI Coupling Network (See Fig. 5.17)

The simplified approximate design is

$$\frac{R_L}{R_{LO}} = \left(\frac{C_2}{C_1}\right)^2 \tag{5.108}$$

$$L = \frac{25{,}300}{f^2 C_1 C_2 / (C_1 + C_2)} \tag{5.109}$$

$$\frac{159{,}000}{fC_2} < \frac{R_{LO}}{5} \tag{5.110}$$

This circuit has the advantage that it reduces the harmonic content.

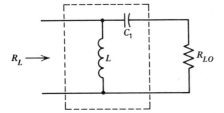

Figure 5.16 Capacitive coupling network.

5.7.5 General Remarks

In the above circuits X is in Ω, L is in μH, C is in pF, and f is in MHz. Many of the components in the above circuits may be combined with components already in the oscillator circuitry so that the coupling networks will actually require fewer components than shown.

5.8 f_T AND β_o REQUIREMENTS

In Section 5.5.2, assumption and condition 6 states that the design in this chapter is based upon the condition that the effect of $r_{bb'}$ is negligible. This condition imposes values for f_T and β_o.

5.8.1 Minimum Value of f_T

It is easily shown from Eqs. (2.63) and (2.75) that $X_{C_{be'}}$, which is the reactance of $C_{be'}$ in Fig. 5.9a, is

$$X_{C_{be'}} = \frac{f_T}{g_m f} \qquad (5.111)$$

Assuming that, for this discussion,

$$C_{bed} \gg C_{bet}$$

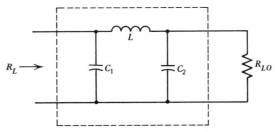

Figure 5.17 PI coupling network.

if

$$X_{C_{be}} > 10 r_{bb'} \qquad (5.112)$$

$r_{bb'}$ will have a negligible effect.

Assuming that $r_{bb'} = 60\ \Omega$, which is pessimistic, and solving for f_T

$$f_T > 600 g_m f \qquad (5.113)$$

5.8.2 Minimum Value of β_o

Again, it is easily shown for $r_{be'}$ in Fig. 5.9a

$$r_{be'} = \frac{\beta_o}{g_m} \qquad (5.114)$$

if

$$r_{be'} > 20 r_{bb'} \qquad (5.115)$$

$r_{bb'}$ will have a negligible effect.

Solving for β_o and assuming $r_{bb'} = 60\ \Omega$

$$\beta_o > 1200 g_m \qquad (5.116)$$

It should be noted that if β_o does not satisfy Eq. (5.116), it can be compensated for by a relationship similar to Eq. (2.53).

$$g_m = \frac{1}{1/g'_m + r_{bb'}/\beta_o} \qquad (5.117)$$

or else it can be compensated for in the trimming procedure, if necessary.

5.9 FREQUENCY STABILITY ANALYSIS

5.9.1 Introduction

As stated in Eq. (5.46), the oscillator will operate at the frequency at which, to a first approximation,

$$X_3 = -(X_1 + X_2) \qquad (5.46)$$

Any factor that causes one of the quantities in Eq. (5.46) to change will cause a change in frequency. It is therefore appropriate to investigate more closely what these quantities are and what may cause frequency changes. For the purpose of analysis

$$X_3 = F(f) \tag{5.118}$$

and must be described as invariable with respect to all parameters except f. X_3 may also have an inherent frequency, as defined at specified conditions. The overall circuit frequency stability is deteriorated by X_1 and X_2, and one of the major goals of good circuit design procedure is to minimize this deterioration.

From Eqs. (5.46) and (5.118) one obtains

$$\Delta f = \frac{-\Delta X_1 - \Delta X_2}{dX_3/df} \tag{5.119}$$

$$= \frac{-\Delta(X_1 + X_2)}{dX_3/df} \tag{5.119a}$$

also,

$$\Delta X_n = \sum_{m=1}^{m} \frac{\partial X_n}{\partial U_m} \Delta U_m \tag{5.120}$$

where $m = 1, 2,$ and so on $\tag{5.121}$
and where U_m could be any parameter, such as temperature, frequency, time, voltage, and so on.

X_1 and X_2 are contributed by the circuit and are determined by the power output, power in Z_3, power conversion efficiency, transistor characteristics, and output load characteristics. In general, strong efforts are made to make X_1 and X_2 small so that the transistor and load will make only small contributions to the total X_1 and X_2 and, therefore, can produce only small variations in X_1 and X_2. However, there are limits as to how small X_1 and X_3 can be made. In addition, X_3 according to Eq. (5.46) also becomes smaller as $X_1 + X_2$ decreases and this becomes undesirable.

5.9.2 The Stabilizing Capacitor C_L'

In order to reduce the effect of $\Delta(X_1 + X_2)$ on the frequency, a stabilizing capacitor is inserted as shown in Fig. 5.18. This capacitor has the following characteristics:

1. It is of the most stable type.
2. Its value is as small as is consistent with the stray elements across it, which should be negligible.
3. Its value is that required to make X_3 the correct value.

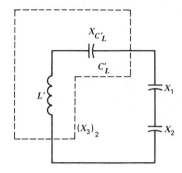

Figure 5.18 Circuit with stabilizing capacitor C'_L.

Equation (5.46) now becomes

$$(X_3)_2 = X'_3 + X_{C'_L} = -(X_1 + X_2) \tag{5.122}$$

where

$$X'_3 = X_3 - X_{C'_L} \tag{5.123}$$

and Eq. (5.119a) becomes

$$\Delta f = \frac{-\Delta(X_1 + X_2)}{d(X'_3 + X_{C'_L})/df} \tag{5.124}$$

If $(X_3)_1$ is an inductor, L_3, then

$$\frac{d(X_3)_1}{df} = (2\pi L_3) = \frac{-(X_1 + X_2)}{f} \tag{5.125}$$

If $(X_3)_2$ is the network of Fig. 5.18,

$$\frac{d(X_3)_2}{df} = 2\pi L'_3 + \frac{1}{2\pi C'_L f^2}$$

$$= \frac{-(X_1 + X_2 + 2X_{C'_L})}{f} \tag{5.126}$$

from Eq. (5.122) and it will be noted that

$$a_{dx} = \frac{d(X_3)_1/df}{d(X_3)_2/df} = 1 + \frac{2X_{C'_L}}{X_1 + X_2} \tag{5.127}$$

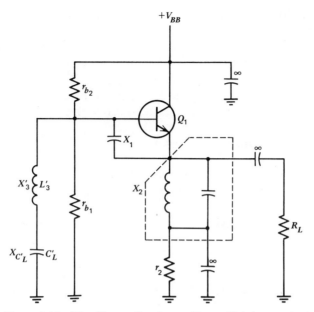

Figure 5.19 The Clapp–Gouriet oscillator (Colpitts version).

which means that

$$\frac{\Delta f_2}{\Delta f_1} = \frac{1}{a_{dx}} \tag{5.128}$$

which may be a substantial improvement in the stability.

This technique of reducing Δf is used in the Clapp–Gouriet oscillator shown in Fig. 5.19.

If X_3 is a crystal, C'_L has no significant effect in stabilizing the frequency, but it does control the operating reactance of the crystal as determined by Eq. (5.122). The calculation of the required value of C'_L is presented in Section 7.2.6.4.

5.9.3 Stability Analysis in Crystal Oscillators

From Equation (3.19)

$$\frac{dX_3}{df} \approx \frac{1}{\pi C_1 f^2} = \frac{2 X_{C_1}}{f} \tag{5.129}$$

so that

$$\frac{\Delta f}{f} = \frac{-\Delta(X_1 + X_2)}{2 X_{C_1}} \tag{5.130}$$

so that the crystal C_1 should be minimum in order for the circuit to have minimum effect upon the frequency stability.

It should be noted that Eq. (5.129) is only approximate since it does not take into account the effect of C_0, which increases the value of dX_3/df as f approaches the crystal antiresonant frequency f_a. One can take advantage of this effect, but it is unwise to operate too close to f_a because X_3 then becomes a critical function of the stray circuit elements across the crystal.

5.9.4 Causes of Variations in X_1 and X_2

Figure 5.9b shows the composition of X_1 and X_2. In this figure, the components, shown in solid lines, are physical wired-in components, while those shown in dashed lines are contributions of the transistor.

Variations in the capacitative elements will of course produce directly variations in X_1 and X_2, while variations in the resistive elements will indirectly produce variations in X_1 or X_2 in accordance with Eq. (1.8). One cause of variation is aging.

Obviously, all the components are temperature sensitive, and temperature changes will produce changes in the component values. Some of the components, particularly the transistor capacitances, are voltage sensitive so that changes in power supply voltages will produce changes in these components.

Again $C_{be'}$ and $r_{be'}$ are transistor current sensitive, so that changes in the transistor characteristic or changes in the power supply which cause the transistor current to change will also cause $C_{be'}$ and $r_{be'}$ to change. Another aspect of this effect is discussed in Section 7.3.2.

R_L is determined by conditions external to the oscillator. If R_L is changed for any reason, X_2 will change, which in turn will cause a frequency shift, Δf. The magnitude of Δf is determined by ΔR_L and the rest of the circuit. $\Delta f/f$ for a specified ΔR_L is one measure of the goodness of the oscillator and is sometimes called the oscillator isolation.

6
Limiting

6.1 INTRODUCTION

In Chapter 1 (Section 1.1.1) we stated that an oscillator must include a nonlinearity called *limiting* which maintains the oscillation amplitude in stable equilibrium. This section presents an explanation of the limiting processes and the various types of limiters. This has been only sketchily treated in the literature, and an attempt is therefore made to develop a formal logical treatment of this rather difficult and somewhat vague subject.

By limiting is meant the reduction of the effective transconductance or loop gain of an oscillator as the amplitude increases. The limiter may take the following basic forms:

1 Self-limiter—which means that the amplifier producing the oscillations limits its gain by its own nonlinear action. Most oscillators use a single transistor as the amplifier and limiter and are, therefore, the self-limiting type. All the oscillators for which algorithms are supplied in this book fall into this category. Because of the simplicity of the circuitry, this type of limiting is also the most difficult to analyze and some phases have not as yet been completely and satisfactorily quantitatively explained.

In general, oscillators using this type of limiting do not have as good short-term performance as oscillators with other limiter types, but surprisingly good performance is obtainable, even with crystals operating at very low drive power.

2 External limiter—which means that limiting is performed by auxiliary devices or circuits such as diodes, symmetrical clippers, and transistors. In two-transistor oscillators, one transistor is used specifically for limiting purposes. The external limiter is discussed in Chapter 13, except that diode limiting is treated in Section 6.5.

3 Automatic Level Control Limiting (ALC)—in which the rectified output voltage is compared to a dc reference voltage. The difference voltage is then used to adjust the amplifier gain to the value demanded by the oscillator circuit. This type of limiting produces the best short- and long-term performance. It is discussed in Chapter 14.

6.2 THE SELF-LIMITING SINGLE BIPOLAR TRANSISTOR OSCILLATOR, SINUSOIDAL v_{BE} OR $v_{BE'}$

6.2.1 Types of Limiting Actions

There are two distinct types of limiting actions in this oscillator.

1 Base emitter cutoff, or current limiting in which the fundamental Fourier component of the collector current is reduced by swinging the base emitter voltage below the contact potential. The minimum instantaneous collector voltage then exceeds the peak base voltage at equilibrium. The latter condition exists where

$$V_{CE} > 1.4(V_{ce} + V_{be}) + 700 \text{ mV}$$

2 Collector voltage limiting in which the fundamental component of the collector current is reduced by the potential of the collector swinging close to the base potential, thus drastically reducing the collector emitter resistance r_{ce}. This type of limiting is sometimes called collector voltage bottoming.

It should be noted that the material in this section is applicable to both the common emitter and common base transistor applications.

6.2.2 Advantages and Disadvantages of Each Type of Limiting

1 In a large production run of the same design, if the resonators, Z_3, have approximately the same resistance, R_3, the *be* cutoff limiter is definitely superior. However, it is difficult to economically maintain the resistance of supposedly identical quartz crystals equal. In fact, military specifications only specify the maximum resistance for a given crystal type and therefore the acceptable resistance is permitted to cover a wide range. This results in a wide range of crystal drive and power output in the same oscillator when the *be* cutoff limiter is used, and each unit will have to be individually adjusted if constant power output or crystal drive is desired. On the other hand, when the base collector limiter is used, the same crystals will produce essentially constant power output although the crystal drive will vary as the crystal resistance varies.

2 The *be* cutoff limiter has the advantage that its frequency versus power supply voltage characteristic is better than when the collector voltage limiter is used. The effect is explained in Section 7.3.2.

6.2 The Self-Limiting Single Bipolar Transistor Oscillator, Sinusoidal v_{BE} or $v_{BE'}$

This advantage may not be very important now since closely regulated, economical power supplies are very readily available.

3 The *be* cutoff limiter has the further advantage that often it produces smaller operating Q degradation than the collector voltage limiter.

4 On the other hand, the oscillator circuit using collector voltage limiting requires less labor to design and analyze, as will be evident in later chapters during the presentation of the design and analysis procedures.

5 The *be* limiter will permit higher-frequency operation of the oscillator since V_{CE} is higher and therefore f_T is larger.

6 When the crystal drive in crystal oscillators is very small, the base emitter limiter (or the auxiliary diode limiter described in Section 6.5) must be used for the reasons given in Section 6.5.

6.2.3 The Base Emitter Cutoff Limiter

The theory of operation of this type of limiter has already been presented in Section 2.4. However, it is interesting and instructive to demonstrate the limiting action. This is done, using the circuit of Fig. 6.1, as follows:

1   r_{b_2}, r_{b_1}, and r_2 are selected in accordance with the procedure described in Section 2.5.

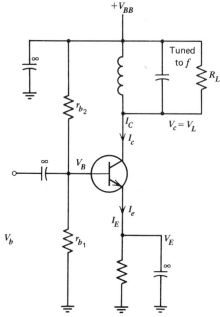

Figure 6.1   Circuit for analyzing the base emitter cutoff limiter.

172 Limiting

2. V_{BB} and R_L are selected so that $V_{BB} - 1.4V_{c_{max}} > V_B + 1.4V_{b_{max}}$ to make certain that only base emitter cutoff limiting will take place.
3. V_b is varied, and for each value of V_b, V_c is recorded.
4. Then

$$I_c = \frac{V_c}{R_L} \tag{6.1}$$

$$g_m = \frac{I_c}{V_b} = \frac{\gamma_1 I_C}{V_b} \tag{6.2}$$

$$\alpha = \frac{g_m}{g_{m_0}} \tag{6.3}$$

where g_{m_0} is the value of g_m for small V_b (small-signal condition). γ_1 in Eq. (6.2) and α in Eq. (6.3) versus V_b are already plotted in Fig. 2.12.

6.2.4 The Collector Base Voltage Limiter

If the precautions specified in Section 6.2.3, item 2, are not taken and R_L is made too large, or V_{BB} too small, v_C approaches v_B and the limiting condition exists when $v_B \approx v_C$ because transistor action cannot exist when $v_B > v_C$.

The theory of collector base voltage limiting has not been extensively developed and very little information on a quantitative basis is available in the literature. This section will therefore develop a theoretical procedure which offers some quantitative treatment and is in reasonably good agreement with experiment.

Figure 6.2a shows the schematic of a circuit suitable for analyzing the base collector voltage limiter. Figure 6.2b shows the voltage relations that exist when limiting starts. Figure 6.2c shows the voltage relations that exist when the input signal V_b exceeds the limiting starting point. Note the broadening of the v_C curve at the bottom, leading to the name *bottoming effect*. The departure of the v_C wave shape from a sine wave is, of course, a function of the R_L circuit effective Q.

In both Figs. 6.2b and 6.2c the dc voltage difference between base and emitter is shown as approximately 700 mV although it varies with signal level due to bias shifts, but the voltage change has little effect on the total performance because of the biasing circuitry used. From Fig. 6.2 one obtains

$$V_{BB} - 1.4V_L = V_B + 1.4V_B - 700$$

and

$$V_B = V_E + 700$$

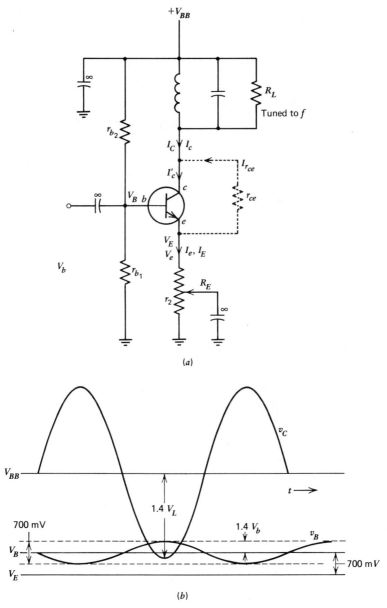

Figure 6.2 Circuit for analyzing the collector base voltage limiter. (a) Schematic. Notes: $I_C \approx I_E$. $I_o \approx I_e$. (b) Wave shapes at the limiting starting point. (c) Wave shapes past the limiting starting point.

174 Limiting

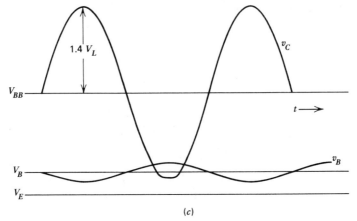

(c)

Figure 6.2 (Continued).

from which

$$V_E = V_{BB} - 1.4(V_L + V_b) \tag{6.4}$$

Figure 6.2a shows the transistor output resistance r_{ce}, drawn in dashed lines, which normally approaches ∞ before limiting takes place, but becomes progressively smaller as the input signal is increased. It will be seen that as the signal starts from 0 and increases, all the I'_c current goes to the load circuit. Past the limiting point, part of the total current I'_c goes to the load circuit as I_c and the remainder goes into r_{ce} as $I_{r_{ce}}$. The division of the total current changes as a function of the input signal V_b and the dc current I_E.

Also, past the limiting point, I'_c and $I_{r_{ce}}$ increase and I_c remains essentially constant.

By definition,

$$g_m = \frac{I_c}{V_b} \tag{6.5}$$

$$g_{m_L} = \frac{I'_c}{V_b} \tag{6.6}$$

and

$$A_L = \frac{g_{m_L}}{g_m} = \frac{I'_c}{I_c} \tag{6.7}$$

From Eqs. (2.65) and (2.50)

$$g_{m_{L_0}} = \frac{1}{26/I_E + R_E} \quad \text{at } T = 300°\text{K} \tag{6.8}$$

6.2 The Self-Limiting Single Bipolar Transistor Oscillator, Sinusoidal v_{BE} or $v_{BE'}$

assuming β is reasonably large and $r'_e = r_{bb'} = 0$. From Figs. 2.12b and 2.13c, past the limiting point,

$$\gamma_1 = \frac{I'_{c_{max}}}{I_C} = 1.4 \tag{6.9}$$

6.2.4.1 Limiting when $R_E = 0$

Assume that, for maximum efficiency, $I'_c = I'_{c_{max}}$ and $r_{bb'} = 0$, $r'_e = 0$, and $R_E = 0$. Then

$$V_b = V_{be} \tag{6.10}$$

and from Eq. (6.8)

$$g_{m_{L_0}} = \frac{I_C}{26} \tag{6.11}$$

From Eqs. (2.70), (6.6), (6.10), and (6.11),

$$\alpha V_b \frac{I_C}{26} = I'_c \tag{6.12}$$

and from Eq. (2.79) for $V_{be} \gtrsim 113$ mV

$$36 \frac{I_C}{26} = I'_c$$

or

$$I_C \approx 0.7 I'_c \tag{6.13}$$

which is consistent with Eq. (6.9), or, from Eq. (6.7)

$$I_C = 0.7 A_L I_c \tag{6.14}$$

$V_b \equiv V_{be}$ will now be computed from Eqs. (6.5) and (6.6).

$$V_b = \frac{I'_c}{g_{m_L}} = \frac{I_c}{g_m} \tag{6.15}$$

$$= \frac{I'_c}{\alpha g_{m_{L_0}}}$$

$$\approx \frac{36}{\alpha} \text{ mV} \tag{6.16}$$

from Eqs. (6.11) and (6.13) where $\alpha < 0.3$ to make certain that Eq. (6.14) is valid.

Consider the following example:

$V_{BB} = 10$ V
$R_L = 570 \ \Omega$
$V_L = 2$ V
$R_E = 0, \ r'_e = 0$
$A_L = 2$

Then

$$I_c = \tfrac{2}{570} = 3.5 \text{ mA}$$

$$I'_c = 2(3.5) = 7.0 \text{ mA}$$

$$I_C = 0.7(7) = 4.9 \text{ mA}$$

$$I_{r_{ce}} = 3.5 \text{ mA}$$

$$V_b \approx \frac{0.36}{\alpha}$$

For $\alpha = 0.3$, $V_b \approx 120$ mV

$$g_{m_L} \approx \tfrac{7}{120} \approx 0.058$$

$$g_m \approx 0.029$$

For $\alpha = 0.1$, $V_b \approx 360$ mV

$$g_{m_L} \approx 0.0194, \qquad g_m \approx 0.0097$$

6.2.4.2 Limiting When $R_E \gg r_{e_0}$

It is desirable to stabilize the characteristics of the transistor. This is possible by making R_E a value large compared to r_{e_0}. This is done at the expense of decreasing $I_{r_{ce}}$, as shown in the following example:

$$V_{BB} = 10 \text{ V}$$

$$R_L = 570 \ \Omega$$

$$V_L = 2 \text{ V}$$

$$R_E = 5 r_{e_0}, \qquad r'_e = 1 \ \Omega$$

$$A_L = \ ?$$

then $I_c = 2/570 = 3.5$ mA.

6.2 The Self-Limiting Single Bipolar Transistor Oscillator, Sinusoidal v_{BE} or $v_{BE'}$

I_C will be kept at 4.9 mA = $1.4I_c$ since a computer analysis showed that very little is gained by increasing I_C.

$$r_{e_0} = \frac{26}{4.9} + 1 = 6.3 \, \Omega$$

$$R_E = 5(6.3) = 31.5 \, \Omega$$

$$g_{m_{L_0}} = \frac{1}{6(6.3)} = 0.026 \, \mho$$

Assume

$$g_m = \frac{g_{m_{L_0}}}{2} = 0.013 \, \mho$$

then

$$V_b = \frac{3.5}{0.013} = 269 \text{ mV}$$

From Fig. 2.13c, $\gamma_1 = 1.2$ for $\rho = 5$ so that $I'_c = 1.2(4.9) = 5.88$ mA $= 1.7I_c$ and $I_{r_{ce}} = 2.38$ mA $= 0.7I_c$.

This design is still considered satisfactory because there is a substantial $I_{r_{ce}}$, although not as much as described in Section 6.4.1. In the algorithms for oscillators using this type of limiting, this design will be used. It should be further noted that the design is based on crystals having the maximum resistance. When the resistance decreases, I_c decreases, I'_c remains constant, and $I_{r_{ce}}$ increases, which implies that this design is safe.

To formalize the design, the following equations are developed:

$$I_C = 0.7 A_{L_0} I_c \tag{6.17}$$

$$A_{L_0} = \frac{g_{m_{L_0}}}{g_m} \tag{6.18}$$

$$g_{m_{L_0}} = \frac{1}{r_{e_0} + R_E} \tag{6.19}$$

$$= \frac{1}{R_E + 26/I_E + 1} \tag{6.19a}$$

178 Limiting

for the special case where

$$R_E = 5r_{e_0} = 5\left(\frac{26}{I_E} + 1\right) \quad (6.20)$$

$$g_{m_{L_0}} = \frac{1}{R_E(1 + \frac{1}{5})} = \frac{0.83}{R_E} \quad (6.20a)$$

$$= \frac{1}{6(26/I_E + 1)} \quad (6.20b)$$

$$R_{in} = \frac{X_1^2}{r_{be}}, \quad = g_m \frac{X_1^2}{\beta_o} \quad (6.21)$$

$$C_{bed} = g_m \frac{159{,}000}{f_T} \quad (6.22)$$

6.2.4.3 The Uses of the Design Procedures of Sections 6.2.4.1 and 6.2.4.2

This section will discuss when each of the above design procedures should be used.

The method in Section 6.2.4.2 is generally preferred because it minimizes transistor variations and it reduces phase noise. However, it requires a specific V_b which at high frequencies may not be realizable for a given crystal current. It will be noted from the example that V_b in the example of Section 6.2.4.1 is 120 mV while the equivalent V_b in the example of Section 6.2.4.2 is 259 mV. Also, for a given crystal drive the procedure of Section 6.2.4.1 permits a wide range of V_b for the same A_{L_0}, while the V_b for the procedure of Section 6.2.4.2 is fixed.

6.3 THE SELF-LIMITING SINGLE BIPOLAR TRANSISTOR OSCILLATOR, SINUSOIDAL i_E

6.3.1 Types of Limiting Actions

Again in this oscillator there are two distinct types of limiting actions:

1 R_{IN} variation as a function of I_e, as described in Section 2.5.2.2.
2 Collector voltage limiting which is identical to that described in Section 6.2.

6.3.2 Advantages of Each Type of Limiting

Section 6.2.2 applies except that when be cutoff limiting is mentioned; read R_{IN} variation limiting.

6.3.3 The R_{IN} Limiter

The theory of this type of limiter is presented in Section 2.5.2.2. The following additional material is presented to adapt this theory to oscillator design.

1. To ensure that this type of limiting is operative, $v_{C_{min}}$ must always exceed $v_{B_{max}}$.
2. For the most effective limiting, it is advisable that Eq. (2.97) be modified to

$$I_E \approx 1.4 I_e \qquad (6.23)$$

so that

$$\gamma \approx 1 \qquad (6.24)$$

It will be seen from the extrapolated dotted portion of Fig. 2.15 that, at this operating point, R_{IN} can assume any value larger than $2/g_{m_0}$ as demanded by the circuit equilibrium conditions. (See Chapter 11 for the effect on the operating Q.)

3. The requirements of items 1 and 2 fix the other operating points and the associated component values.

6.3.4 The Collector Base Voltage Limiter

Essentially, all the material presented in Section 6.2.4 is applicable.

6.4 THE SELF-LIMITING SINGLE JUNCTION FIELD EFFECT TRANSISTOR OSCILLATOR, SINUSOIDAL v_{GS}

Again, in this oscillator, there are two distinct types of limiting.

1. Clamped bias limiting, described in Section 2.6.4. This type of limiting is analogous to the base emitter cutoff limiting for the bipolar transistor and uses the same design procedure.
2. Drain gate voltage limiting, which is almost identical to the collector base voltage limiting for the bipolar transistor described in Section 6.2.4.

6.5 DIODE LIMITER (SEE ALSO SECTION 13.5.2)

Diode limiting is really a special form of collector limiting.

Equation (6.4) and Figs. 6.2b and 6.2c show that when V_L is small, setting the various bias points to the necessary precision becomes very difficult as V_{CB}, which is now required to be very small, is the difference of two large quantities,

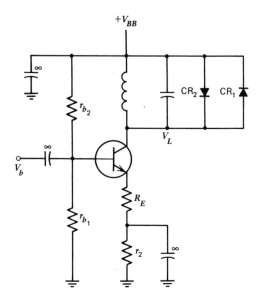

Figure 6.3 Diode limiting circuit.

V_C and V_B. This is further complicated by the bias differences that exist between the nonoscillating and the oscillating states, and the bias shifts that are caused by temperature variations.

One method of resolving these difficulties is to use auxiliary diodes as shown in Fig. 6.3. In this circuit V_L is limited by the nonlinear characteristics of CR_1 and CR_2 which tend to maintain V_L constant over a wide range of the current in the diodes. A good approximation is that V_L will be 0.5 V for silicon diodes.

Sometimes CR_2 is omitted with approximately the same results. However, the output waveform is more symmetrical when both diodes are present and the frequency voltage sensitivity is smaller.

This circuit does not completely keep V_L constant. This is due to the fact that the diodes do not maintain a strictly constant voltage as the current varies and, in addition, the diode voltage varies with temperature. The temperature variation can be compensated by a thermistor-controlled circuit, but additional complications in design and reliability are introduced.

6.6 PRINCIPAL REFERENCE

Reference 2.3 contains extensive material on nonlinear processes suitable for limiting, including complete exposition of the large-signal transistor characteristics treated in Chapter 2.

7

The Normal Pierce Oscillator

7.1 INTRODUCTION

This chapter develops the additional and/or modified relations of Chapter 5 required to design this type of oscillator. It also contains comments about the practical aspects of the design. Finally, it formulates two design algorithms—one for each of the major types of limiting of Section 6.2. The algorithms are specifically applicable to the self-limiting types of crystal oscillator, but they are also useful for non-self-limiting types and oscillators with other two-terminal resonators.

By the normal Pierce oscillator is meant the oscillator first shown in Fig. 5.1a and now repeated in a modified form in Fig. 7.1. All the components in this figure are physical; that is, they are all installed components. However, there are additional resistors and capacitors which form part of the transistor. This chapter will show how to compute the values of the physical components which when combined with the transistor properties will produce a prespecified oscillator performance.

7.2 OSCILLATOR CIRCUIT ANALYSIS

7.2.1 Introduction

As stated in Section 5.5.1, the Pierce circuit is basically the same as the Colpitts oscillator and all the theory developed in Chapter 5 for the Colpitts oscillator

182 The Normal Pierce Oscillator

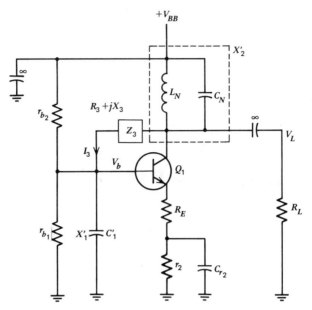

Figure 7.1 Pierce oscillator detailed schematic.

applies, with slight simplifications, to the Pierce oscillator. Since the analysis of the Colpitts oscillator was based upon the detailed ac circuit schematics of Fig. 5.9, it is instructive to show similar schematics for the Pierce oscillator.

Figures 7.2a, 7.2b, and 7.2c are analogous schematics for the Pierce oscillator corresponding to Figs. 5.9a, 5.9b, and 5.9c for the Colpitts oscillator, and Section 5.5 should be consulted for the derivation of these schematics. It will be noted that the only difference is, apart from the change of datum point (ground), that r_b is across X_1 in the Pierce oscillator, while r_b is across $X_1 + X_2$ in the Colpitts oscillator. This seemingly trivial difference has the following important effects:

1. r_b has a considerably smaller loading effect and results in a higher operating Q in the Pierce oscillator.
2. The Pierce oscillator is easier to design.

7.2.2 Modified Equations

Some of the equations derived in Chapter 5 have to be modified as follows:
Equation (5.40) now becomes

$$R_T = R_3 + \frac{X_1^2}{r_{be'}} + \frac{X_1^2}{r_b} + \frac{X_2^2}{R_L} = g_m X_1 X_2 \qquad (7.1)$$

7.2 Oscillator Circuit Analysis

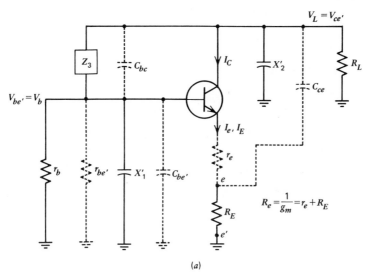

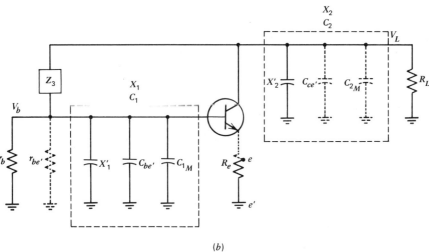

Figure 7.2 Pierce oscillator. (a) Detailed schematic of the ac circuitry. *Note:* $i_C \approx i_E$. (b) Modified schematic, Step 1. (c) Modified schematic, Step 2.

where by definition (see Section 5.5.3) and Eq. (5.40)

$$\frac{X_1^2 g_m}{\beta_o} = \frac{X_1^2}{r_{be'}} \equiv R_{in} \qquad (7.2)$$

$$\frac{X_1^2}{r_b} \equiv R_b \qquad (7.3)$$

$$\frac{X_2^2}{R_L} \equiv R_2 \qquad (7.4)$$

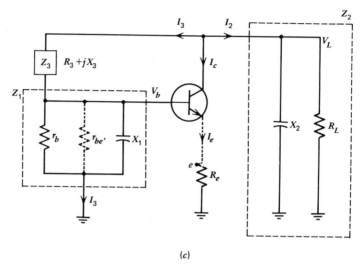

(c)

Figure 7.2 (*Continued*).

Also,

$$V_b = I_3 Z_1 \approx -jI_3 X_1 \tag{7.5}$$

and

$$|V_b| \approx |I_3||X_1| \tag{7.5a}$$

Equation (5.59) becomes

$$\frac{|I_3|}{|I_e|} = \frac{|X_2|}{R_T} \tag{7.6}$$

where R_T is defined in Eq. (7.1). Equation (5.65) becomes

$$V_L = I_3[R_3 + R_{in} + R_b + jX_2] \tag{7.7}$$

or

$$|V_L| = |I_3|\sqrt{(R_3 + R_{in} + R_b)^2 + X_2^2} \tag{7.8}$$

The relationships presented in this chapter deserve greater elaboration, but an adequate presentation is prohibited by space limitations. Equations (7.1) to (7.7) particularly merit close examination, as their significance changes substantially with the type of limiting. Some of these aspects will be treated in the following sections.

7.2.3 Fundamental Design Equations

By definition

$$|I_c| = g_m|V_b| = g_m|I_3|X_1 \tag{7.8}$$

from Eq. (7.5a). By inspection

$$1000 P_L = \frac{V_L^2}{R_L} \tag{7.9}$$

$$1000 P_{Z_3} = I_3^2 R_3 \tag{7.9a}$$

Let

$$\eta \equiv \frac{P_L}{P_{Z_3}} = \frac{V_L^2}{I_3^2 R_3 R_L} \tag{7.10}$$

or

$$\eta = \frac{(R_3 + R_{in} + Rb)^2 + X_2^2}{R_3 R_L} \quad \text{from Eq. (7.7)} \tag{7.11}$$

$$= \frac{(R_3 + R_{in} + R_b)^2 + R_2 R_L}{R_L R_3} \quad \text{from Eq. (7.4)} \tag{7.12}$$

solving for R_2

$$R_2 = \frac{\eta R_L R_3 - (R_3 + R_{in} + R_b)^2}{R_L} \tag{7.13}$$

$$= \eta R_3 - \frac{(R_3 + R_{in} + R_b)^2}{R_L} \tag{7.14}$$

When a design is first started, the values of R_{in} and R_b are unknown. If it is assumed that

$$R_3 \gg R_{in} + R_b \tag{7.15}$$

then Eq. (7.14a) becomes

$$R_2 \approx \eta R_3 - \frac{R_3^2}{R_L} \tag{7.16}$$

and from Eq. (7.1),

$$R_T \approx R_3 + R_2 \tag{7.17}$$

Also from Eq. (5.77),

$$Q_{op} \approx \frac{Q_{Z_3} R_3}{R_3 + R_2} = \frac{Q_{Z_3}}{1 + (R_2/R_3)} \tag{7.18}$$

which becomes, from Eq. (7.16),

$$Q_{op} \approx \frac{Q_{Z_3}}{1 + \eta - R_3/R_L} \tag{7.19}$$

Usually,

$$1 + \eta \gg \frac{R_3}{R_L} \tag{7.20}$$

therefore

$$Q_{op} \approx \frac{Q_{Z_3}}{1 + \eta} \tag{7.21}$$

which states that for small degradations of Q, η must be small.

If a degradation of Q_{Z_3}, of approximately 25% is allowed, then

$$\eta \approx 0.33$$

which is a recommended value.

Equations (7.10) and (7.21) state that for small Q degradation, the output power is much smaller than the Z_3 power.

Equation (7.16) should be further investigated. This equation states

$$R_2 \approx \eta R_3 - \frac{R_3^2}{R_L}$$

Occasionally, R_2 calculates as being a negative quantity, which is obviously physically impossible. What it does mean is that the design is unsound. The only way of correcting this defect is to increase R_L, which in turn requires increasing V_L, which in turn requires increasing the BV_{CE} of the transistor and, possibly, V_{BB}.

7.2.4 Discussion of $V_{L_{max}}$ and R_L

Obviously, the larger V_L is, the greater is the power output, but $V_{L_{max}}$ is limited by the BV_{CE} of the transistor; and the relationship of BV_{CE} to $V_{L_{max}}$ will now be explored. BV_{CE} is emphasized over BV_{CB} because in every silicon transistor BV_{CB} is considerably larger than BV_{CE}.

The relationships are different for the collector limited and the *be* cutoff oscillators. This difference is caused by the fact that in the collector base voltage limited oscillator, V_L is relatively independent of R_3 for a given oscillator. However, Eqs. (2.79), (7.1), and (7.5) state that as R_3 decreases in an oscillator designed for *be* cutoff limiting, I_3 increases. Equation (7.7), in turn, states that V_L increases, although not as fast as I_3 increases. Particularly, when Z_3 is a crystal, R_3, of necessity, varies from unit to unit and provision must be made in the circuit to allow for the variation.

7.2.4.1 Calculation of $V_{L_{max}}$ for the Collector Base Limiting Oscillator

Figure 7.3 shows the voltage relationships in this type of limiting. V_e is shown approximately equal to V_b to provide a factor of safety, although V_e is appreciably smaller than V_b.

By inspection it is seen that

$$v_{CE_{max}} = v_{C_{max}} - v_{E_{min}}$$

$$\approx 2.8(V_{L_{max}} + V_b) \qquad (7.22)$$

assuming $V_b - V_e$.

But

$$v_{CE_{max}} \leqq BV_{CB}$$

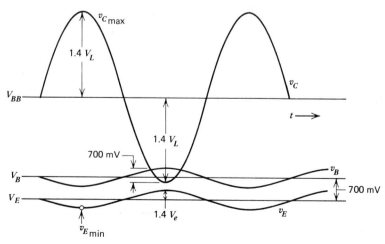

Figure 7.3 Voltage relationships for collector base limiting in the Pierce oscillator.

188   The Normal Pierce Oscillator

therefore,

$$V_{L_{max}} \leq \frac{BV_{CE} - 2.8V_b}{2.8}$$

$$\leq 0.357 BV_{CE} - V_b \qquad (7.23)$$

To compensate for $-V_b$,

$$V_{L_{max}} \approx 0.33 BV_{CE} \qquad (7.23a)$$

7.2.4.2 Calculation of $V_{L_{max}}$ for the be Cutoff Limiting Oscillator

Figure 7.4 shows the voltage relationships in this type of limiting. In this oscillator $v_{B_{max}}$ must be less than $v_{C_{min}}$ for Z_3 having the minimum R_3. Since the oscillator is designed for $R_{3_{max}}$, when V_L is minimum, provision must be made for the increase of V_L as R_3 decreases. Let

$$\frac{R_{3_{max}}}{R_{3_{min}}} = 2 \qquad (7.24)$$

then

$$\frac{V_{L_{max}}}{V_{L_{min}}} \approx 1.5 \qquad (7.25)$$

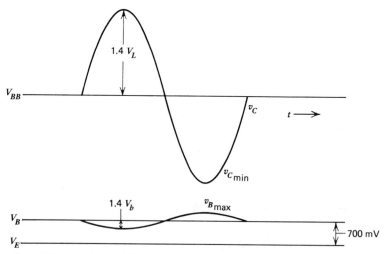

Figure 7.4 Voltage relationships for base emitter cutoff limiting in the Pierce oscillator.

and from Eq. (7.23a)

$$V_{L_{max}} = \frac{0.33}{1.5} BV_{CE} = 0.22 BV_{CE} \qquad (7.26)$$

7.2.4.3 Calculation of V_E, Which Should Be Greater Than 2 V

7.2.4.3.1

It is evident from Fig. 7.3 that for collector limiting,

$$V_E = V_{BB} - 1.4(V_L + V_b) \qquad (7.27)$$

7.2.4.3.2

Similarly, from Fig. 7.4, for *be* cutoff limiting,

$$V_E = V_{BB} - (\geq 2.1)(V_L + V_B) \qquad (7.28)$$

It is very desirable that $v_{CE_{min}} = v_{C_{min}} - v_{E_{max}}$ in Fig. 7.4 exceed 1700 mV to ensure proper operation of the transistor at high frequencies. Therefore, Eq. (7.28) becomes

$$V_E = V_{BB} - (\geq 2.1)(V_L + V_B) - 1700 \qquad (7.28a)$$

Equation (7.28a) sets a value of 1700 mV for $v_{CE_{min}}$. However, it is desirable that $v_{CE_{min}}$ be larger than 1700 mV. This is usually possible for the common values of V_{BB} and for small V_L.

Consider the following example (all units in mV):

$$V_{BB} = 10{,}000, \qquad V_L = 200, \qquad V_b = 113$$

Then, from Eq. (7.28a), $V_E = 7642$.

This results in the following situation:

1. $v_{CE_{min}}$ is calculated to be 1700.
2. r_2 will dissipate an excessive part of the available V_{BB}.
3. V_{CE} will be difficult to maintain because of the tolerances and available standard values of components.

This situation is not tolerable because it can be remedied. It is therefore recommended that the procedure below be followed:

1. Calculate V_{E_1} by Eq. (7.28a).
2. If $V_{E_1} > V_{BB}/2$, make

$$V_E = \frac{V_{BB}}{2} \qquad (7.28b)$$

3 If $V_{E_1} \leq V_{BB}/2$, make

$$V_E = V_{E_1} \tag{7.28c}$$

In the above example, $V_E = 5000$, which now makes $v_{CE_{min}} = 4342$, a preferable value.

7.2.4.4 Calculation of R_L and Its Effect upon V_L

Once $V_{L_{max}}$ is known, R_L is found from

$$R_L = \frac{V_{L_{max}}^2}{1000 P_L} \tag{7.29}$$

Often, R_L computed from Eq. (7.29) is too large and cannot be physically realized at the operating frequency f.

Equation (1.16) states that

$$R_L \leq \frac{10{,}000}{\sqrt{f}}\ \Omega$$

If this value of R_L is substituted into Eq. (7.29),

$$V_L = \sqrt{\frac{10^7 P_L}{\sqrt{f}}} \tag{7.30}$$

for P_L in mW, V_L in mV, and f in MHz. Obviously, V_L must then be made the smaller of the values obtained from Eq. (7.30) and Eqs. (7.23a) or (7.26) as applicable.

7.2.5 Calculation of the Approximate g_m When R_2, X_1, and X_2 Have Already Been Computed

From Eq. (7.1),

$$R_T = R_3 + R_2 + R_{in} + R_b = g_m X_1 X_2$$

and neglecting R_b because it is unknown:

$$R_T \approx R_3 + R_2 + R_{in} \tag{7.31}$$

From Eq. (7.2),

$$R_{in} = \frac{X_1^2}{r_{be'}} \tag{7.32}$$

7.2 Oscillator Circuit Analysis

and from Eq. (5.114)

$$r_{be'} = \frac{\beta_o}{g_m}$$

Therefore,

$$R_T = R_3 + R_2 + \frac{X_1^2 g_m}{\beta_o} = g_m X_1 X_2 \qquad (7.33)$$

Solving for g_m

$$g_m = \frac{(R_3 + R_2)}{X_1(X_2 - X_1/\beta_o)} \qquad (7.34)$$

7.2.6 Some Miscellaneous Reactance Formulas

7.2.6.1

$$C = \frac{1}{2\pi f X_C} \quad \text{(general formula)} \qquad (7.35)$$

or

$$C = \frac{159,000}{f X_C} \qquad (7.36)$$

where C is in pF, f is in MHz, and X_C is in Ω.

7.2.6.2

$$L = \frac{X_L}{2\pi f} \qquad (7.37)$$

where L is in μH, f is in MHz, and X_L is in Ω.

7.2.6.3

$$X_{C_{r_2}} = \frac{0.05}{g_m} \qquad (7.38)$$

$X_{C_{r_2}}$ is made very small compared to $1/g_m$, the dynamic emitter resistance.

7.2.6.4 Calculation of C_L'

A crystal resonator rated at antiresonance is required to work into a load capacitance $C_{L_{df}}$ to be at its rated frequency f. Normally the capacitance equivalent of $X_1 + X_2$ is much larger than $C_{L_{df}}$. Therefore a capacitance C_L' has

192 The Normal Pierce Oscillator

to be connected in series, and it is

$$X_{C_L'} = \left[\frac{159{,}000}{C_{L_{df}} f} - (X_1 + X_2)\right] \qquad (7.39)$$

and

$$C_L' = \frac{159{,}000}{X_{C_L'} f} \qquad (7.40)$$

7.3 FREQUENCY VARIATIONS DUE TO EXTERNAL FACTORS

The frequency variation, caused by changes in load impedance and power supply voltages, is treated in this section.

7.3.1 Frequency Variation Due to Changes in the Load R_L

The treatment will consider changes only of the magnitude of R_L and will assume that the load remains resistive. Extension of the treatment to other types of load variation will be obvious to the reader.

In Fig. 7.2c, the load R_L is across the reactance X_2. Normally, since $R_L \gg X_2$, Z_2 can be written

$$Z_2 = \frac{X_2^2}{R_L} + jX_2 \qquad (7.41)$$

but, in this case, the small changes in X_2 are important, so the complete impedance expression must be used.

$$Z_{2_n} = R_{2_n} + jX_2 = \frac{R_{L_n}(jX_2)}{R_{L_n} + jX_2} = \frac{R_{L_n} X_2^2}{R_{L_n}^2 + X_2^2} + j\frac{R_{L_n}^2 X_2}{R_{L_n}^2 + X_2^2} \qquad (7.42)$$

It is seen that when R_L is changed to R_{L_2} from R_{L_1}, then

$$\Delta X_2 = \frac{R_{L_2}^2 X_2}{R_{L_2}^2 + X_2^2} - \frac{R_{L_1}^2 X_2}{R_{L_1}^2 + X_2^2} \qquad (7.43)$$

Then the fractional frequency shift due to the change in R_L will be

$$\frac{\Delta f}{f} = \frac{\Delta X_2}{\left(\Delta X_3 (\Delta f/f)^{-1}\right)} \qquad (7.44)$$

where $(\Delta X_3 (\Delta f/f)^{-1})$ is defined in Step b10 of the algorithms.

Example

$$X_2 = 120, \quad R_{L_1} = 2000, \quad R_{L_2} = \infty \text{ (open circuit)},$$

$$\left(\Delta X_3 (\Delta f / f)^{-1}\right)_{\text{crystal}} = 2 \times 10^6 \, \Omega$$

then

$$\frac{\Delta f}{f} = \frac{120 - 119.57}{2 \times 10^6} = 2.1 \times 10^{-7}$$

7.3.2 Frequency and Amplitude Variation Due to Change in the Supply Voltage V_{BB}

The frequency variation is very difficult to calculate since it is influenced mainly by minute variations in the transistor characteristics with supply voltage change. However, some useful qualitative conclusions can be obtained.

7.3.2.1 Frequency Variations Independent of Type of Limiting

In general, when the voltage V_{BB} is increased, C_{bet}, C_{ce}, and C_{cb} decrease and f_T increases. All of these changes produce an increase in frequency.

7.3.2.2 Frequency and Amplitude Variations in Oscillators with Emitter Base Cutoff Limiting

The oscillation equation (7.1) is rewritten here with some slight change in notation applicable to the Pierce oscillator.

$$R_T = \frac{X_2^2}{R_L} + R_3 + \frac{X_1^2}{r_b} = g_m \left[X_1 X_2 - \frac{X_1^2}{\beta_o} \right] \quad (7.45)$$

Examination of the terms, X_2^2/R_L, R_3, X_1^2/r_b, X_1, and X_2 discloses that they are independent of signal amplitudes. If it is assumed that β_o is also independent of signal amplitude, then it follows from Eq. (7.45) that g_m is also independent of signal amplitude. If g_m is independent of signal amplitude, all effects, excluding those that are listed in Section 7.3.2.1, that may influence the frequency are also independent of signal amplitude. Therefore, there is no frequency change due to change of V_{BB}.

Equation (7.1) gives no information on the amplitude change due to the change in V_{BB}. To calculate the new signal strength V_L the following procedure is followed:

1 Calculate the new I_E using the equations in Section 2.5.
2 Calculate $I_e = 1.4 I_E$. This is because $\gamma_1 = 1.4$.
3 Calculate $|V_L| = |I_2| X_2$ from Eq. (5.59a).

When the above steps are followed, it will be found that V_L is proportional to I_E, which in turn is proportional to V_{BB}, when V_E is much larger than 700 mV.

7.3.2.3 Frequency and Amplitude Variation with Collector Base Voltage Limiting

As V_{BB} increases, the output increases almost proportionally since V_{CB} increases proportionally and $V_{CB} \approx 1.4(V_L + V_b)$.

The tendency at the same time is for g_{mL} to increase. This produces the following effects:

1. C_{bed} increases.
2. r_{be} decreases.
3. r_{ce} decreases.

The effect of point 1 is to decrease the frequency, while the effects of points 2 and 3 are to increase the frequency, so that it is difficult to be certain whether the frequency will increase or decrease. However, these effects, combined with that of Section 7.3.2.1, usually will cause the frequency to increase as the voltage supply is increased.

7.4 THE DESIGN PROCEDURE

The design procedure can best be described as a series of levels of successive approximations. At each approximation level, the known information is used to calculate additional quantities which are then incorporated into the original information to form the basis for the next approximation level.

The rather arbitrary criterion, for termination of the design procedure, is that it is terminated at the end of the level where

$$R_T < 1.1 R'_T \tag{7.46}$$

where R'_T is the value of R_T at the beginning of the level and R_T is the value at the end. R_T is defined in Eq. (7.1).

The number of levels never exceeds two, so that the design procedure is not excessively long and tedious. In general, the Pierce oscillator design requires only one level of approximation. After the levels of approximation described above have been completed, the resulting quantities are translated into the real physical components which make up the oscillator. This translation is performed taking into account the mode of operation of the resonator.

The procedures described above have been applied to several specific single transistor self-limiting crystal oscillators and are presented in the form of algorithms to render the complete design a formal, logical, and readily executable process (see Chapter 12).

7.5 THE PIERCE OSCILLATOR, COLLECTOR BASE LIMITING

7.5.1 The Design Algorithm for This Oscillator (Algorithm 12.1)

Algorithm 12.1 is contained in Chapter 12. It is recommended that Section 12.2 be read before using the algorithm.

The user has to supply the information requested in Sections a, b, c, and d. It will be noted that items c7, c9, c10, and c11 are not used in the algorithm but checks should be made, after completion of the design, to make sure that these values are not exceeded.

As noted in Eq. (6.18)

$$A_{L_0} = \frac{g_{m_{L_0}}}{g_m}$$

Since the loop gain is proportional to g_m and the loop gain, at stable oscillation, equals 1, it follows that A_{L_0} is the *small-signal* loop gain.

The algorithm has been prepared for oscillators with crystal networks operating in the inductive mode; however, it is easily adaptable for replacement of the crystal network by another two-terminal network, such as an inductor of reactance value $\approx -(X_1 + X_2)$ at the desired frequency.

Section 1.2.1.6 states that the algorithms have been prepared on the basis of maximum practical conversion efficiency. This means that the load resistances are the maximum practical. If for any reason a smaller load resistance, R_L, is desired in the normal Pierce and Colpitts oscillators, it is necessary to

1. Insert the desired R_L in Step 3.
2. Calculate the new V_L corresponding to the new P_L and R_L.
3. Insert the new V_L in Step 2.
4. Continue with the execution of the algorithm.

The algorithms in general, although there are some obvious exceptions, include the design of an SC crystal mode selector making up X_1 (see Section 5.6.5.3). For very small I_{x}, the physical realization of the mode selector becomes impractical; for example, C_b becomes too small. In that case, X_1 should be the equivalent C_1' and the mode selector should be one of the types shown in Fig. 5.13 (see Section 5.6.5.4).

7.5.2 Design Examples for Algorithm 12.1

Design Examples 7.1 and 7.2 are included as examples of designs calculated from the algorithm. The important data is stated in the table and the calculated circuit components values are those on the circuit diagram.

Examples are given for 2 frequencies. Examination of the data on Design Example 7.1 shows that the design is unsatisfactory in that V_{CE} is too small to

	Item	Value	All units in Ω MHz Ω mW pF μH mA, mV dc or rms
Oscillator Performance	f	20	
	P_L	1.66	
	P_x	5	
	V_{BB}	10,000	
Principal Crystal Data	R_{df}	20	
	I_x	15.8	
	Cut	AT	
	N	1	
	C_L	32	
Transistor Data	β_o	30	
	f_T	700	
	BV_{CE}	15,000	
	C_{cb}	1	
	C_{bet}	2	
	C_{ce}	2	
	P_{dis}		
	Type	2N918	
Circuit Parameters	A_{L_0}	2	
	η	0.33	
Calculated Data	I_{BB}	5.14	
	g_m	0.0125	
	V_E	6914	

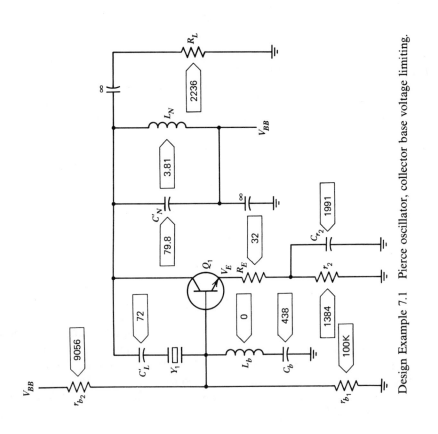

Design Example 7.1   Pierce oscillator, collector base voltage limiting.

	Item	Value	All units in Ʊ MHz Ω mW pF μH mA, mV dc or rms
Oscillator Performance	f	0.8	
	P_L	3.3	
	P_x	10	
	V_{BB}	10,000	
Principal Crystal Data	R_{df}	625	
	I_x	4	
	Cut	AT	
	N	1	
	C_L	32	
Transistor Data	B_o	20	
	f_T	300	
	BV_{CE}	15,000	
	C_{cb}	1	
	C_{bet}	2	
	C_{ce}	2	
	P_{dis}		
	Type	2N2369A	
Circuit Parameters	A_{L_o}	2	
	η	0.33	
Calculated Data	I_{BB}	5.11	
	g_m	.011	
	V_E	2634	

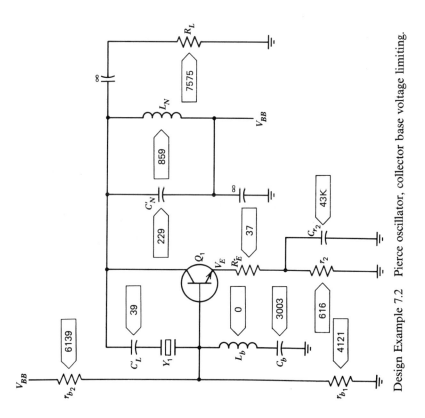

Design Example 7.2 Pierce oscillator, collector base voltage limiting.

198   The Normal Pierce Oscillator

be closely controlled and excessive variation in output will result due to variations in transistor characteristics and component values. It may be stated, in general, that this type of design is unsatisfactory for low P_x and/or very high frequencies.

Normally,

$$\frac{R_2^2}{R_L} \approx \eta R_{df}$$

but Design Example 7.2 is a case where this is not true. This is because R_{df}^2/R_L is not negligible compared to ηR_{df}.

Line c3 of the algorithm requires that $f_{T_{min}}$ be 150 and 5.3 MHz, and line c2 requires $\beta_{o_{min}}$ be 15 and 13 for Design Examples 7.1 and 7.2, respectively. It will be noted that these requirements are met by the transistors specified. Also, at the same frequency, the required $f_{T_{min}}$ is approximately proportional to the square root of the output power.

7.5.3   Trimming for Algorithm 12.1 (see Fig. 7.1)

7.5.3.1   Introduction

After the prototype is constructed, using the nearest standard values for the components, it is tested. Some adjustment may then be found necessary to make the measured performance identical to the specification performance. The adjustment process is called *trimming* and this section discusses the trimming procedure.

7.5.3.2   Basis of the Trimming Procedure

The trimming is based upon the following approximate relations:

1. $|V_L| \propto V_{CE}$ (7.47)
     from Eq. (7.23a).

2. $|I|_3 \approx V_L / \sqrt{R_3^2 + X_2^2}$ (7.48)
     from Eq. (7.8).

3. $|V_b| \approx |I_3||X_1|$ (7.49)
     from Eq. (7.5a).

4. $P_L \propto V_L^2 / R_L$ (7.50)

5. $|V_b|$ and I_E must be maintained so that the proper type of limiting is in effect. This is a rather broad requirement and will permit a large range of $|V_b|$ and I_E, the values of which must be optimized for lowest X_1 and highest conversion efficiency consistent with the variations in transistor properties. This may involve changes in the values of r_2, R_E, r_{b1}, and r_{b2}.

Note: For a crystal resonator $I_3 \equiv I_x$.

7.5.3.3 Typical Trimming Steps

The following describes the action required to increase a given characteristic. Obviously, the opposite action decreases the same characteristic.

7.5.3.3.1 Output Power P_L

Equation (7.50) states that increasing V_L or decreasing R_L increases the power. V_L, in turn, can be increased by increasing V_{CE}, as required by Eq. (7.47), by adjusting r_{b1} and r_{b2}. At the same time, it may be necessary to decrease r_2 for proper limiting action. Similarly, decreasing R_L may require adjustments of the dc biasing.

7.5.3.3.2 Current I_3 (I_X in Crystal Oscillators)

Equation (7.48) states that I_3 can be increased by increasing V_L and/or decreasing X_2. Any action that will increase V_L also increases I_3 (see Section 7.5.3.3.1). Decreasing X_2 is accomplished by increasing C'_N. Of course, adjusting V_L will also affect the power output.

It is interesting to note that Eq. (7.48) also implies that changing C'_1 will have little effect on I_3. The same equation states that the current increases slightly as R_3 decreases so that if a new crystal with lower $R_{df} \equiv R_3$ is used, the current increases but at a slow rate, so that in the case where $X_2 > R_3$, I_3 may be considered constant as R_3 varies.

7.5.4 Frequency Instability due to Variations in Components, Other Than Z_3

Section 5.9 presents a stability analysis of this type of oscillator. Study of this analysis discloses that for maximum stability, X_1 and X_2 should be minimized. However, the requirements of power output, Z_3 power loss, operating Q, and conversion efficiency dictate the values of X_1 and X_2 as outlined in the algorithms. Accordingly, not much can be done for increasing the stability without changing the other performance characteristics. For example, increasing the power input substantially will permit a reduction in X_2. Also, increasing the power supply voltage will permit greater stabilization of the dc bias circuit and thus increase the oscillator stability.

7.6 THE PIERCE OSCILLATOR, *be* CUTOFF LIMITING

7.6.1 The Design Algorithm for This Oscillator (Algorithm 12.2)

The user has to supply the information requested in Sections a, b, and c. It will be noted that items c7, c9, c10, and c11 are not in the algorithm but checks should be made after completion of the design to make sure that these values are not exceeded.

In Section d, 0.3 has been chosen as the recommended value of α because it is comfortably within the straight-line portion of Fig. 2.12b. This value of α also corresponds to the saturation value of γ_1 and to relatively small values of γ_2 and γ_3 as shown in Fig. 2.12c. This value of α also produces a small-signal loop gain of 3.3 which is considered satisfactory.

The comments in Section 7.5.1 about replacement of the crystal networks by another network, about replacing the calculated R_L by another R_L, and about the crystal mode selector apply equally well here.

7.6.2 Design Examples for Algorithm 12.2

Design Examples 7.3, 7.4, 7.5, and 7.6 are included as examples of designs calculated from this algorithm. The input data is stated in the table and the calculated circuit component values are shown on the circuit diagram.

Examples are given for two frequencies and two power levels.

Examination of Design Example 7.4 shows the design is satisfactory for low power. The conversion efficiency is very poor and can be improved by increasing the power supply voltage.

Design Example 7.5 and 7.6 are the same design except for power supply voltage. It will be noted that the higher voltage yields the higher efficiency. This is due to the fact that the higher voltage permits a higher load impedance. Design Example 7.5 is defective in that it requires a β_o min of 55, which means that a higher β transistor is necessary.

Comparison of these examples with Design Examples 7.1 and 7.2 discloses that the *be* limiting designs are more efficient than the collector voltage limiting designs. Also the values of C_1' are considerably different in the two designs. This again confirms the fact that satisfactory designs are possible over a wide range of values.

7.6.3 Trimming for Algorithm 12.2

See Fig. 7.1.

7.6.3.1 *Introduction*

See Section 7.5.3.1.

7.6.3.2 *Basis of the Trimming Procedure*

The trimming is based upon the following relations:

1   from Fig. 2.12b.

$$I_e \approx 1.4 I_E \tag{7.51}$$

2   from Eq. (7.8).

$$|V_L| \approx |I_3|\sqrt{R_T^2 + X_2^2} \tag{7.52}$$

$$\approx |I_e|\frac{|X_2|}{R_T}\sqrt{R_T^2 + X_2^2} \tag{7.53}$$

All units in Ω MHz Ω mW pF μH mA, mV dc or rms	Item	Value			
Oscillator Performance	f	20			
	P_L	1.66			
	P_x	5			
	V_{BB}	10,000			
Principal Crystal Data	R_{df}	20			
	I_x	15.8			
	Cut	AT			
	N	1			
Transistor Data	β_o	30			
	f_T	700			
	BV_{CE}	15,000			
	C_{cb}	1			
	C_{bet}	2			
	C_{ce}	2			
	P_{dis}				
	Type	2N918			
Circuit Parameters	α	0.3			
	γ_1	1.4			
	V_{be}	113			
	η	0.33			
Calculated Data	I_{BB}	2.84			
	g_m	0.030			
	V_E	4018			

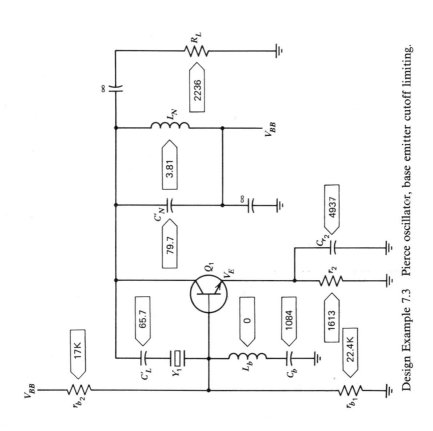

Design Example 7.3 Pierce oscillator, base emitter cutoff limiting.

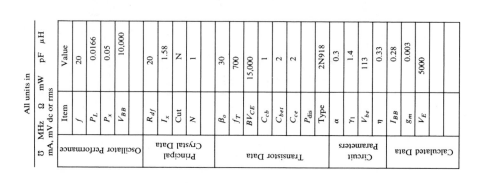

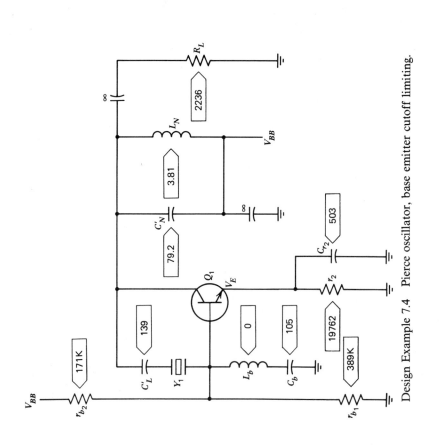

Design Example 7.4 Pierce oscillator, base emitter cutoff limiting.

	Item	Value
Oscillator Performance	f (MHz)	0.8
	P_L (mW)	3.3
	P_x (mW)	10
	V_{BB} (mV)	10,000
Principal Crystal Data	R_{d1} (Ω)	625
	I_x (mA)	4
	Cut	AT
	N	1
Transistor Data	B_o	20
	f_T (MHz)	250
	BV_{CE}	15,000
	C_{cb} (pF)	1
	C_{be1} (pF)	2
	C_{ce} (pF)	2
	P_{dis}	
	Type	2N2369A
Circuit Parameters	α	0.3
	γ_1	1.4
	V_{be} (mV)	113
	η	0.33
Calculated Data	I_{BB} (mA)	4.69
	g_m	0.046
	V_E (mV)	1132

All units in Ʊ MHz Ω mW pF μH mA, mV dc or rms

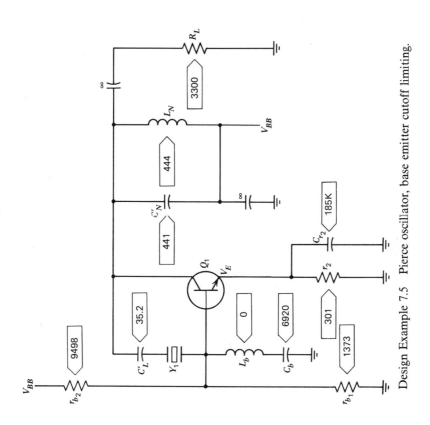

Design Example 7.5 Pierce oscillator, base emitter cutoff limiting.

203

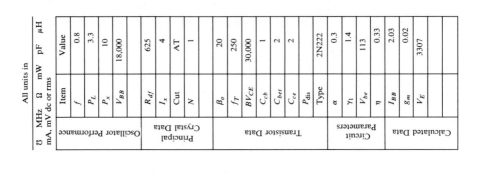

	Item	Value	All units in Ω MHz Ω mW pF μH mA, mV dc or rms
Principal Oscillator Performance	f	0.8	
	P_L	3.3	
	P_x	10	
	V_{BB}	18,000	
Crystal Data	R_{df}	625	
	I_x	4	
	Cut	AT	
	N	1	
Transistor Data	β_o	20	
	f_T	250	
	BV_{CE}	30,000	
	C_{cb}	1	
	C_{bet}	2	
	C_{ce}	2	
	P_{dis}		
	Type	2N222	
Circuit Parameters	α	0.3	
	γ_1	1.4	
	V_{be}	113	
	η	0.33	
Calculated Data	I_{BB}	2.03	
	g_m	0.02	
	V_E	3307	

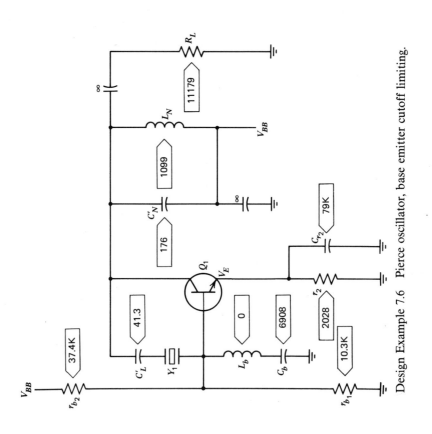

Design Example 7.6 Pierce oscillator, base emitter cutoff limiting.

204

from Eqs. (7.6) and (7.52)

3
$$|I_3| = |I_e|\frac{|X_2|}{R_T} \tag{7.54}$$

from Eq. (7.6).

4
$$V_b \approx I_3 X_1 \tag{7.49}$$

from Eq. (7.5a).

5
$$P_L \propto V_L^2/R_L \tag{7.50}$$

6 Section 7.5.3.2(5) is applicable here.

Note: For a crystal resonator, $I_3 = I_x$.

7.6.3.3 Typical Trimming Steps

The following describes the action required to increase a given characteristic. Obviously, the opposite action decreases the same characteristic.

7.6.3.3.1 Output Power P_L

Equation (7.50) states that increasing V_L or decreasing R_L will increase the power. V_L, in turn, can be increased by increasing I_E or X_2, as stated in Eq. (7.53). I_e is increased by increasing I_E, as stated in Eq. (7.51). I_E is increased by adjusting r_{b_1}, r_{b_2}, and/or r_2.

7.6.3.3.2 Current I_3 (I_x in Crystal Oscillators)

Equation (7.54) states that I_3 increases as I_e and/or X_2 are increased. Any action that will increase V_L will also increase I_3. (See Section 7.6.3.3.1.) Increasing X_2 is accomplished by decreasing C'_N. If it is desired to change I_3 without changing V_L, it is necessary to adjust R_L for the required power, after I_3 has been adjusted.

It is interesting to note that Eq. (7.54) implies that changing C'_1 will have little effect on I_3. The same equation states that I_3 increases as R_3 decreases so that if a new crystal with lower R_{df} is used, the current will increase but at a much more rapid rate than in Section 7.5.3.3.2 because of the difference in Eqs. (7.48) and (7.54).

7.6.4 Frequency Instability Due to Variations in Components Other Than Z_3

See Section 7.5.4.

7.7 MEASUREMENT TECHNIQUES FOR THE PIERCE OSCILLATOR (SEE ALSO CHAPTER 19)

The Pierce oscillator is unique in that it facilitates meaningful measurements since the emitter is ac grounded and since most measuring instruments have one of their terminals grounded. This section discusses some of the types of measurements, which are relatively easy in the Pierce oscillator but much more difficult in other oscillator configurations.

7.7.1 Measurement of I_3 in Fig. 7.1

In a crystal oscillator Z_3 is the crystal and therefore $I_3 = I_x$. I_3 is determined by measuring or calculating $|Z_1|$, which is the impedance from the base to ground, and then measuring V_b with a high-impedance RF voltmeter. Then

$$|I_3| = \frac{|V_b|}{|Z_1|} \tag{7.55}$$

For greater accuracy, the voltmeter loading should be included in $|Z_1|$.

7.7.2 Measurement of Q_{op} in Crystal Oscillators

There are many procedures by which Q_{op} can be measured. This section describes a relatively easy procedure for measuring the approximate Q_{op}.

From Eqs. (7.5) and (7.7)

$$\frac{V_L}{V_b} \approx \frac{R_T + jX_2}{-jX_1} \approx A \angle \theta \tag{7.56}$$

where

$$R_T \approx R_3 + R_{in} + R_b \tag{7.57}$$

assuming $R_2 \to 0$. Solving for θ

$$\theta \approx \tan^{-1}\frac{X_2}{R_T} + \frac{\pi}{2} \tag{7.58}$$

and

$$\frac{\partial \theta}{\partial f} \approx \frac{(1/R_T)(\partial X_2/\partial f)}{1 + (X_2/R_T)^2} \tag{7.59}$$

assuming R_T is constant over the range of f.

$$\approx \frac{(1/R_T)(\partial X_2/\partial f)}{1 + \tan^2(\theta - \pi/2)} \tag{7.59a}$$

7.7 Measurement Techniques for the Pierce Oscillator

from Eq. (7.58). Since X_1 is constant over the range of f,

$$Q_{op} = \frac{f}{2}\left|\frac{\partial \theta}{\partial f}\right|\left[1 + \tan^2\left(\theta - \frac{\pi}{2}\right)\right] \qquad (7.60)$$

from Eqs. (5.77) and (7.59a).

Equation (7.60) shows how Q_{op} is measured, which may be summarized as follows:

1. A small shift in frequency, ∂f, is produced by changing X_2 slightly.
2. θ and $\partial \theta$ are measured by means of a vector voltmeter, set up to measure V_b and V_L. (Note: The meter reads θ in degrees. This should be converted to radians.)
3. Q_{op} is computed with Eq. (7.60).

8

Isolated Pierce Oscillator

8.1 INTRODUCTION

The Pierce oscillator discussed in the previous chapter has the following serious disadvantages.

1. The power output for stable oscillator design is only a small fraction of the loss in Z_3. (In crystal oscillators, Z_3 is the crystal.)
2. The isolation, meaning the effect of variation in load upon the frequency, is quite poor.

The isolated Pierce oscillator is a modification of the normal Pierce oscillator which does not have the above disadvantages. It does possess the disadvantage of being relatively difficult to design. This chapter adapts the design procedure of the previous chapter to be applicable to the isolated Pierce oscillator.

8.2 OSCILLATOR CIRCUIT ANALYSIS

8.2.1 Introduction

The *isolated Pierce oscillator* means the oscillator, the circuit schematic diagram of which is shown in Fig. 8.1. It will be noted that it is the same as that for the normal Pierce oscillator of Chapter 7 except for resistor R_s.

8.2 Oscillator Circuit Analysis

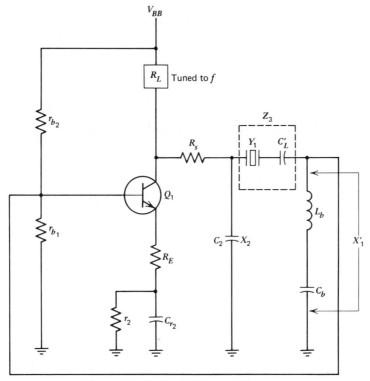

Figure 8.1 Isolated Pierce oscillator.

Figure 8.2a shows the ac equivalent circuit of Fig. 8.1. In this circuit, the transistor has been replaced by a current generator having transconductance g_m. Also the ac loading of the transistor input and bias circuits is shown as the equivalent resistances R_b and R_{in}, respectively, which have the same significance as that for the normal Pierce oscillator.

In this circuit

$$I_c = g_m V_b \tag{8.1}$$

If it is assumed that the Q of the PI circuit to the right of the dashed line exceeds 5, then

$$X_3 \approx -(X_1 + X_2) \tag{8.2}$$

$$R_n \simeq \frac{X_2^2}{R_3 + R_{in} + R_b} \tag{8.3}$$

$$I_3 \simeq I_s \frac{X_2}{R_3 + R_{in} + R_b} \tag{8.4}$$

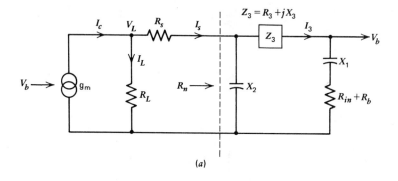

(a)

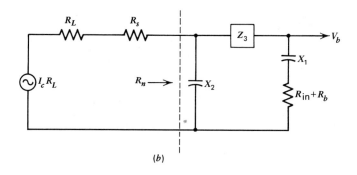

(b)

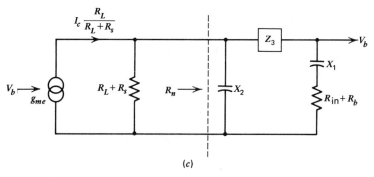

(c)

Figure 8.2 Equivalent ac circuits of isolated Pierce oscillator. (*a*) Equivalent circuit of Fig. 8.1. (*b*) Thévenin's equivalent of (*a*). (*c*) Norton's equivalent of (*b*).

210

8.2 Oscillator Circuit Analysis

$$I_s \simeq \frac{V_L}{R_s + R_n} \tag{8.5}$$

$$I_c = \left(1 + \frac{R_L}{R_s + R_n}\right) I_L \tag{8.6}$$

$$\simeq \left(1 + \frac{1}{m_r}\right) I_L \tag{8.6a}$$

assuming $R_s \gg R_n$ and

$$m_r \equiv \frac{R_s}{R_L} \tag{8.7}$$

$$V_b \simeq I_3 X_1 \tag{8.8}$$

$$g_m = \frac{I_c}{V_b} \simeq \frac{I_c}{I_3 X_1} \tag{8.9}$$

or

$$g_m X_1 \simeq \frac{I_c}{I_3} \tag{8.9a}$$

The conditions existing in that part of the circuit to the left of the dotted line are now known. It is very desirable to be able to use the theory developed for the Pierce oscillator in Chapters 5 and 7 to determine the conditions in the rest of the circuit.

To do so, the circuit of Fig. 8.2a is converted into the circuit of Fig. 8.2c, which is the normal Pierce oscillator circuit. This is accomplished as follows: The circuit of Fig. 8.2a is first converted into the Thévenin equivalent circuit of Fig. 8.2b by replacing the current generator with the equivalent voltage generator. Norton's theorem is then used to convert the circuit of Fig. 8.2b into the equivalent circuit of Fig. 8.2c.

8.2.2 Additional Design Equations

By definition of g_m, from Fig. 8.2c,

$$g_{me} = \frac{I_c R_L}{V_b(R_L + R_s)} = \frac{R_L}{R_L + R_s} g_m \tag{8.10}$$

$$= \frac{1}{1 + m_r} g_m \tag{8.10a}$$

Isolated Pierce Oscillator

Assuming

$$R_3 + R_{in} + R_2 \gg R_b \qquad (8.11)$$

then

$$R_T \simeq R_2 + R_{in} + R_3 \qquad (8.12)$$

Let

$$R_t = R_3 + R_{in} \qquad (8.13)$$

and let

$$X_2 = nR_t \qquad (8.14)$$

It is recommended that

$$n \geqq 5 \qquad (8.14a)$$

then

$$R_n \simeq \frac{X_2^2}{R_t} \qquad (8.15)$$

$$= 25R_t \quad \text{for } n = 5 \qquad (8.15a)$$

From Eq. (8.5),

$$I_s = \frac{V_L}{R_s + 25R_t} \quad \text{for } n = 5 \qquad (8.16)$$

From Eq. (8.4)

$$I_3 \simeq \frac{V_L}{R_s + R_n} \frac{X_2}{R_t} \qquad (8.17)$$

$$\simeq \frac{I_L R_L}{R_s + R_n} \frac{X_2}{R_t} \qquad (8.17a)$$

$$\simeq \frac{I_L R_L}{R_s} \frac{X_2}{R_t} \quad \text{when } R_n \ll R_s \qquad (8.17b)$$

The last equation may be written, from Eqs. (8.6) and (8.9), as

$$I_3 \equiv \frac{1}{1 + m_r} I_c \frac{X_2}{R_t} \qquad (8.17c)$$

It is interesting to formulate the relations for

$$\eta = \frac{P_L}{P_3} \tag{8.18}$$

$$\eta = \frac{V_L^2}{R_L}(I_3^2 R_3)^{-1} \tag{8.18a}$$

$$= \frac{(R_s + R_n)^2}{R_L R_3 (X_2/R_t)^2} \quad \text{from Eq. (8.17)} \tag{8.18b}$$

$$= \frac{R_s^2}{25 R_L R_3} \quad \text{for } n = 5 \text{ and } R_s \gg R_n \tag{8.18c}$$

Equation (8.18b) is very useful for determining the power ratio for given values of R_s and R_L. It is very interesting that large values of R_n and large values of R_s/R_L yield large η. However, the maximum value of R_s is limited to the largest *practical* value of R_s which at high frequencies is not very large.

If Eq. (8.18b) is solved for R_s, then

$$R_s = \frac{X_2}{R_t}\sqrt{\eta R_3 R_L} - R_n \tag{8.19}$$

$$\approx \frac{X_2}{R_t}\sqrt{\eta R_3 R_L} \quad \text{when } R_s \gg R_n \tag{8.19a}$$

Equation (8.19) may be used for determining R_s for a given η.

Equation (8.9a) states that

$$g_m X_1 = \frac{I_c}{I_3}$$

which, from Eqs. (8.6) and (8.17a), is

$$= \frac{R_s + R_n + R_L}{R_L X_2/R_t} \tag{8.20}$$

$$\approx \frac{1 + m_r}{X_2/R_t} \tag{8.20a}$$

when $R_s + R_L \gg R_n$.

The general basic oscillatory equation discussed at great length in Chapters 5 and 7 states that

$$g_{me} X_1 X_2 = R_T \tag{8.21}$$

It will be found after calculating the values of the components that Eq. (8.21) will not be completely satisfied because of the approximations made during the derivations. For example, R_T, from Eq. (8.21), using the values for X_1, X_2, and g_{me} calculated in the algorithm, is always less than that given by Step 17 in the algorithm.

The relationship between the two can be shown to be

$$\frac{R_T(\text{of Step 17})}{R_T(\text{of Eq. (8.21)})} \approx \frac{R_s + R_n}{R_s} \tag{8.22}$$

This discrepancy can be corrected by replacing X_2 by X_2' where

$$X_2' = nR_T \tag{8.23}$$

Therefore, Eq. (8.21) becomes

$$g_{me} X_1 X_2' = R_T \tag{8.24}$$

and

$$C_2 = \frac{159{,}000}{fX_2'} \tag{8.25}$$

8.2.3 Practical Values of R_s

Section 1.2.1.7 states that the maximum realizable resistor value is

$$R_{\max} = \frac{32{,}000}{f}$$

This is based upon $X_{C_p} > 10R$.

In this application, it is sufficient to make

$$X_{C_p} > 5R;$$

therefore,

$$R_{\max} = 64{,}000/f \tag{8.26}$$

8.2.4 Oscillator Design for Minimum Power Consumption

It is obvious that, for minimum power consumption, R_L should be maximum. R_L is determined by the following:

1. The BV_{CE} of the transistor.
2. The maximum physically realizable R_L.
3. The value of $\eta = P_L/P_x$.

8.5 Algorithm and Design Examples

Items 1 and 2 are fully treated in Section 7.2.4. Item 3 will now be discussed.

Equation (8.18b) may be rewritten

$$R_L = \frac{R_s^2}{n^2 R_3 \eta} \quad \text{for } R_s \gg R_n \tag{8.27}$$

It is seen that, for R_L to be maximum, R_s must be maximum, which however is limited as stated in Eq. (8.26). Therefore, from Eqs. (8.27) and (8.26)

$$R_{L_{\max}} = \frac{(64{,}000)^2}{f^2 n^2 R_3 \eta} \tag{8.28}$$

The value of R_L to be used is the smaller of the values as determined by item 1, item 2, and/or Eq. (8.28).

8.3 LIMITING

Since it is desirable to maintain the output power independent of the resistance, R_3, and since the isolating characteristic of this oscillator renders the frequency much less sensitive to changes in supply voltage than in the normal Pierce oscillator, it is in order to use collector base limiting. Accordingly, the material in Section 7.2.4.1 is applicable to this design. However, for very small output power, Section 7.2.4.2 is applicable, as base emitter cutoff limiting is then in order. Often, the algorithm will alert the user that a different type of limiting is called for by calculating a negative value for r_{b_2}. The preparation of the algorithm for base emitter cutoff limiting is left to the reader as an exercise, using the other algorithms as a guide.

8.4 THE DESIGN PROCEDURE

The design procedure is similar to that for the normal Pierce oscillator described in Section 7.4, except for the modification demanded by the conversion from the isolated Pierce oscillation to the normal Pierce oscillation as described in Section 8.2. The procedure is surprisingly straightforward for the relatively complex oscillator circuit.

8.5 ALGORITHM AND DESIGN EXAMPLES

8.5.1 The Design Algorithm for the Isolated Pierce Oscillator (Algorithm 12.3)

See Section 7.5.1 for some pertinent remarks which are also applicable to this algorithm.

216 Isolated Pierce Oscillator

The algorithm has been prepared for maximum conversion efficiency consistent with the transistor characteristics, maximum practical load resistance, and the desired $\eta = P_L/P_x$.

The algorithm has been prepared on the basis that the g_m design procedure is the same as that for the normal Pierce oscillator. This basis allows the operating Q to vary widely depending on P_L/P_x. Another basis could have been to attempt to make the ratio of Q_{op}/Q_x a predetermined value, as is done in the normal Pierce oscillator, but this often results in unrealizable circuits; that is, some of the components are calculated to be negative. The basis used renders the algorithm very simple and always results in realizable circuits.

8.5.2 Design Examples for Algorithm 12.3

Design Examples 8.1 and 8.2 are the design examples for this algorithm. They have been chosen to illustrate the effects of large frequency differences and large crystal resistance differences. Both examples have the same P_L and η.

R_L in Design Example 8.1 was fixed by the maximum realizable value of R_s, while R_L in Design Example 8.2 was fixed by the BV_{CE} of the transistor. It will be noted that the conversion efficiency of Design Example 8.2 is considerably greater than that of Design Example 8.1. This is due to the larger R_L.

C_2 in Design Example 8.2 is rather small for an oscillator at 0.8 MHz. This is due to the very large crystal resistance. R_{df}/R_T, cited in the tables, equals Q_{op}/Q_x.

In both examples, η is 100, which is much larger than that in the normal Pierce circuit, typically $\frac{1}{3}$. The major advantage of this oscillator is thereby demonstrated.

8.5.3 Trimming for Algorithm 12.3

8.5.3.1 Introduction

See Section 7.5.3.1.

8.5.3.2 Basis of the Trimming Procedure

The trimming is based upon the following approximate relations, many of which are similar to those for the normal Pierce oscillator, but are repeated here for convenience.

1      $|V_L| \propto V_{CE}$ (7.47)

2      $P_L \propto \dfrac{V_L^2}{R_L}$ (7.50)

3      $|V_b| \approx |I_3||X_1|$ (7.49)

4      $P_x \approx P_L \dfrac{25 R_L R_x}{R_s^2}$ from Eq. (8.18c)   (8.29)

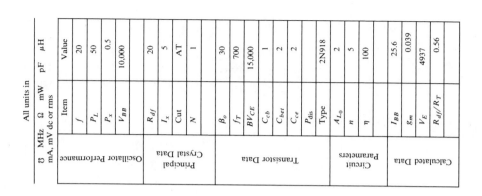

All units in Ω MHz Ω mW pF μH mA, mV dc or rms	Item	Value
Oscillator Performance	f	20
	P_L	50
	P_x	0.5
	V_{BB}	10,000
Principal Crystal Data	R_{df}	20
	I_x	5
	Cut	AT
	N	1
Transistor Data	β_o	30
	f_T	700
	BV_{CE}	15,000
	C_{cb}	1
	C_{bet}	2
	C_{ce}	2
	P_{dis}	
	Type	2N918
Circuit Parameters	A_{L_0}	2
	n	5
	η	100
Calculated Data	I_{BB}	25.6
	g_m	0.039
	V_E	4937
	R_{df}/R_T	0.56

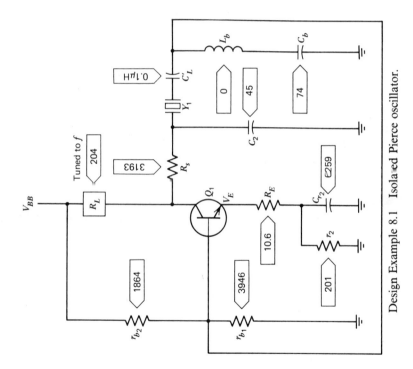

Design Example 8.1   Isolated Pierce oscillator.

All units in mA, mV dc or rms	G MHz Ω mW pF μH		
		Item	Value
Oscillator Performance	Principal	f	0.8
		P_L	50
		P_x	0.5
		V_{BB}	10,000
Crystal Data		R_{df}	625
		I_x	.89
		Cut	AT
		N	1
Transistor Data		β_o	30
		f_T	700
		BV_{CE}	15,000
		C_{cb}	1
		C_{bet}	2
		C_{ce}	2
		P_{dis}	
		Type	2N918
Circuit Parameters		A_{L_0}	2
		n	5
		η	100
Calculated Data		I_{BB}	16.5
		g_m	0.04
		V_E	2644
		R_{df}/R_T	0.518

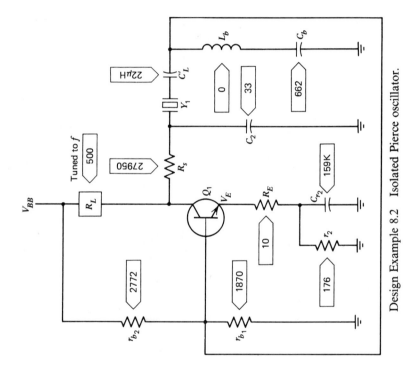

Design Example 8.2 Isolated Pierce oscillator.

8.5.3.3 Typical Trimming Steps

The following describes the action required to increase a given characteristic. Obviously, the opposite action decreases the same characteristic.

8.5.3.3.1 Output Power P_L

Equation (7.50) states that increasing V_L or decreasing R_L will increase the power. V_L, in turn, can be increased by increasing V_{CE}, as required by Eq. (7.47), by adjusting r_{b_1} and r_{b_2}. At the same time, it may be necessary to decrease r_2 for proper limiting action. Similarly, decreasing R_L may require adjustments of the dc biasing.

8.5.3.3.2 Crystal Power P_x

Equation (8.27) states that decreasing R_s will increase P_x. Also, any action that will increase P_L will also increase P_x. Decreasing R_x will also tend to decrease P_x, as stated in the same equation.

8.5.4 Frequency Instability Due to Variation in Components Other than Z_3

Obviously, variations in components due to time, environment, and component defects will produce variations in frequency. What is particularly interesting is the effect of variations caused by the operator such as changes in R_L and tuning of the output circuit.

Changes in R_L will produce frequency changes smaller than those in the equivalent normal Pierce oscillator because of the isolation provided by R_s as is evident from Fig. 8.2c. The isolation increases as the ratio R_s/R_L increases and as the ratio R_s/X_2 increases.

A problem not present in the normal Pierce oscillator is the tuning of the load R_L. The entire analysis assumes that R_L is resistive, but if it is improperly tuned, R_L becomes Z_L, which is a complex impedance of magnitude and phase angle, depending upon the tuning. Equation (8.10) more correctly should be

$$g_{me} = \frac{Z_L}{Z_L + R_s} g_m \qquad (8.30)$$

which in turn is a complex transconductance, even when g_m is real. Therefore, one would expect a change in frequency as Z_L is tuned. The extent of this change can be computed as described in Section 5.4 and may be substantial. This effect can be used constructively as a fine frequency tuning adjustment.

9

The Normal Colpitts Oscillator

9.1 INTRODUCTION

This chapter develops the additional and/or modified relations of Chapters 5 and 7 required to design this type of oscillator. It also contains comments about the practical aspects of the design. Finally, it formulates two design algorithms—one for each of the major types of limiting of Section 6.2. The algorithms are specifically applicable to the self-limiting types of crystal oscillator but they are also useful for non-self-limiting types and for other types of two-terminal resonators.

By the normal Colpitts oscillator is meant the oscillator whose circuit is shown in Fig. 5.1b and repeated in a modified form in Fig. 5.8. All the components in that figure are physical; that is, they are all installed components.

The theory of the oscillator is fully covered in Chapter 5, except for the limiting considerations. It is therefore recommended that Chapter 5 be read in its entirety, beginning with Section 5.3.4, before proceeding with this chapter.

Many of the equations developed in Chapter 7 for the Pierce oscillator will also be used for formulating the design algorithms and will be referenced therein in the "Text Equation" column. In using these equations, V_{be} or $V_{be'}$ as appropriate, must be substituted for V_b in the equations of Chapter 7.

9.2 OSCILLATOR CIRCUIT ANALYSIS

9.2.1 Introduction

Most of the circuit analysis is already presented in Chapter 5. Additional material developed in Section 7.2.3 is also applicable. The reader is referred to Chapter 5 and Section 7.2.3, as the material is not being repeated.

9.2.2 Discussion of $V_{L_{max}}$ and R_L

Section 7.2.4 presents the discussion of V_L and R_L for the normal Pierce oscillator. The discussion for the Colpitts oscillator is almost identical, and for that reason the reader is urged to read Section 7.2.4, as the identical material will not be repeated here.

9.2.2.1 Calculation of $V_{L_{max}}$ for the Collector Base Limiting Oscillator

The results are identical to those for the Pierce oscillator but that fact is not self-evident and the results will therefore be derived.

Figure 9.1 shows the voltage relationships in this type of limiting. By inspection it is seen to be

$$v_{CE_{max}} = V_{BB} - v_{E_{min}} \tag{9.1}$$

$$\approx 2.8V_L + 1.4V_{be'}$$

But

$$v_{CE_{max}} \leqq BV_{CE} \tag{9.2}$$

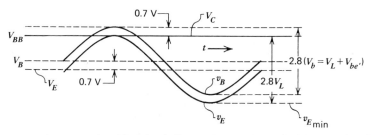

Figure 9.1 Voltage relationships for collector base voltage limiting in the Colpitts oscillator.

therefore,

$$V_L \leq \frac{BV_{CE} - 1.4V_{be'}}{2.8}$$

$$\leq 0.357 BV_{CE} - 0.5 V_{be'}$$

$$\approx 0.33 BV_{CE} \tag{9.3}$$

to compensate for the $V_{be'}$ term.

9.2.2.2 Calculation of $V_{L_{max}}$ for the be Cutoff Limiting Oscillator

Figure 9.2 shows the voltage relationship for this type of limiting. In this oscillator $v_{B_{max}}$ must be less than $v_C = V_{BB}$ for Z_3 having the minimum R_3. Since the oscillator is designed for $R_{3_{max}}$, when V_L is minimum, provision must be made for the increase of V_L as R_3 decreases. Let

$$\frac{R_{3_{max}}}{R_{3_{min}}} = 2 \tag{9.4}$$

then

$$\frac{V_{L_{max}}}{V_{L_{min}}} \approx 1.5 \tag{9.5}$$

and from Eq. (9.3)

$$V_L = \frac{0.33}{1.5} BV_{CE} = 0.22 BV_{CE} \tag{9.6}$$

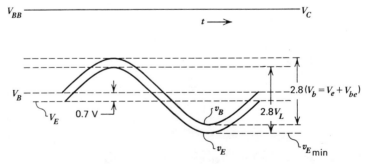

Figure 9.2 Voltage relationships for the base emitter cutoff limiting in the Colpitts oscillator.

9.2.2.3 Calculation of V_E, Which Should Be Greater Than 2 V

It is evident from Fig. 9.1 that, for collector limiting,

$$V_E = V_{BB} - [1.4(V_L + V_{be'}) - 700] - 700$$

$$= V_{BB} - 1.4(V_L + V_{be'}) \tag{9.7}$$

Similarly, from Fig. 9.2, for the be cutoff limiting oscillator

$$V_E = V_{BB} - 2.1(V_L + V_{be}) - 1700 \tag{9.8}$$

or, when

$$V_{E_1} > \frac{V_{BB}}{2}, \quad V_E = \frac{V_{BB}}{2} \tag{9.8a}$$

which is derived in the same manner as that described in Section 7.2.4.3.2, to which the reader is referred.

9.2.3 Calculation of C_{r_2}

It is desirable that

$$X_{C_{r_2}} \leqq 0.02 X_{L_N} \tag{9.9}$$

in order not to affect the stability of X_{L_N} and

$$C_{r_2} = \frac{159{,}000}{X_{C_{r_2}} f} \tag{9.10}$$

9.3 FREQUENCY VARIATIONS DUE TO EXTERNAL FACTORS

The material in Section 7.3 is completely applicable.

9.4 THE DESIGN PROCEDURE

The design procedure outlined in Section 7.4 is applicable except that there are two levels of approximation. It is therefore seen that the design for the Colpitts oscillator is more difficult than that for the Pierce oscillator.

9.5 THE COLPITTS OSCILLATOR, COLLECTOR BASE LIMITING

9.5.1 The Design Algorithm for This Oscillator (Algorithm 12.4)

See Section 7.5.1 for pertinent remarks which are also applicable to this algorithm.

It will be noted that Steps 6 to 20 constitute the first level of approximation, and Steps 22 to 28, the second level. The remaining steps constitute the translation of the calculated quantities into the physical components which make up the oscillator.

9.5.2 Design Examples for Algorithm 12.4

Design Examples 9.1 and 9.2 are design examples for this algorithm. The results are quite similar to those for the design examples for Algorithm 12.1 discussed in Section 7.5.2. The major differences are that I_{BB} is larger because of the greater losses, and the operating Q is somewhat smaller.

9.5.3 Trimming for Algorithm 12.4

See Fig. 5.8.

9.5.3.1 Introduction

See Section 7.5.3.1.

9.5.3.2 Basis of the Trimming Procedure

The trimming is based upon the following approximate relations:

1. $$V_L \propto V_{CE} \tag{9.11}$$
    from Eq. (9.3).

2. $$|I_3| \approx \frac{V_L}{\sqrt{R_3^2 + X_2^2}} \tag{9.12}$$
    from Eq. (5.65).

3. $$V_{be} \approx |I_3||X_1| \tag{9.13}$$
    from Eq. (5.52a).

4. $$P_L \propto \frac{V_L^2}{R_L} \tag{9.14}$$

5. $|V_{be'}|$ and I_E must be maintained so that the proper type of limiting is in effect. This is a rather broad requirement and will permit large ranges of $|V_{be'}|$ and I_E, the values of which must be optimized for lowest X_1 and

All units in µH, MHz, Ω, mW, pF, mA, mV dc or rms		Item	Value		
Oscillator Performance		f	20		
		P_L	1.66		
		P_x	5.0		
		V_{BB}	10,000		
Principal Crystal Data		R_{df}	20		
		I_x	15.8		
		Cut	AT		
		N	1		
Transistor Data		β_o	30		
		f_T	700		
		BV_{CE}	15,000		
		C_{cb}	1		
		C_{bet}	2		
		C_{ce}	2		
		P_{dis}			
		Type	2N918		
Circuit Parameters		A_{Lo}	2		
		η	33		
Calculated Data		I_{BB}	5.5		
		V_E	6929		
		g_m	0.14		

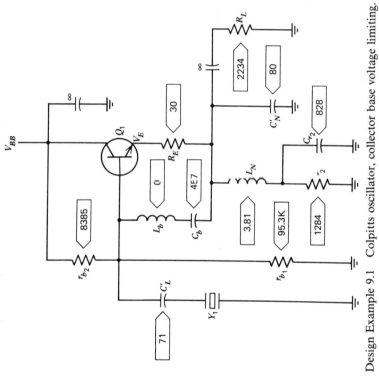

Design Example 9.1 Colpitts oscillator, collector base voltage limiting.

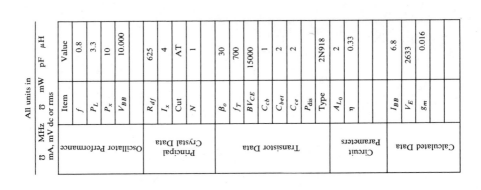

All units in G MHz U mW pF μH mA, mV dc or rms			
	Item	Value	
Oscillator Performance	f	0.8	
	P_L	3.3	
	P_x	10	
	V_{BB}	10.000	
Principal Crystal Data	R_{df}	625	
	I_x	4	
	Cut	AT	
	N	1	
Transistor Data	β_o	30	
	f_T	700	
	BV_{CE}	15000	
	C_{cb}	1	
	C_{bef}	2	
	C_{ce}	2	
	P_{dis}		
	Type	2N918	
Circuit Parameters	A_{L_0}	2	
	η	0.33	
Calculated Data	I_{BB}	6.8	
	V_E	2633	
	g_m	0.016	

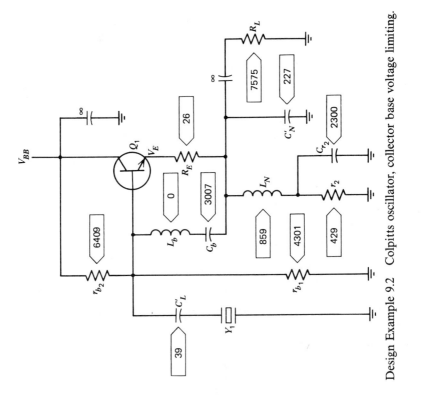

Design Example 9.2 Colpitts oscillator, collector base voltage limiting.

226

highest conversion efficiency, consistent with the variations in transistor properties. This may involve adjustments in the values of r_2, r_{b_1}, and r_{b_2}.

Note: For a crystal resonator, $I_3 \equiv I_x$.

9.5.3.3 Typical Trimming Steps

Section 7.5.3.3 is applicable, taking into account the number changes of the equivalent equations.

9.5.4 Frequency Instability Due to Variations in Components, Other Than Z_3

Section 7.5.4 is applicable.

9.6 THE COLPITTS OSCILLATOR, *be* CUTOFF LIMITING

9.6.1 The design algorithm for the oscillator is Algorithm 12.5. The comments in Section 7.6.1 are applicable here.

9.6.2 Design Examples for Algorithm 12.5

Design Examples 9.3 and 9.4 pertain to Algorithm 12.5. The results are quite similar to those for the design examples for Algorithm 12.2 discussed in Section 7.6.2. The major differences are that I_{BB} is larger because of the greater losses, and the operating Q is somewhat smaller.

9.6.3 Trimming for Algorithm 12.5

See Fig. 5.8.

9.6.3.1 Introduction

Same as Section 7.5.3.1.

9.6.3.2 Basis of the Trimming Procedure

The trimming is based upon the following relations:

1 from Fig. 2.12*b*.
$$I_3 = 1.4 I_e \tag{9.15}$$

2 from Eq. (5.66).
$$|V_L| = |I_3|\sqrt{R_T^2 + X_2^2} \tag{9.16}$$

$$= |I_e|\frac{|X_2|}{R_T}\sqrt{R_T^2 + X_2^2} \tag{9.17}$$

from Eq. (5.59a).

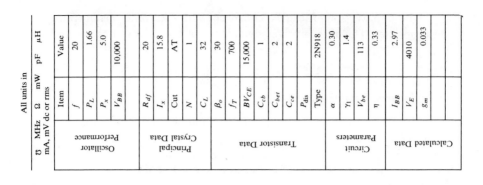

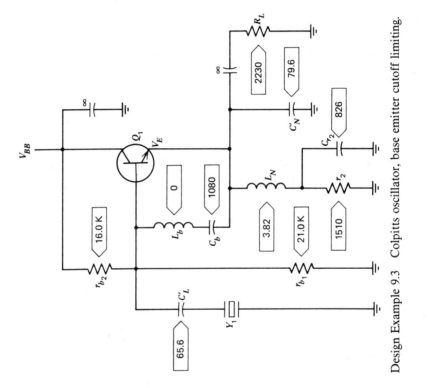

Design Example 9.3 Colpitts oscillator, base emitter cutoff limiting.

All units in ℧ MHz Ω mW pF μH mA, mV dc or rms	Item	Value
Oscillator Performance	f	20
	P_L	0.0166
	P_x	0.05
	V_{BB}	10,000
Principal Crystal Data	R_{df}	20
	I_x	1.53
	Cut	AT
	N	1
Transistor Data	β_o	30
	f_T	700
	BV_{CE}	15,000
	C_{cb}	1
	C_{bet}	2
	C_{ce}	2
	P_{dis}	
	Type	2N918
Circuit Parameters	α	0.3
	γ_1	1.4
	V_{be}	113
	η	0.33
Calculated Data	I_{BB}	0.28
	V_E	5000
	g_m	0.0032

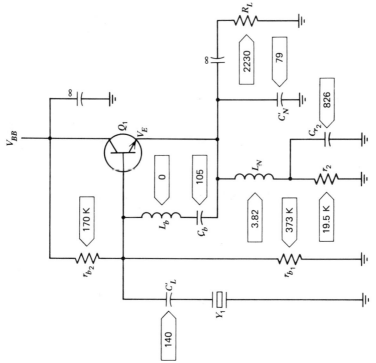

Design Example 9.4 Colpitts oscillator, base emitter cutoff limiting.

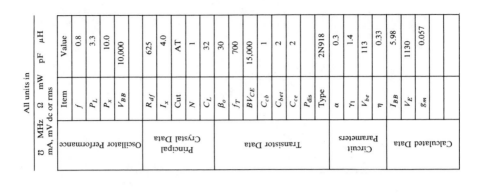

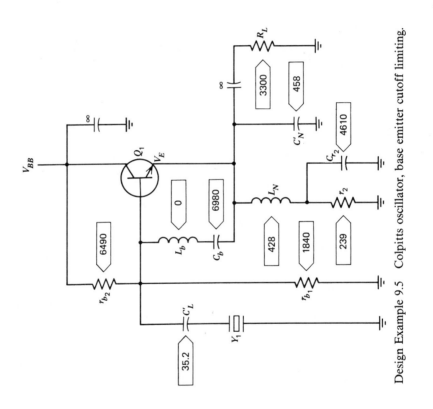

Design Example 9.5 Colpitts oscillator, base emitter cutoff limiting.

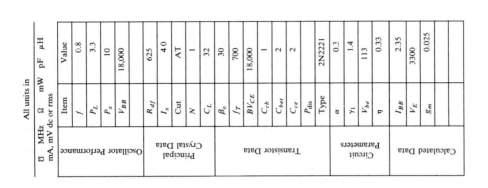

All units in Ω, MHz, Ω, mW, pF, μH, mA, mV dc or rms		Item	Value
Oscillator Performance		f	0.8
		P_L	3.3
		P_x	10
		V_{BB}	18,000
Principal Crystal Data		R_{df}	625
		I_x	4.0
		Cut	AT
		N	1
		C_L	32
Transistor Data		β_o	30
		f_T	700
		BV_{CE}	18,000
		C_{cb}	1
		C_{bet}	2
		C_{ce}	2
		P_{dis}	
		Type	2N2221
Circuit Parameters		α	0.3
		γ_1	1.4
		V_{be}	113
		η	0.33
Calculated Data		I_{BB}	2.35
		V_E	3300
		g_m	0.025

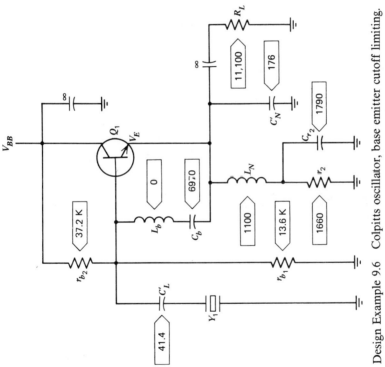

Design Example 9.6 Colpitts oscillator, base emitter cutoff limiting.

3 $$|I_3| = |I_e|\frac{|X_2|}{R_T} \tag{9.18}$$
from Eq. (5.59a).

4 $$|V_{be}| = |I_3||X_1| \tag{9.19}$$
from Eq. (5.52a).

5 $$P_L \propto \frac{V_L^2}{R_L} \tag{9.20}$$

6 Section 7.5.3.2(5) is applicable here.

9.6.3.3 Typical Trimming Steps

Section 7.6.3.3 is applicable, taking into account the number changes of the equivalent equations.

9.6.4 Frequency Instability Due to Variations in Components Other Than Z_3

See Section 7.5.4.

10
The Semi-isolated Colpitts Oscillator

10.1 INTRODUCTION

By the semi-isolated Colpitts oscillator is meant the oscillator whose circuit is shown in Fig. 10.1. All the components shown in the circuit are physical; that is, they are all installed components.

The theory of the oscillator is extensively covered in Chapter 5. Some additional theory is presented in Chapters 7 and 9, and also in this chapter.

This chapter develops the additional and/or modified relations of Chapters 5, 7, and 9. It also contains comments about the practical aspects of the design. Finally, it formulates a design algorithm for *be* cutoff limiting. The algorithm is specifically applicable to the self-limiting crystal oscillator, but it is also useful for non-self-limiting types and other types of resonators.

10.2 OSCILLATOR CIRCUIT ANALYSIS

10.2.1 General Oscillator Description

This oscillator is a variation of the normal Colpitts oscillator discussed in Chapter 9. It has several important properties:

1 It is capable of large values of P_L/P_3.
2 The output frequency can be the same as the frequency of oscillation, f, or a harmonic, Mf.

234 The Semi-isolated Colpitts Oscillator

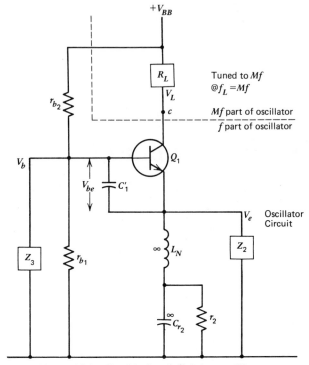

Figure 10.1 Semi-isolated Colpitts oscillator.

3 Moderate isolation is provided at the fundamental output frequency, which is the reason why it is called *semi-isolated*. At the harmonic output frequencies it provides excellent isolation.

4 The output power is relatively independent of the values of R_3, which is very important in crystal oscillators, where considerable variation in R_3 is normal.

5 Production units tend to be very much alike.

It will be seen from the above that the properties are very desirable. It is, however, rather difficult to design, but it is well worth the effort.

Because of property 1, the circuit is particularly useful when substantial power output at low crystal power dissipation is required.

10.2.2 Discussion of Fig. 10.1

If one compares Fig. 10.1 to Fig. 5.8, one is struck with the similarity. The only major difference is the location of the power output point. One would therefore expect the theory to be similar, which it is. Another point worth noting is that at harmonic power output the fundamental voltage at point c is close to zero, so that the two circuits below point c are identical and can be analyzed identically. However, when the output is at the fundamental frequency the two circuits differ in Miller effects, which seriously deteriorate the isolation in the circuit of Fig. 10.1. This aspect is discussed later.

10.2.3 Isolation Characteristics

One of the important assumptions, which is fairly realistic, made in this book is that the currents and voltages at the different harmonic frequencies do not interact. As a result of the assumption, it follows that the isolation at harmonic output frequencies is excellent since the oscillator operates at the fundamental frequency. However, the isolation for fundamental output frequency is much worse, primarily because of Miller effects, which are very serious for high ratios of P_L/P_x because of the high values of V_L/V_b and V_b/V_e.

10.2.4 The Miller Effects

Figure 10.2a shows the various ac voltages existing at each electrode to ground and the capacitances between the electrodes. It should be noted that V_{L_1} is the fundamental component of V_L.

In Fig. 10.2b, C_{ce} has been replaced by C_{cem}. In Fig. 10.2c, C_{cb} has been replaced by C_M. In Fig. 10.2d, C_M in turn is replaced by C_{1M} and C_{2M}.

The method of performing these replacements is similar to that described in Section 5.5.4.2.5. The results are the following:

$$C_{cem} \approx M_{ce} C_{ce} \qquad (10.1)$$

where

$$M_{ce} \approx 1 + \frac{V_{L_1}}{V_e} \qquad (10.2)$$

$$C_M \approx M_{cb} C_{cb} \qquad (10.3)$$

where

$$M_{cb} \approx 1 + \frac{V_{L_1}}{V_b} \qquad (10.4)$$

$$C_{1M} \approx \left(1 + \frac{1}{M_M}\right) C_M \qquad (10.5)$$

where

$$M_M \approx \frac{V_{be}}{V_e} \qquad (10.6)$$

$$C_{2M} \approx (1 + M_M) C_M \qquad (10.7)$$

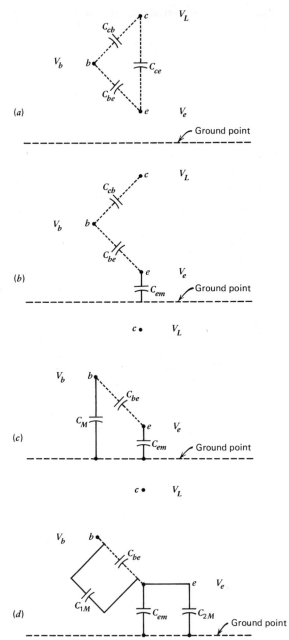

Figure 10.2 Miller effects. (a) Schematic circuit for Miller effects. (b) Replacement of C_{ce} by C_{em}. (c) Replacement of C_{be} by C_M. (d) Apportionment of C_M.

It should be noted that the above relationships are only approximate and do not take into account the Miller resistance contributions as pointed out in Section 5.5.4.2.5.

In practice, when M_{cb} is very large which, in general, is true when P_L/P_3 is very large, the Miller contributions are very important. This is particularly true when R_L is a tuned load. If R_L is changed by the environment or during the tuning procedure, large frequency shifts and amplitude variations may result and the oscillation may even be killed by the resistance contributions. Therefore, for fundamental output circuits it is necessary that M_{cb} be made very small. This results in the output being very small or the conversion efficiency being very poor. Fundamental output is not recommended for this circuit except when R_L consists of a low-value resistor.

10.2.5 Limiting

Section 6.2.2 indicates that *be* cutoff limiting is the preferred type of limiting except where the output power is a function of the resonator resistance and the application demands constancy of power output versus the resonator resistance. However, in this circuit, the power output does not depend upon the crystal resistance. Therefore *be* cutoff limiting is used.

Section 2.4 presents the theory of this type of limiting and the results are summarized in Figs. 2.12 and 2.13. Since it is very difficult to use curves in computer-aided design procedures, equations for the applicable parts of these curves will now be derived.

It is very desirable to operate in the saturated region of the γ_M curves. Therefore, α should decrease as M increases. A suitable relationship is

$$\alpha = \frac{0.3}{M} \tag{10.8}$$

From Figs. 2.12 and 2.13, it then follows that

$$\gamma_M = 1.6 - 0.2M \tag{10.9}$$

and as a special case

$$\gamma_1 = 1.4 \tag{10.9a}$$

From Eqs. (2.79) and (10.8)

$$V_{be} = 120M = \frac{36}{\alpha} \tag{10.10}$$

10.2.6 Discussion of $V_{L_{max}}$ and V_E

Since V_L is only a weak function of R_3, Eq. (9.6) can be modified to

$$V_L = 0.27 B V_{CE} \tag{10.11}$$

The suitable relationship for V_E is

$$V_E = V_{BB} - 1.7 V_L - 1700 \tag{10.11a}$$

The factor 1.7 is used instead of 1.4 to account for the RF voltage in the oscillator circuit.

When the calculated $V_E > V_{BB}/2$,

$$V_E = \frac{V_{BB}}{2} \tag{10.11b}$$

which is derived in the same manner as that described in Section 7.2.4.3.2.

10.2.7 Calculation of R_L

The discussion of Section 7.2.4.4 applies. However, Eq. (1.16) must be modified to

$$R_L \leq \frac{10{,}000}{\sqrt{Mf}} \tag{10.12}$$

as the load is tuned to the harmonic of the oscillator frequency f.

Similarly, Eq. (7.30) becomes

$$V_L = \sqrt{\frac{10^7 P_L}{\sqrt{Mf}}} \tag{10.13}$$

When the output frequency is f, then $R_L \equiv R_{L_1}$, which means that it is tuned to f. However, when the output frequency is Mf, then R_L is tuned to Mf and R_{L_1} is very small and is considered to be zero, so that for $M \neq 1$,

$$R_{L_1} = 0 \tag{10.14}$$

for $M = 1$,

$$R_{L_1} = R_L \tag{10.15}$$

Equations (10.14) and (10.15) are important for determining the Miller effects.

10.2.8 The Calculation of r_{2ac}

When Z_2 in Fig. 10.1 is a capacitor, L_N and C_{r_2} become unnecessary, and r_2 now has an ac component, r_{2ac}. Of course, r_{2ac} then contributes an ac circuit loss which must be included in the calculations. This situation exists in crystal oscillators when the crystal is operating at its fundamental overtone, in which case $N = 1$. Therefore, for $N = 1$

$$r_{2ac} = r_2 \tag{10.16}$$

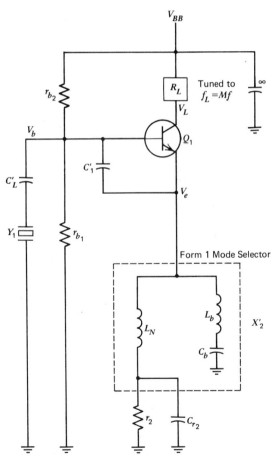

Figure 10.3 Semi-isolated Colpitts oscillator for SC-cut crystals with form 1 mode selector. *Note*: For $N = 1$, $L_N = C_{r_2} = 0$.

240 The Semi-isolated Colpitts Oscillator

for $N \neq 1$

$$r_{2ac} \to \infty \qquad (10.17)$$

since r_2 is then isolated by C_{r_2} and L_N.

10.2.9 Overtone and Mode Selection

Figures 10.3 and 10.4 show the oscillator configured for using the SC-cut crystal with either the form 1 or form 2 mode selector. (See Section 5.6.5.4.)

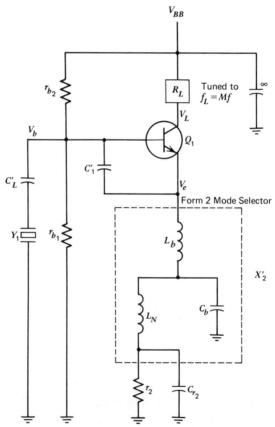

Figure 10.4 Semi-isolated Colpitts oscillator for SC-cut crystals with form 2 mode selector.

10.3 FREQUENCY VARIATIONS DUE TO EXTERNAL FACTORS

The material in Section 7.3 is applicable except that, as pointed out in Section 10.2.3, the effect of changes in external load is much less, especially for harmonic output circuits.

10.4 THE DESIGN PROCEDURE

The design procedure is very straightforward and is principally determined by the output power, the harmonic number M, and the crystal characteristics. Again, the algorithms are prepared for crystal resonators, but other types of resonators may also be used with slight modifications of the algorithms. The algorithms are prepared for all crystal cuts and include both forms of SC mode selectors, as shown in Figs. 10.3 and 10.4. In this connection, it should be noted that this oscillator circuit is normally used for low crystal drive applications where C_1' is very small and mode selection in the base emitter circuit therefore becomes impractical.

10.5 THE DESIGN ALGORITHM AND DESIGN EXAMPLES

10.5.1 The Design Algorithm (Algorithm 12.6)

The user has to supply the information requested in Sections a, b, c, and d. It will be noted that Items c7, c9, c10, and c11 are not in the algorithm, but checks should be made after the design to make sure that these values are not exceeded.

The comments in Section 7.5.1 about replacement of the crystal networks by other types of networks and about replacing the calculated value of R_L by another R_L apply equally well here.

10.5.2 Design Examples 10.1 to 10.6

There are six design examples; three for $M = 1$ and three for $M = 2$.

For $M = 1$, R_L is fixed by the BV_{CE} of the transistor in accordance with Eq. (10.10). For $M = 2$, R_L is fixed by the maximum realizable load in accordance with Eq. (10.12).

$(P_L/P_x)_{max}$ is fixed by the maximum realizable load in accordance with Eq. (10.12).

All the design examples have the same P_L and P_x. P_L/P_x is very high, namely, 200.

A surprising result is the relatively high conversion efficiency of the circuits: 43% for $M = 1$ and 36% for $M = 2$. This, of course, is diminished by the losses

	Item	Value	All units in pF	μH
Oscillator Performance	f MHz	20		
	P_L mW	10		
	P_x mW	0.05		
	V_{BB} mV dc or rms	12,000		
Principal Crystal Data	M	1		
	R_{df} Ω	20		
	I_x mA	1.58		
	Cut	AT		
	N	1		
	C_L pF	32		
Transistor Data	β_o	30		
	f_T	700		
	BV_{CE}	15,000		
	C_{cb}	2		
	C_{bet}	2		
	C_{ce}			
	P_{dis}			
	Type	2N918		
Circuit Parameters	α	0.3		
	γ_1	1.4		
	γ_m			
	V_{be} mV	120		
Calculated Data	I_{BB} mA	2.04		
	g_m	0.020		
	V_E	3410		

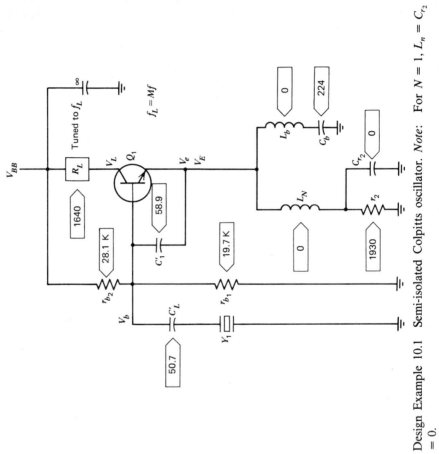

Design Example 10.1 Semi-isolated Colpitts oscillator. *Note:* For $N = 1$, $L_n = C_{r_2} = 0$.

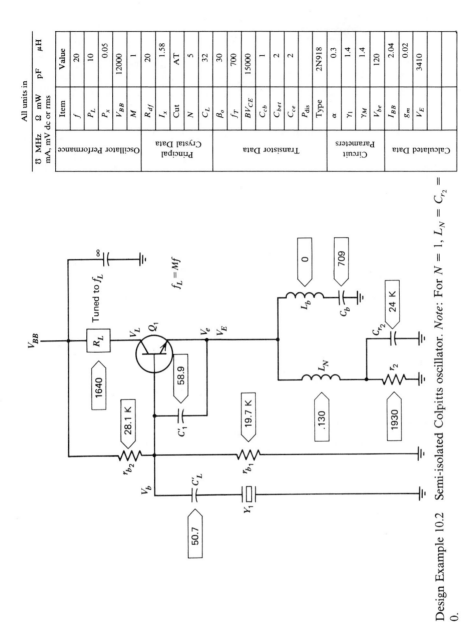

Design Example 10.2 Semi-isolated Colpitts oscillator. *Note:* For $N = 1$, $L_N = C_{r_2} = 0$.

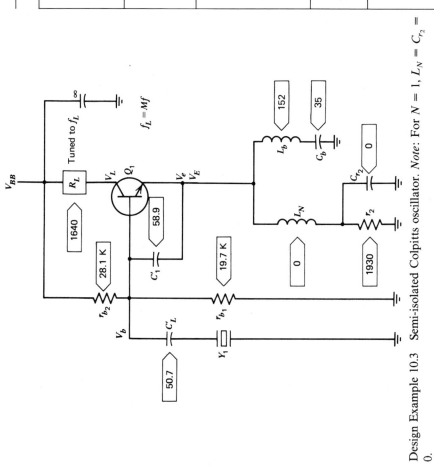

Design Example 10.3 Semi-isolated Colpitts oscillator. *Note:* For $N = 1$, $L_N = C_{r_2} = 0$.

		All units in			
		Ω mW mA, mV dc or rms	MHz	pF	μH
	Item		Value		
Oscillator Performance	f		20		
	P_L		10		
	P_x		0.05		
	V_{BB}		12,000		
Principal Crystal Data	M		2		
	R_{df}		20		
	I_x		1.58		
	Cut		AT		
	N		5		
	C_L		32		
Transistor Data	β_o		30		
	f_T		700		
	BV_{CE}		15,000		
	C_{cb}		1		
	C_{bet}		2		
	C_{ce}		2		
	P_{dis}				
	Type		2N918		
Circuit Parameters	α		0.15		
	γ_1		1.4		
	γ_M		1.2		
	V_{be}		240		
Calculated Data	I_{BB}		2.42		
	g_m		0.012		
	V_E		3540		

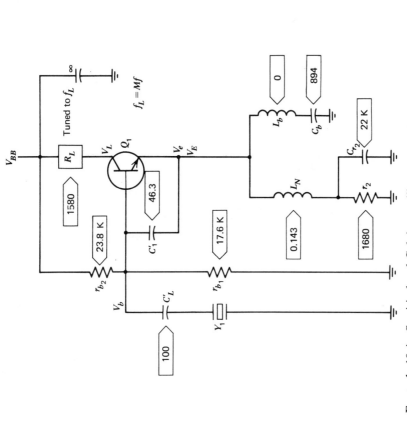

Design Example 10.4 Semi-isolated Colpitts oscillator. *Note:* For $N = 1$, $L_N = C_{r_2} = 0$.

	Item	Value	All units in Ω MHz mW pF μH mA, mV dc or rms
Oscillator Performance	f	20	MHz
	P_L	10	mW
	P_x	0.05	mW
	V_{BB}	12,000	mV
	M	2	
Principal Crystal Data	R_{df}	20	Ω
	L_x	1.58	
	Cut	SC	
	N	5	
	C_L	32	pF
Transistor Data	$β_o$	30	
	f_T	700	MHz
	BV_{CE}	15,000	mV
	C_{cb}	1	pF
	C_{bet}	2	pF
	C_{ce}	2	pF
	P_{dis}		
	Type	2N918	
Circuit Parameters	$α$	0.15	
	$γ_1$	1.4	
	$γ_M$	1.2	
	V_{be}	240	mV
	I_{BB}	2	mA
Calculated Data	g_m	0.012	
	V_E	3540	mV

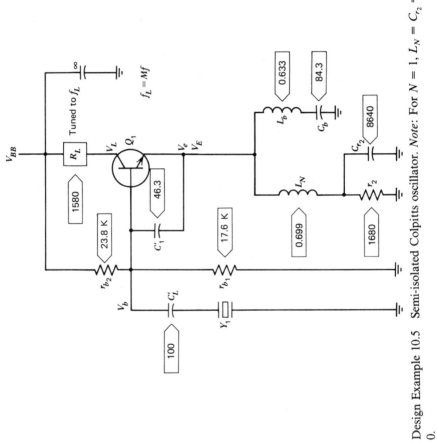

Design Example 10.5 Semi-isolated Colpitts oscillator. *Note*: For $N = 1$, $L_N = C_{r_2} = 0$.

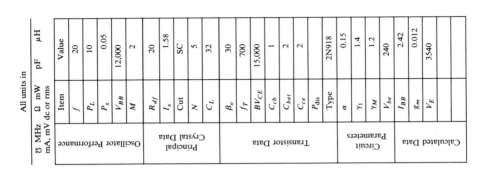

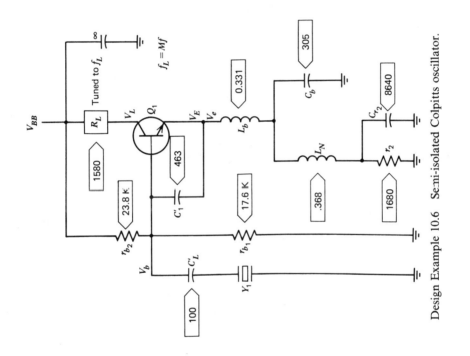

Design Example 10.6 Semi-isolated Colpitts oscillator.

in the various circuit elements, except the load, transistor, and the crystal. The distortion of the output signal is a function of the Q of the load circuit, R_L.

Because of the high value of P_L/P_x, the Miller effects in the $M = 1$ circuits are very large, and considerable difficulty will be encountered in tuning R_L. For example, in Design Example 10.1, C_1' is 59 pF, while the Miller effects provide an additional 45 pF. Similarly C_2' is 709 pF, and the Miller effects provide an additional 289 pF. These examples are, of course, deliberately made to be extreme to show the importance of the Miller effects. It is recommended that, if M must be 1, P_L/P_x be made as small as possible.

For $M = 2$, the total contribution of the Miller effect and the transistor transition and diffusion capacitances is only 6 pF to C_1', which is 46 pF. The similar contribution to C_2, which is effectively 460 pF, is only 8 pF, so that for all practical purposes C_2 may be considered equal to C_2'.

Design Example 10.3 is the same as Design Example 10.1 except for the crystal cut. The emitter circuits are modified accordingly.

Design Example 10.2 is the same as Design Example 10.1 except for the crystal overtone. This also requires modification in the emitter circuitry.

Design Example 10.4 is the same as Design Example 10.2 except for M. The circuits are similar except for the tuning of R_L and the values of the components.

Design Examples 10.5 and 10.6 are identical except for the form of the mode selector circuit, which is necessary because of the SC-cut crystal. In this case both forms 1 and 2 are realizable, but there are many cases where one form is much more realizable than the other, and the choice becomes self-evident.

The values of g_m are given on the design examples to check the adequacy of β_o and f_T in accordance with Steps c2 and c3 of the algorithm.

10.5.3 Trimming for Algorithm 12.6

See Algorithm Figures 12.6.1 and 12.6.2.

10.5.3.1 Introduction

See Section 7.5.3.1.

10.5.3.2 Basis of the Trimming Procedure

The trimming is based upon the following relations:

1. $$I_e \approx 1.4 I_E$$
2. α decreases as X_1 increases.
3. γ_M increases slightly as α decreases ($M \neq 1$).
4. $$I_{e_M} = \gamma_M I_E$$
5. $$V_L = I_{e_M} R_L$$

6	$1000 P_L = \dfrac{V_L^2}{R_L} \approx \dfrac{I_{e_M}^2 R_L}{1000}$
7	$I_3 \approx I_{e_1} \dfrac{(X_2)}{R_T}$
8	$V_{be} \approx I_3 X_1$
9	Section 7.5.3.2, item 5 is applicable here.

Note: For a crystal resonator $I_3 = I_x$.

10.5.3.3 Typical Trimming Steps

The following describes the action required to increase a given characteristic. Obviously, the opposite action decreases the same characteristic.

10.5.3.3.1 Output Power P_L

Item 6 states that increasing I_{e_M} and/or R_L will increase the power. I_{e_M}, in turn, can be increased by increasing I_E and/or γ_M as stated in item 4. I_E is increased by adjusting the bias circuit. γ_M is increased by increasing X_1 but when γ_M reaches 1.2 it has almost reached saturation, and further efforts will result in little improvement.

For $M = 1$ great caution should be exercised in increasing R_L because that will increase frequency instability.

10.5.3.3.2 Current I_3

Item 7 states that I_3 increases as I_{e_1} and/or X_2 are increased. Increasing X_2 is accomplished as described in Sections 5.6.4 and 5.6.5. I_{e_1} is increased by increasing I_E which will also increase P_L.

It is interesting to note that changing X_1 will have little effect upon I_3.

Item 7 also states I_3 increases as R_{df} decreases so that if a new crystal with lower R_{df} is used, the current increases.

11

The Butler Oscillator

11.1 INTRODUCTION

The Butler oscillator, also called the Bridged-T Oscillator, is the most popular member of a family of oscillators wherein the emitter current is the crystal (or other resonator) current. The emitter current is therefore sinusoidal, and the output is relatively free of harmonic content. Also, the crystal (or other resonator) acts as a filter to reduce the noise outside the effective bandwidth of the crystals, f/Q_{op}.

The reason for the popularity of this oscillator is not the noise characteristic mentioned above, but the fact that it can function at very high frequencies because the transistor is used in its common base configuration, which, as pointed out in Section 2.5, is particularly desirable for high-frequency operation.

By the Butler oscillator is meant the oscillator, whose circuit is shown in Fig. 11.1. All the components in that circuit are physical; that is, they are all installed components.

In comparing Fig. 11.1 to Figs. 5.5b and 5.5c, one is struck by their similarity. They are identical except that the circuit of Fig. 11.1 has a crystal in series with the emitter. The principal effect of the crystal is to vary the magnitude and phase of the g_m of the transistor, as the frequency varies, by emitter local degeneration action.

One would therefore expect that the analysis of Fig. 11.1 would be similar to that of Figs. 5.5b and 5.5c. However, adapting the analysis of these circuits to the circuit of Fig. 11.1 turns out to be very tedious and yields little insight into the circuit operation. Therefore, the analysis of Section 11.2 has been developed.

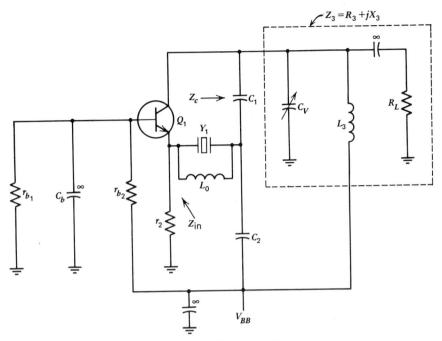

Figure 11.1 Butler oscillator.

11.2 OSCILLATOR CIRCUIT ANALYSIS

11.2.1 Introduction

One could endeavor to use two-port network theory for the oscillator, but because of the large signals existing within the oscillator, the results will not be very useful and will be very complicated.

Instead, it has been found to be more practical to use the feedback and negative resistance oscillator models combined with an approximate hybrid-PI network model for the transistor.

11.2.2 Transistor Approximations

In order to arrive at a reasonably simple oscillator treatment, it is necessary to make many approximations as to the transistor behavior, beginning with the model of Fig. 2.14 and the large-signal characteristics of Section 6.3.

Obviously these approximations are valid only for high-performance transistors of the type referenced in Section 2.5.1. Some of the approximations are questionable even for these high-performance transistors, but they must be made in order to produce a relatively simple analytical model. These approximations will cause the resultant model to be only approximate, but

11.2.2.1 Assumptions and Approximations for Fig. 2.14

1 $\alpha = -1$. (11.1)

To ensure that this assumption is valid, it is required that

$$f_T > 4fI_E \tag{11.1a}$$

2 $r_{bb'}$ is assumed zero.
3 $r_{ce} \to \infty$ (11.2)
4 C_{bet} and C_{bed} can be neglected, as it is in shunt with r_e, which is small.
5 C_{cb} can be lumped into the Z_3 network of Fig. 11.1.

11.2.2.2 Assumptions for Section 6.3

It is assumed that the material in this section is valid at high frequencies up to 200 MHz. Again, this assumption is somewhat questionable, but very useful.

11.2.3 Derivation of the Oscillatory Conditions

Figure 11.2 is the ac equivalent circuit of Fig. 11.1, but in a more general form, since it does not restrict the makeup of Z_1, Z_2, and Z_3 as is done in Fig. 11.1. The restrictions will be applied in conformance with the requirements of the oscillatory equations.

In the following analysis r_2 is neglected as being much larger than the emitter input impedance, so that Z_{IN} is the emitter input impedance.

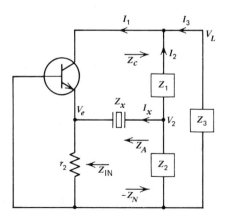

Figure 11.2 Schematic of ac circuitry of the Butler oscillator.

11.2 Oscillator Circuit Analysis

Let

$$Z_A = Z_{IN} + Z_x = R_A + jX_A \tag{11.3}$$

where

$$R_A = R_{IN} + R_x, \quad X_A = X_{IN} + X_x \tag{11.3a}$$

By inspection,

$$I_2 = I_1 \frac{Z_3}{Z_3 + Z_1 + Z_2 Z_A/(Z_2 + Z_A)} \tag{11.4}$$

$$I_x = \frac{-I_2}{Z_2 + Z_A} Z_2 \tag{11.5}$$

$$= \frac{-I_1 Z_3}{Z_3 + Z_1 + Z_2 Z_A/(Z_2 + Z_A)} \cdot \frac{Z_2}{Z_2 + Z_A} \quad \text{from Eq. (11.4)}$$

$$\tag{11.5a}$$

$$= \frac{-I_1 Z_3 Z_2}{(Z_3 + Z_1)(Z_2 + Z_A) + Z_2 Z_A} \tag{11.5b}$$

but at equilibrium oscillation,

$$I_1 = \alpha I_x$$

$$= -I_x \quad \text{from Eq. (11.1)}$$

Therefore, from Eqs. (11.5b) and (2.82)

$$(Z_3 + Z_1)(Z_2 + Z_A) + Z_2 Z_A = Z_3 Z_2 \tag{11.6}$$

or

$$Z_A(Z_1 + Z_2 + Z_3) = -Z_1 Z_2 \tag{11.7}$$

Let

$$Z_1 + Z_2 + Z_3 = Z_s = R_s + jX_s \tag{11.8}$$

then Eq. (11.7) becomes

$$Z_A Z_s = -Z_1 Z_2 \tag{11.9}$$

solving for Z_A

$$Z_A = \frac{-Z_1 Z_2}{Z_s} = \frac{-Z_1 Z_2}{R_s + jX_s} \tag{11.10}$$

If Z_1 and Z_2 are specified to be $-jX_1$ and $-jX_2$, respectively, as in Fig. 11.1, Eq. (11.10) becomes

$$Z_A = \frac{-(-X_1 X_2)}{R_s + jX_s} \tag{11.11}$$

It should be noted that Eq. (11.11) will not change if Z_1 and Z_2 are both made loss-free inductors. The necessary and sufficient condition is that both X_1 and X_2 have the same sign. If Z_1 and Z_2 are lossy inductors, the equation will remain essentially the same, but will be slightly more complicated.

Equation (11.11) is now separated into its real and imaginary parts,

$$R_A = \underbrace{\frac{-(-X_1 X_2)}{R_s^2 + X_s^2} R_s}_{-R_N} \tag{11.12}$$

$$X_A = \underbrace{\frac{-X_1 X_2}{R_s^2 + X_s^2} X_s}_{-X_N} \tag{11.13}$$

Equation (11.12) contains the amplitude-determining information, Eq. (11.13) contains the frequency-determining information.

For convenience in analysis, the right-hand term in Eq. (11.12) is designated as $-R_N$. Similarly, the right-hand term of Eq. (11.13) is designated as $-X_N$. Together, they form $Z_N = R_N + jX_N$. At equilibrium, Z_A is constrained to be equal to $-Z_N$.

Equations (11.12) and (11.13) are called the *oscillatory equations* or the *equations for oscillation* and much can be learned from the study of these equations.

Also, for the case where Z_1 and Z_2 are loss-free capacitors, from Eqs. (11.8), (11.12), and (11.13),

$$R_s = R_3 \tag{11.14}$$

$$X_s = X_3 - (X_1 + X_2) \tag{11.15}$$

R_N will be recognized as the negative resistance R_g in the negative resistance oscillator model discussed in Section 1.3.1. It is seen, from that section, that oscillation cannot exist when $R_A > R_N$. The oscillation amplitude increases as long as $R_A < -R_N$. The oscillation is at amplitude equilibrium when $R_A = -R_N$.

An important consequence of the foregoing is that, since X_s, R_s, X_1, and X_2 are quantities independent of the amplitude, R_A must increase as the amplitude increases in order to attain amplitude equilibrium.

Frequency Stability Relationships

From Eq. (11.3a)

$$X_A = X_{IN} + X_x$$

If X_{IN} is assumed constant, then

$$\partial X_A = \partial X_x$$

which is a measure of the frequency stability of the oscillator since ∂f is almost proportional to ∂X_x.

For any particular independent parameter, U, it is desirable to make $\partial X_A / \partial U \to 0$ to achieve minimum frequency change with variation of that parameter.

11.2.4 Relationships between Z_s and the Physical Components of Fig. 11.1

L_3 and C_V together are equivalent to a variable inductor having a reactance

$$X_L = \frac{2\pi f L_3}{1 - (2\pi f)^2 \times 10^{-6} L_3 C_v} \tag{11.16}$$

where C_V is designed so that

$$(2\pi f)^2 \times 10^{-6} L_3 C_v < 1 \tag{11.16a}$$

It is interesting to note from Eq. (11.16) that X_L increases as C_V increases. As shown in Fig. 11.1, Z_3 consists of X_L and R_L in parallel: or

$$Z_3 = \frac{R_L}{1 + (R_L/X_L)^2} + j\frac{X_L}{1 + (X_L/R_L)^2} \tag{11.17}$$

and from Eq. (11.8)

$$Z_s = \underbrace{\frac{R_L}{1 + (R_L/X_L)^2}}_{R_s} + j\underbrace{\left[\frac{X_L}{1 + (X_L/R_L)^2} - (X_1 + X_2)\right]}_{X_s} \tag{11.18}$$

or

$$Z_s \approx \underbrace{\frac{X_L^2}{R_L}}_{R_s} + j\underbrace{[X_L - (X_1 + X_2)]}_{X_s} \qquad (11.18a)$$

when $R_L \geq 5X_L$.

It should be noted that Eq. (11.18a) introduces only a small error in R_s but may introduce a fairly substantial error in X_s since X_s, which is a relatively small quantity, is the difference of two large quantities.

For any given oscillator, X_1, X_2, and R_L are constants. X_L depends upon the tuning of C_V. Therefore, R_s and X_s are not independent quantities and for every value of X_s there is a corresponding value of R_s as determined by Eq. (11.18).

11.2.5 Calculation of Z_c, V_L, V_e, V_2, P_T, P_A, and P_x

From Fig. 11.2

$$Z_c = \frac{(Z_1 + Z_2 Z_A/(Z_2 + Z_A))Z_3}{Z_3 + Z_1 + Z_2 Z_A/(Z_2 + Z_A)}$$

$$= \frac{Z_1 Z_2 + (Z_1 + Z_2)Z_A}{Z_2}$$

from Eq. (11.7) and

$$Z_c = \frac{Z_A}{Z_2}(Z_1 + Z_2 - Z_s)$$

from Eq. (11.10) or

$$Z_c = \frac{-Z_A Z_3}{Z_2} \qquad (11.19)$$

from Eq. (11.8) and

$$V_L = -I_1 Z_c = -I_1 \left(\frac{-Z_A Z_3}{Z_2}\right)$$

$$= -I_x \frac{Z_A Z_3}{Z_2} \qquad (11.20)$$

from Eqs. (2.83) and (11.19).

$$V_e = I_x Z_{IN} \tag{11.21}$$

$$V_2 = I_x Z_A = -V_L \frac{Z_2}{Z_3} \tag{11.22}$$

Let P_T be the total power dissipated

$$P_T = P_L + P_A \tag{11.23}$$

Then

$$10^3 P_T = I_x^2 \text{Re}\left(\frac{-Z_A Z_3}{Z_2}\right) \tag{11.24}$$

$$P_L = P_T - P_A$$
$$= P_T - I_x^2 R_A \times 10^{-3} \tag{11.25}$$

11.2.6 Relationships for the Special Case Where $X_A = 0$, $Z_1 = -jX_1$, and $Z_2 = -jX_2$

In this case the crystal is operating at series resonance if Z_{IN} is real; if not, $X_x = -X_{IN}$. Also $Z_A = R_A$.
From Eqs. (11.11) and (11.18)

$$X_s = 0 \tag{11.26}$$

at which point

$$X_L = (X_1 + X_2)\left[1 + \left(\frac{X_1 + X_2}{R_L}\right)^2\right] \tag{11.27}$$

$$\approx X_1 + X_2 \tag{11.27a}$$

also,

$$R_s = R_3 = \frac{X_1 X_2}{R_A} \tag{11.28}$$

From Eq. (11.15)

$$X_3 = X_1 + X_2 \tag{11.29}$$

258 The Butler Oscillator

From Eq. (11.22)

$$V_2 = I_x R_A \tag{11.30}$$

From Eq. (11.24)

$$10^3 P_T = I_x^2 \operatorname{Re}\left(\frac{-Z_A Z_3}{-jX_2}\right) \tag{11.31}$$

$$= I_x^2 \frac{X_1 + X_2}{X_2} R_A \tag{11.32}$$

from Eq. (11.29) or

$$10^3 P_T = I_x^2 R_A + I_x^2 \frac{X_1}{X_2} R_A \tag{11.33}$$

The first term is $P_A \times 10^3$.
Therefore, from Eq. (11.25),

$$10^3 P_L = I_x^2 R_A \frac{X_1}{X_2} \tag{11.34}$$

In order to simplify the calculations without, however, incurring large errors, let

$$R_A \geqq 5X_2 \tag{11.35}$$

Then

$$V_L \approx \frac{X_1 + X_2}{X_2} V_2 \tag{11.36}$$

Let

$$\frac{X_1}{X_2} = n \tag{11.37}$$

Then

$$X_1 = nX_2 \tag{11.38}$$

and

$$V_L \approx (n + 1)V_2 \tag{11.39}$$

and

$$P_L = nP_A \tag{11.40}$$

from Eq. (11.34). Obviously,

$$10^3 P_L = \frac{V_L^2}{R_L} \tag{11.41}$$

$$R_L = \frac{[(n+1)I_x R_A]^2}{10^3 P_L} \tag{11.42}$$

from Eqs. (11.39), (11.41), and (11.30).

11.2.7 The Concept of the R_A Gain Factor, A_{L_0}

Under small-signal conditions R_A will have a value

$$R_{A_0} = R_x + R_{IN_0} \tag{11.43}$$

This is measured with the oscillating loop open but properly terminated. When the loop is closed, R_A will assume a value

$$R_A = R_x + R_{IN} \tag{11.44}$$

R_{IN} will, of course, increase to stabilize the amplitude of oscillation.
The ratio A_{L_0} is defined as

$$A_{L_0} = \frac{R_A}{R_{A_0}} \tag{11.45}$$

or

$$R_A = A_{L_0}(R_{IN_0} + R_x) \tag{11.45a}$$

11.2.8 Limiting

In this oscillator also, there are two types of limiting.

1. R_{IN} limiting
2. Collector base limiting, used only when V_L is large.

11.2.8.1 R_{IN} Limiting (See Section 6.3.3)

In this case the crystal current I_x is fixed by the dc emitter current I_E in accordance with Eq. (6.23). R_{IN} will then assume the value required to satisfy Eq. (11.12).

To ensure correct limiting,

$$V_E = V_{BB} - 1.7V_L - 1700 \tag{11.46}$$

or when V_E calculates to be $> V_{BB}/2$, then

$$V_E = \frac{V_{BB}}{2} \tag{11.46a}$$

11.2.8.2 Collector Base Limiting

This type of limiting is similar to that described in Section 6.2.4. However, modifications must be made to the material in that section because of the sinusoidal i_E.

In this form of limiting:

I_E is determined by Eq. (2.97).

$$V_E = V_{BB} - 1.4V_L \tag{11.47}$$

as V_e is usually very small and can be neglected.

11.2.8.3 Type of Limiting Used in the Algorithms for the Butler Oscillator

R_{IN} limiting is used in the circuits designed in the algorithms. This is done because of the superior high-frequency transistor performance. Also, when properly designed, the output is almost independent of the crystal resistance, as the output is dependent upon I_x which is essentially fixed by the emitter current I_E, as stated in Eq. (6.23).

11.2.9 Calculation of C_b in Fig. 11.1

The function of C_b is to effectively connect the base to ground at *ac*. This is calculated by making its reactance much smaller than that of the emitter to ground, which is $R_A - R_{df}$; then

$$C_b \gg \frac{1.59 \times 10^6}{(R_A - R_{df})f} \tag{11.48}$$

11.2.10 Calculation of L_0 in Fig. 11.1

The function of L_0 is to neutralize the C_0', which is equal to the sum of the C_0 of the crystal and the parallel wiring capacitance. L_0 is necessary at high frequencies where C_0' would tend to short out the crystal action and oscillations uncontrolled by the crystal may be permitted. It is recommended that L_0 be supplied at $f \geq 50$ MHz; the value of L_0 is

$$\frac{10^6}{C_0'(6.28f)^2} \qquad (11.49)$$

11.2.11 Calculation of the Effect of the Q of L_3 in Fig. 11.1

Often L_3 has a power loss that is appreciable compared to that in the load R_L and therefore should be taken into account.

Assume that L_3 has a $Q = Q_{L_3}$. Then the associated equivalent parallel resistance

$$r_{L_3} = Q_{L_3} X_{L_3} \qquad (11.50)$$

This resistance is in shunt with R_L. Therefore, its power loss is

$$P_L \cdot \frac{R_L}{r_{L_3}} \qquad (11.51)$$

The total power dissipated in R_L and L_3 will be

$$P_L' = P_L \left(1 + \frac{R_L}{r_{L_3}}\right)$$

$$= P_L \left(1 + \frac{R_L}{Q_{L_3} X_{L_3}}\right) \qquad (11.52)$$

from Eq. (11.50).

11.2.12 Calculation of the Oscillator Operating Q

Obviously

$$Q_{op} = \frac{Q_x \cdot R_{df}}{-R_N} \qquad (11.53)$$

where R_N is the negative resistance in the negative resistance oscillator model. For the case where $X_A = 0$,

$$-R_N = R_A \qquad (11.54)$$

Therefore, in this case

$$Q_{op} \approx \frac{Q_x \cdot R_{df}}{R_A} \qquad (11.55)$$

11.2.13 Calculation of R_N, X_N, and X_c in Terms of the Physical Component Values

From Eqs. (11.12) and (11.18)

$$-R_N \approx \frac{X_1 X_2 \dfrac{R_L}{1 + (R_L/X_L)^2}}{\left[\dfrac{R_L}{1 + (R_L/X_L)^2}\right]^2 + \left[\dfrac{X_L}{1 + (X_L/R_L)^2} - (X_1 + X_2)\right]^2} \qquad (11.56)$$

From Eqs. (11.13) and (11.18)

$$-X_N \approx \frac{-X_1 X_2 \left(\dfrac{X_L}{1 + (X_L/R_L)^2} - (X_1 + X_2)\right)}{\left[\dfrac{R_L}{1 + (R_L/X_L)^2}\right]^2 + \left[\dfrac{X_L}{1 + (X_L/R_L)^2} - (X_1 + X_2)\right]^2} \qquad (11.57)$$

From Eqs. (11.19) and (11.17)

$$|Z_c| = \frac{\sqrt{(R_N^2 + X_N^2)}\, R_L X_L}{X_2 \sqrt{R_L^2 + X_L^2}} \qquad (11.58)$$

11.2.14 Discussion of Design Procedures

In order to arrive at quantitative design procedures which are reasonably simple, it is necessary to make additional assumptions that are often incorrect. One assumption is that the transistor input impedance Z_{IN} is independent of the load in the collector circuit. Furthermore, the R_{IN} is calculable as described in Section 2.5.2.2. As implied in Eq. (2.85) and explained in Section 11.4, the latter assumptions are valid only when the effects of C_{ce} can be neglected, which is very often not true at high frequencies. The net result is that X_{IN} may be substantial, and more importantly, it varies with the tuning of the collector load. In addition, R_{IN} is also caused to vary when the collector load changes. Taking into account these variations in any design procedure renders the procedure incredibly complex. It was therefore opted to neglect these variations and to make the above-mentioned assumptions. However, an attempt will be made to explain the consequences of these assumptions during the development of the design procedures. In doing so, these assumptions will be denoted as assumptions $A \equiv \text{Asmp}\,A$.

It is noteworthy that most of the above variations in R_{IN} can be rendered harmless by shunting the emitter with a low-value resistor, or capacitor, but insufficient loop gain may result.

At this point, it is worthwhile examining the influence of the above assumptions upon the theory developed thus far.

The basic oscillatory equations (11.12) and (11.13) are correct since they were derived without Asmp A. Also, Z_N can be rendered easily calculable by making $X_{c_2} \ll R_A$ as stated in Eq. (11.35), so that the contribution of Z_A to Z_N becomes very small. It should be noted, as stated previously, that once Z_N is known, Z_A can be determined from Eqs. (11.12) and (11.13).

All other equations involving Z_A are equally correct. However, the contributions of the transistor and the crystal to Z_A are somewhat indeterminate, and these are now examined in detail. Z_A is known from Eqs. (11.12) and (11.13). Therefore, R_A is known. R_{IN} is known because $R_{IN} = R_A - R_x$, and R_x is known. X_A is known from Eq. (11.13).

$X_A = X_{IN} + X_x$, but the relative contributions of X_{IN} and X_x are unknown because of the dependence of X_{IN} upon the collector load as described above. Also, with X_x unknown, little is known about the frequency stability because it is directly related to the behavior of X_x.

Another important aspect of Asmp A is that R_{IN_0}, and hence A_{L_0}, are assumed to be known. In fact, R_{IN_0} is not known. Again, this is very unfortunate because oscillations can only exist when $R_N > R_{IN_0}$, and therefore one does not know when oscillations will cease.

As a result of the above, Sections 11.2.3.1 and 11.2.7 are meaningless except in special situations which will be described later. Although quantitative information on X_x and R_{IN_0} cannot be obtained from the design procedures in the general case, some useful qualitative information will be supplied, as will be demonstrated later.

Two design procedures will now be developed. Both procedures will include Asmp A. The designs obtained from these procedures will be somewhat incorrect, as pointed out above, but the required performance will be obtained from these designs combined with the trimming procedures.

In the development of these design procedures, additional assumptions may be made to simplify the execution of the procedure. However, errors introduced by these additional assumptions and the true performance can be determined by means of the exact relationships derived in Section 11.2.

For convenience, the substance of Asmp A is repeated here:

1. Z_{IN} is independent of the collector load.
2. X_{IN} = constant, and is equivalent to a simple linear reactance.
3. $R_{IN_0} = \dfrac{26}{I_E}$ (2.87)

neglecting the 1 Ω and $r_{bb'}$.

11.3 THE $X_A = 0$ DESIGN PROCEDURE

11.3.1 Introduction

The Butler oscillator is primarily used because of its ability to operate at much higher frequencies than the other oscillators thus far discussed. Accordingly, one is forced to accept its shortcomings, such as low operating Q.

One of the most popular designs requires that C_1, C_2, and X_3 be tuned to resonance at the operating frequency f. This implies that $X_A = 0$, and for the case where $X_{IN} = 0$, the crystal is operating at series resonance. This operating state is experimentally attained by replacing the crystal with an equivalent resistor and tuning the oscillator so that the resulting operating frequency approximates the crystal frequency. (See, for example, page 94 of Ref. 1.10.) This operating mode has the advantage that its amplitude stability is optimized since the negative resistance term, and hence R_A, in Eq. (11.12), is maximized. Also, the amplitude is maximized, which is a convenient condition to observe.

It will be seen later that this mode of operation does not have the maximum frequency stability. However, because of its popularity, an algorithm for the design of the Butler oscillator operating in this mode, called the $X_A = 0$ mode, is developed using the theory presented in Section 11.2, including the $X_A = 0$ relationships of Section 11.2.6.

11.3.2 The $X_A = 0$ Algorithm

This algorithm is principally prepared using the relationships in Section 1.2.6. It is presented in Algorithm 12.7.

The user has to supply the information in Sections a, b, c, and d. It will be noted that items c7 to c10 are not in the algorithm, but checks should be made after the design to ensure that they are not exceeded. Also, because of the approximations in the algorithms, the influence of Items c4 to c6 must be included in the trimming procedure.

11.3.3 The $X_A = 0$ Design Example

Design Example 11.1 is the design example computed from the algorithm. The ratio $n = C_2/C_1$ is very large because of the large P_L/P_x. The Q_{L_3} has been chosen to be sufficiently high so that the losses of L_3 contribute little to P_L'.

		All units in		
		ʊ MHz mA, mV dc or rms	Ω mW	pF µH
	Item	Value		
Oscillator Performance	f	20		
	P_L	5		
	P_x	0.1		
	V_{BB}	10,000		
Principal Crystal Data	R_{df}	20		
	I_x	2.24		
	Cut	AT		
	N	3		
	C_0'	6		
Transistor Data	β_o	30		
	f_T	700		
	BV_{ce}	15,000		
	C_{cb}	—		
	C_{bet}	—		
	C_{ce}	—		
	P_{dis}	—		
	Type	2N918		
Circuit Parameters	A_{L_0}	2		
	Q_{L_3}	100		
	R_A/X_2	5		
Calculated Data	I_{BB}	3.49		
	$|Z_c|$	1038		
	R_N	56.6		
	X_A	0.7		
	X_L	218.5		

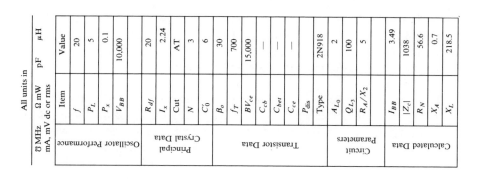

Design Example 11.1 The $X_A = 0$ Butler oscillator.

Some of the important performance characteristics of the example are plotted in Fig. 11.3. The data were calculated from Eqs. (11.56) to (11.58). (See Section 11.4.1 for the significance of X_L.)

The following useful information may be gathered from this figure:

1. From Fig. 11.3*b* it is seen that the circuit will oscillate for values of $X_L \approx 180$ to $250\ \Omega$ since $R_N > [(R_{IN_0} + R_{df}) = 29]\ \Omega$ for these values of X_L. Also, from the same figure, the intersection of the curve with the $R_N = 38\ \Omega$ line determines the X_L region where $|Z_c|$, plotted in Fig. 11.3*c*, is proportional to $|V_L|$. This is so because $|V_L| = I_x|Z_c|$ and, in this region, $[R_N = R_{df} + R_{IN}] \geq [(R_{df} + 2R_{IN_0}) = 38]\ \Omega$, and, as shown in Fig. 2.15, I_x remains essentially constant at $I_E/\sqrt{2}$ when $R_{IN} \geq 2R_{IN_0}$. Outside of this region, V_L varies much more rapidly than $|Z_c|$.

2. Figure 11.3*a* shows that at the design value of X_L the slope of the curve is maximum. This means that the design is *very poor with respect to frequency stability* as any variation in X_L will cause a relatively large shift in X_A, which is directly related to the frequency. (For further discussion of Fig. 11.3*a*, see Section 11.4.2.)

3. Figure 11.3*d* shows that the frequency stability of this design, with respect to variations in R_L, is quite good.

   It should be noted that the above curves are theoretical. In practice, the curves will be similar, but skewed because of the effects of C_{ce}. (See Section 11.4.4.)

11.3.4 Trimming for Algorithm 12.7

See Algorithm Figure 12.7.

11.3.4.1 Introduction

See Section 7.5.3.1.

11.3.4.2 Basis of the Trimming Procedure

The trimming is based upon the following relationships.

1.
$$P_x = \frac{I_x^2 R_{df}}{1000}$$

2.
$$I_x = \frac{I_E}{\sqrt{2}}$$

   from Eq. (6.23)

3.
$$P_L = \left(\frac{n}{1+n}\right)^2 \frac{R_L I_x^2}{1000}$$

   from Eqs. (11.40), (11.56), and (11.57 and assuming $(R_L/X_L)^2 \gg 1$.

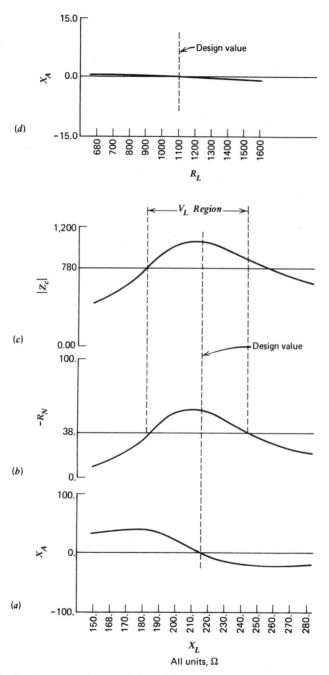

Figure 11.3 Performance characteristics of Design Example 11.1. (a) X_A versus X_L. (b) R_N versus X_L. (c) $|Z_c|$ versus X_L. (d) X_a versus R_L.

11.3.4.3 *Typical Trimming Steps*

The following describes the action required to increase a given characteristic. Obviously, the opposite action decreases the same characteristic.

11.3.4.3.1 *Crystal Power P_x*

Relation 1 states that to increase P_x, I_x must be increased. Relation 2 states that to increase I_x, I_E must be increased by adjusting r_2, r_{b_2}, and r_{b_1} as necessary. Relation 3 also states that the output power is also increased as I_x increases.

11.3.4.3.2 *Output Power P_L*

Relation 3 states that increasing I_x, n, and R_L will increase P_L. However, increasing I_x will increase P_x and if it is desired to maintain P_x constant, only n and R_L may be increased. Increasing R_L will also increase V_L so that readjustment of r_2, r_{b_1}, and r_{b_2} may be necessary to obtain the proper limiting conditions. It should be noted that increasing R_L reduces the share of the total generated power delivered to the external load. In addition, if n or R_L is changed, X_L must be readjusted for maximum output.

11.4 THEORY AND DESIGN OF THE STABLE BUTLER OSCILLATOR

11.4.1 Introduction

Section 11.2.3.1 states that for any independent parameter U, $\partial X_A/\partial U$ should approach zero to achieve minimum frequency change with variation of U.

It is desirable, if possible, to select a physical parameter U_1, the effective value of which is influenced by variations in the other physical parameters in the circuit. It follows that if $\partial X_A/\partial U_1 \to 0$, variations in the other parameters will also have a small effect upon the frequency. It is then sufficient to investigate only the behavior of $\partial X/\partial U_1$ to determine the frequency stability of the oscillator. If $\partial X_A/\partial U_1$ proves to be small, it follows that $\partial X_A/\partial U_n$ will also be small.

It is seen from Eqs. (11.13) and (11.18) that a suitable parameter for U_1 is X_L, which is a direct strong function of C_v and L_3. In addition, the effective value of X_L is a function of the transistor parameters. Equation (11.18) also states that variations in $X_1 + X_2$ can be accounted for by equivalent variations in X_L.

The only parameter which is independent of X_L is R_L, which must be treated separately. It will be seen later that $\partial X_A/\partial X_L$ can be much greater than $\partial X_A/\partial R_L$.

In view of the above, an investigation will be made of the behavior of $\partial X_A/\partial X_L$ and $\partial X_A/\partial R_L$. The results of the investigation will be used to prepare an algorithm for circuits in which $\partial X_A/\partial X_L$ is small. Oscillators in which $\partial X_A/\partial X_L$ is small are called *stable oscillators*.

11.4.2 Basis of the Stable Oscillator Designs

Figure 11.3a shows that there are two values of X_L at which $\partial X_A / \partial X_L \to 0$. One of these points at which X_A is plus is called the X_{A+} point. The point, at larger X_L, where X_A is minus, is called the X_{A-} point. Similarly, the value of X_L at these points is called X_{L+} and X_{L-}.

The X_{A+} point has been experimentally observed and often the oscillator is set to this point by first tuning to the maximum output point (see Fig. 11.3c) in the direction of increasing X_L and then backing up a bit until the output is about 3 dB less. However, the reason for the increased stability has not been understood.

The X_{A-} point is difficult to observe experimentally because the value of R_{IN_0} is so large at this point, for the reasons discussed in Section 11.4.4, that unless the oscillator is specifically designed to operate at this point it ceases oscillating before the point is reached.

An analysis, which is very long and tedious and therefore not repeated here, shows that these points exist when $|Z_c|$, in Fig. 11.1, $\approx 0.7R_L$. The values of X_L for these points are computed below.

11.4.3 The Stable Butler Oscillator Relationships

Assume that the circuit is so designed that Z_A is very large compared to X_2. Z_A therefore contributes very little to Z_c. With this assumption,

$$|Z_c| \approx R_L \| X_p = 0.7 R_L \quad (11.59)$$

where X_p is the parallel combination of $X_1 + X_2$ and X_L.
Obviously, when $|X_c| \approx 0.7 R_L$

$$X_p \approx R_L \quad (11.60)$$

and

$$\theta_{Z_c} = 45° \quad (11.60a)$$

Therefore,

$$R_L \approx |X_p| = \left| \frac{X_L(X_1 + X_2)}{X_L - (X_1 + X_2)} \right| \quad (11.61)$$

Solving Eq. (11.61) for X_L

$$X_{L+} = \left| \frac{R_L(X_1 + X_2)}{R_L + (X_1 + X_2)} \right| \quad (11.62)$$

270 The Butler Oscillator

and

$$X_{L-} = \left| \frac{R_L(X_1 + X_2)}{R_L - (X_1 + X_2)} \right| \tag{11.63}$$

R_L is now computed as follows:

$$1000 P_L = \frac{(I_x |Z_c|)^2}{R_L} \tag{11.64}$$

$$\approx \frac{I_x^2 (0.7 R_L)^2}{R_L}$$

from Eq. (11.59), so that

$$R_L \approx \frac{2000 P_L}{I_x^2} \tag{11.65}$$

Continuing the assumption that Z_A is very large compared to X_2, then, in Fig. 11.2,

$$V_2 \approx \frac{X_2}{X_2 + X_1} V_L \approx \frac{X_2}{X_2 + X_1} I_x Z_c \tag{11.66}$$

and

$$I_x \approx \frac{V_2}{Z_A} \equiv \frac{X_2}{X_2 + X_1} \frac{I_x Z_c}{Z_A} \tag{11.67}$$

from which

$$Z_A \approx \frac{X_2 Z_c}{X_2 + X_1} \tag{11.68}$$

From Eq. (11.60a)

$$|\theta_{Z_c}| \approx 45°$$

Therefore, from Eq. (11.68)

$$|\theta_{Z_A}| \approx 45° \tag{11.69}$$

or

$$R_A \approx X_A \approx 0.7 |Z_A| \tag{11.70}$$

Let
$$X_1 = nX_2 \tag{11.71}$$

Then, from Eqs. (11.68), (11.70), and (11.71),
$$1.4 R_A \approx \frac{|Z_c|}{1+n} \tag{11.72}$$

Solving for n,
$$n = \frac{0.7|Z_c|}{R_A} - 1$$

which becomes from Eq. (11.59)
$$n \approx \frac{0.5 R_L}{R_A} - 1 \tag{11.73}$$

Final Relationships for X_{L+} and X_{L-}

From Eqs. (11.62), (11.63), and (11.73)
$$X_{L+} \approx \frac{R_L}{2 R_A / X_2 + 1} \tag{11.74}$$

$$X_{L-} \approx \frac{R_L}{2 R_A / X_2 - 1} \tag{11.75}$$

and
$$\frac{X_{L-}}{X_{L+}} = \frac{2 R_A / X_2 + 1}{2 R_A / X_2 - 1} \tag{11.76}$$

which increases as R_A/X_2 decreases. For example,

$$\frac{R_A}{X_2} = 5, \qquad \frac{X_{L-}}{X_{L+}} = 1.11$$

$$\frac{R_A}{X_2} = 2.5, \qquad \frac{X_{L-}}{X_{L+}} = 1.5$$

As will be demonstrated later, it is desirable to have larger values of X_{L-}/X_{L+} because the slope of the X_A versus X_L curves then generally decreases. Therefore, C_1 and C_2 may decrease to the point where they may be unrealizably small at very high frequencies. In addition, the transistor will contribute more to X_L, which is undesirable.

11.4.4 The Effect of C_{ce}

As pointed out in Eq. (2.85)

$$Y_{IN} = y_{11b} - \frac{y_{12b} y_{21b}}{y_{22b} + Y_L} \tag{2.85}$$

at the usual frequency of operation,

$$y_{22b} \to 0$$

and

$$y_{11b} \approx -y_{21b}$$

therefore

$$Y_{IN} \approx y_{11b}\left[1 + \frac{y_{12b}}{Y_L}\right] \tag{11.77}$$

For $Y_L = G_L + jB_L$,

$$Y_{IN} \approx y_{11b}\left[1 + \frac{y_{12b}[G_L - jB_L]}{G_L^2 + B_L^2}\right] \tag{11.78}$$

but

$$y_{12b} \approx -j|y_{12b}| \tag{11.79}$$

therefore

$$Y_{IN} \approx y_{11b}\left[1 - \frac{|y_{12b}|B_L}{G_L^2 + B_L^2} - j\frac{|y_{12b}|G_L}{G_L^2 + B_L^2}\right] \tag{11.80}$$

It is evident that for

$$X_{L+}, \text{ the sign of } B_L = -$$
$$X_{L-}, \text{ the sign of } B_L = +$$

Therefore

$$Y_{IN} \text{ tends to increase for } X_{L+}$$
$$\text{decrease for } X_{L-}$$

or

$$R_{IN} \text{ tends to decrease for } X_{L+}$$
$$\text{increase for } X_{L-}$$

Therefore, the oscillator will cease operation sooner as X_L increases on the X_{L-} side.

It should be noted that the above analysis is for the small-signal case. Its quantitative validity for the large signals present in oscillators is therefore questionable, but it is certainly qualitatively valid. If Y_{12b} (including the contributions of the layout) is known, it is worthwhile calculating the effect as described above. If it turns out to be very small, it is in order to design for the X_{L-} point because of its superior frequency stability.

11.4.5 The Stable Oscillator Design Algorithm

This algorithm is prepared for the X_{L+} point because of the reasons explained in Section 11.4. However, the algorithm also lays the groundwork for the design procedure for the X_{L-} point.

All the equations used in preparing the algorithm have already been derived except for the relationship for Q_{op}, which is

$$Q_{op} \approx Q_x \frac{R_{df}}{-R_N} \qquad (11.81)$$

The algorithm is presented in Algorithm 12.8. It gives in detail the design procedure for the X_{L+} point and indicates how the design procedure is modified for the X_{L-} point. Since crystals at very high frequencies are normally rated for operation at series resonance, the algorithm includes inductor, L_A, in Step 23, which permits the crystal to operate at series resonance and facilitates the correlation of the crystal specification with the circuit operation. For X_{L-}, X_A will consist of a capacitor.

11.4.6 The Stable Oscillator Design Example

Design Example 11.2 is the design example computed from the algorithm at X_{L+}. This design example is based upon the assumption that $X_2 = R_A/5$. The ratio $n = C_2/C_1$ is large because of the large P_L/P_x.

Some of the important performance characteristics of Design Example 11.2 are plotted in Fig. 11.4. The data was calculated from precise Eqs. (11.56) to (11.58). (For the explanation of the V_L region see Section 11.3.3, item 1.)

The following useful information may be gathered from this figure:

1 From Fig. 11.4b it is seen that the circuit will oscillate for values of $X_L \approx 170$ to $240 \, \Omega$.

2 From Fig. 11.4c, it is seen that V_L at X_{L+} and X_{L-} is about 3 dB below the peak.

3 From Fig. 11.4d, it is seen that the frequency stability versus R_L of this design is considerably worse than the design of Fig. 11.3d. This, however, is

		All units in	
		Ω mW pF μH	
		Ω MHz mA, mV dc or rms	
	Item	Value	
Oscillator Performance	f	20	
	P_L	5	
	V_{BB}	10000	
Principal Crystal Data	R_{df}	20	
	I_x	2.24	
	Cut	AT	
	N	3	
	C_0'	6	
Transistor Data	β_o	30	
	f_T	700	
	BV_{ce}	15000	
	C_{cb}	—	
	C_{bet}	—	
	C_{ce}	—	
	P_{dis}		
	Type	2N918	
Circuit Parameters	A_{L_0}	2	
	Q_{L_3}	100	
	R_A/X_2	5	
Calculated Data	I_{BB}	3.65	
	$|Z_c|$	1324	
	R_N	54.8	
	X_A	66.2	
	X_L	181	

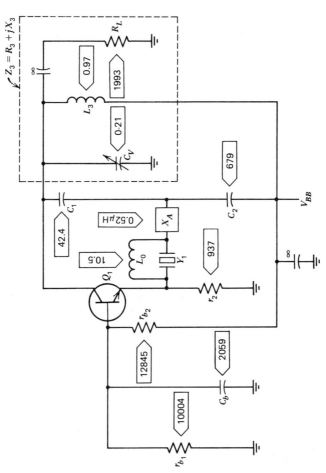

Design Example 11.2   The stable Butler oscillator.

274

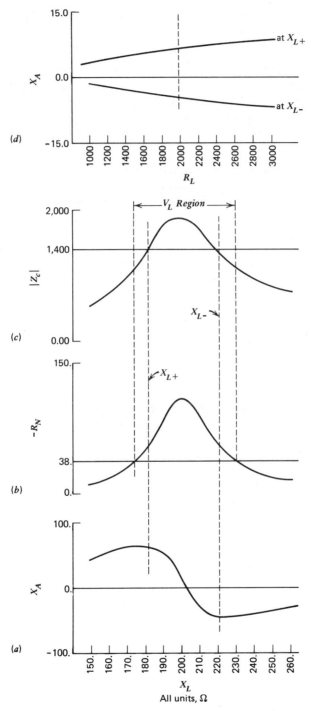

Figure 11.4 Performance characteristics of Design Example 11.2. (a) X_A versus X_L. (b) R_N versus X_L. (c) $|Z_c|$ versus X_L. (d) X_A versus R_L.

not important in those cases where the oscillator feeds an amplifier and R_L is therefore relatively constant.

4 Figure 11.4a shows that the slope of the curve is indeed 0 at X_{L+} and X_{L-} as called for by the design procedure. The figure also shows that the curve is flat for only relatively small ranges of X_L. It is desirable that the X_L range of X_A flatness be increased. A clue as to how to achieve this increase is obtained by noting that the ratio of X_{L-}/X_{L+} is rather small, and a larger ratio may be necessary.

Equation (11.76) states that a larger ratio may be obtained by decreasing R_A/X_2.

In the design of Fig. 11.4, $R_A/X_2 = 5$. A second design was therefore made with $R_A/X_2 = 2.5$.

The circuit values are given in Design Example 11.3 and the performance is that in Fig. 11.5. It will be noted that the X_A versus X_L curves are considerably flatter. However, this design has the disadvantage that the X_L values are approximately double the values of those in Fig. 11.4 and may be unrealizable at the higher frequencies.

Both Figs. 11.4 and 11.5 show that the slope of the X_A versus X_L curve is considerably smaller at X_{L-} than at X_{L+}, and therefore the X_{L-} point is preferable except for the C_{ce} effects discussed in Section 11.4.4.

Also, both Figs. 11.4 and 11.5 prove that the approximate design procedure derived in Section 11.4.3 yields results which are consistent with exact Eqs. (11.56) to (11.58), except for the value of $|X_A|$, which is 15% higher than the design value at X_{L+} and 15% lower at X_{L-}.

Again, the curves of Figs. 11.4 and 11.5 are theoretical. In practice, the curves will be similar but skewed because of the effect of C_{ce} (see Section 11.4.4).

11.4.7 Trimming for Algorithm 12.8

11.4.7.1 Introduction

See Section 7.5.3.1.

11.4.7.2 Basis of the Trimming Procedure

1
$$P_x = \frac{I_x^2 R_{df}}{1000}$$

2
$$I_x = \frac{I_E}{\sqrt{2}}$$

from Eq. (6.23)

3
$$P_L = \frac{I_x^2 R_L}{2000}$$

from Eq. (11.65)

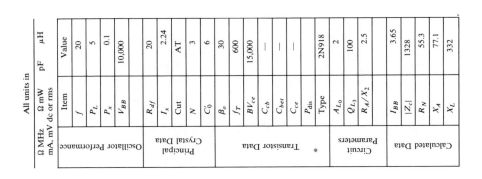

	All units in			
	Ω MHz mA, mV dc or rms	Ω mW	pF	μH
	Item		Value	
Oscillator Performance	f		20	
	P_L	5		
	P_x		0.1	
	V_{BB}	10,000		
Principal Crystal Data	R_{df}	20		
	I_x	2.24		
	Cut	AT		
	N	3		
	C'_0		6	
Transistor Data	β_o	30		
	f_T	600		
	BV_{ce}	15,000		
	C_{cb}	—		
	C_{bet}	—		
	C_{ce}	—		
	P_{dis}			
	Type	2N918		
Circuit Parameters	A_{L_0}	2		
	Q_{L_3}	100		
	R_A/X_2		2.5	
Calculated Data	I_{BB}			3.65
	$|Z_c|$	1328		
	R_N	55.3		
	X_A	77.1		
	X_L	332		

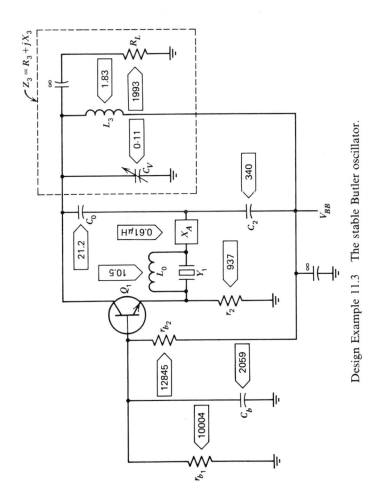

Design Example 11.3 The stable Butler oscillator.

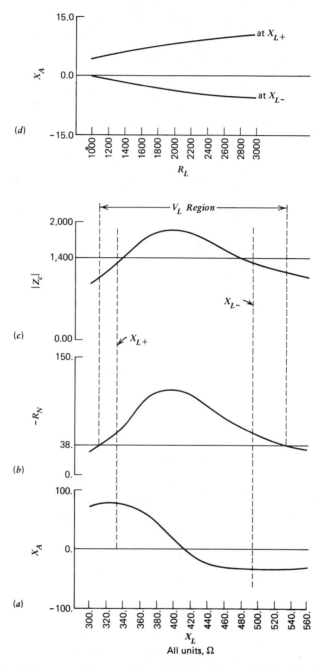

Figure 11.5 Performance characteristics of Design Example 11.3. (a) X_A versus X_L. (b) R_N versus X_L. (c) $|Z_c|$ versus X_L. (d) X_A versus R_L.

3b $$n_2 = \frac{R_{L_2}}{R_{L_1}}(n_1 + 1) - 1$$

from Eq. (11.73)

3c $$X_{L+_2} = \frac{R_{L_2}}{R_{L_1}} X_{L+_1}$$

from Eq. (11.74)

For the significance of the subscripts 1 and 2, see Section 11.4.7.3.2.

11.4.7.3 Typical Trimming Steps

The following describes the action required to increase a given characteristic. Obviously, the opposite action describes the same characteristic.

11.4.7.3.1 Crystal Power P_x

Relation 1 states that to increase P_x, I_x must be increased. Relation 2 states that to increase I_x, I_E must be increased by adjusting r_2, r_b, and r_{b_2} as necessary. Also, Relation 3a states that the output power also increases as I_x increases.

11.4.7.3.2 Output Power P_L

Relation 3a states that increasing I_x and R_L increases P_L. However, increasing I_x will increase P_x and, if one wants to maintain P_x constant, only R_L may be increased. At the same time, if R_L is changed from R_{L_1} to R_{L_2} n must be changed from n_1 to n_2 as per Section 11.4.7.2, relation 3b. X_{L+} must be changed from X_{L+_1} to X_{L+_2} as per Section 11.4.7.2, relation 3c. X_{L+} may also be adjusted experimentally by setting it to maximum P_L and then backing off 3 dB down.

12

Algorithms

12.1 INTRODUCTION

This chapter presents the algorithms for the single bipolar transistor self-limiting oscillators described in Chapters 7 to 11. Obviously, the reader must first select the circuit most suitable for a particular application. Table 12.1 has been prepared to aid in making the selection. This table lists the relative performance characteristics of each circuit.

Strong efforts have been exerted to list quantitative data but, unfortunately, this has been found impractical in the cases of isolation and frequency stability because of the complex nature of these characteristics.

The table shows that, in most of the circuits, the isolation is poor, particularly for reactive changes in the load. In those cases where good or excellent isolation is required, the oscillator must feed an isolating amplifier. This subject is further discussed in Chapter 18. Although the characteristics in the table have already been fully defined, for convenience the table column headings are explained below.

12.1.1 Explanation of Column Headings for Table 12.1

12.1.1.1 Frequency Range

By frequency range is meant the frequency f at which the circuit will operate with a suitable crystal. As indicated by the notes, the frequency range is influenced by many factors.

12.1.1.2 Relative Frequency Stability

This indicates the contribution of the circuit to the stability of the oscillator, assuming that the same crystal is used and the circuit is properly designed for that crystal.

Table 12.1 Self-Limiting Bipolar Transistor Oscillator Circuit Selection

Oscillator Type		f Freq. Range MHz	Relative Frequency Stability	$\dfrac{P_L}{P_x}$	Isolation	Alg.	Chap.	Notes a
Normal Pierce	Coll. lim.	0.5 to 50	Medium	$< \dfrac{1}{2}$	Poor	12.1	7	2, 4
	be lim.	0.5 to 75	Highest	$< \dfrac{1}{2}$	Poor	12.2	7	1, 2, 4
Isolated Pierce		1 to 30	Medium	to 200	Good for R Poor for X	12.3	8	5
Normal Colpitts	Coll. lim.	1 to 40	Medium	$< \dfrac{1}{2}$	Poor	12.4	9	2, 4
	be lim.	1 to 60	High	$< \dfrac{1}{2}$	Poor	12.5	9	1, 2, 4
Semi-Isolated Colpitts	$M = 1$	1 to 30	Medium High for Lo $\dfrac{P_L}{P_x}$	to 20	Poor to Good	12.6	10	2, 3, 4, 5
	$M \neq 1$	1 to 60	High	to 100	Very Good	12.6	10	4, 5, 6
Butler	$X_A = 0$	20 to 200	Medium	to 100	Poor	12.7	11	
	Stable	20 to 200 MHz	Medium	to 100	Poor	12.8	11	

a Please note the following: (1) P_L dependent on R_{df}. (2) Suitable for a range of frequencies without tuning, for $N = 1$. P_L dependent on f. (3) Upper frequency increases as P_L/P_x decreases. (4) Upper frequency decreases as I_x increases. (5) P_L/P_x decreases as f increases. (6) $f_L = Mf$.

12.1.1.3 P_L / P_x

This signifies the ratio of the output power P_L to the crystal dissipated power P_x. The reader is cautioned that, while high values of P_L / P_x appear desirable, as P_x decreases, in general, the long-term stability improves and the short-term stability deteriorates.

12.1.1.4 Isolation

By isolation is meant the effect of changes in the load impedance upon the frequency f.

12.1.1.5 f and Mf

By f is meant the frequency actually generated in the oscillator. Mf is the frequency available at the output terminals of the semi-isolated Colpitts oscillator. It is derived from f by multiplying action with the multiplication factor M.

12.2 GENERAL DISCUSSION OF THE ALGORITHMS

12.2.1 Resonator Description

The algorithms have been prepared for oscillators with crystals and associated components, operating in all the useful overtones and modes. However, the algorithms are easily adaptable for replacement of the crystal network by another two-terminal network as described in the discussion chapter (see Chapters 7 to 11) for the particular algorithm. Of course all the material in the algorithm concerning overtone and mode operation should be then disregarded.

12.2.2 Form of the Algorithm

12.2.2.1 A Common Input Section

Each algorithm consists of a common input section and additional pages as required by the specific oscillator being designed.

The input section contains the input information which falls into the following four categories:

a Oscillator performance requirements and the power supply voltage.

b Crystal resonator characteristics both directly specified and calculated as described in Chapter 3. Much of this information is not strictly necessary for the basic oscillator design but will be necessary for computing the frequency changes due to variations in parameters. The $R_{df_{max}}$ is always used to ensure that the oscillator loop gain always increases with other crystals of smaller R_{df}.

c Transistor characteristics. The information in this category is most difficult to obtain and very often estimates must be made from whatever data are available. At times one must even resort to guesswork, particularly as to the values of the various capacitances. For completeness, this category includes several parameters, BV_{CB}, BV_{EB}, and P_{dis} not mentioned in the design algorithm, but which may be exceeded, so one should be aware of them.

d Circuit Parameters. These are parameters usually associated with the type of limiting used in the specific circuit.

The algorithms do not allow for component losses except where specifically stated in the algorithm. Therefore, when highly lossy components are used, the calculated circuitry may be substantially in error, unless they are included in R_L.

12.2.2.2 Schematic Diagram

Each algorithm includes a schematic diagram which shows all the physical components, except R_L, exactly as they will be installed in the oscillator. R_L is the load that the transistor sees and must be converted into a form suitable for the user, as described in Section 5.7.

One should be aware that, as in all schematics, the stray elements which are a function of the physical layout are not shown; these stray elements can strongly influence the oscillator performance, especially at higher frequencies.

The schematic diagram may also be used as a form for recording design data.

12.2.2.3 Calculation of Component Values

This part of the algorithm presents the step-by-step procedure for transforming the input information given on Page 1 into the specific oscillator design. The format has been planned to be useful both to the novice who is merely interested in obtaining the final design and to the person who may desire to learn why each step is executed in the manner shown.

12.2.3 Supplementary Information Contained in the Algorithm

1 The algorithm steps are frequently annotated with notes, guides, restrictions, references, and comments. These should be read and observed carefully, as they will lead to greater accuracy and fuller understanding.
2 The reader who is interested in learning the derivation of, and/or reason for, any step may consult the applicable text equation referenced for

each step. In some cases the notation SE is shown in lieu of an equation number. SE signifies Self-Explanatory.

12.2.4 Design Examples

Each chapter discussing circuit configurations for which algorithms are supplied includes several design examples, which were prepared using the algorithms. The input information is given in the table and the output information is shown on the schematic. The component values shown are those computed. In practice, the nearest standard values will be used.

12.2.5 Use of the Algorithms

The algorithms can be used for the following purposes:

1. For designing an oscillator to obtain a specified performance as shown in the design examples of Section 12.2.4.
2. To investigate the effects of variations in parameters.

It is recommended that the algorithms be used for programming a computer or one of the more powerful hand-held calculators in order to facilitate their use for design and investigation. To help in the programming, each step includes references to the applicable preceding steps.

12.2.6 Conversion Efficiency in the Algorithm Designs

It will be noted that almost all the algorithms have been prepared on the basis of the maximum practical conversion efficiency. Each chapter describes the procedure to be followed when characteristics other than maximum conversion efficiency are preferred, especially in the isolated Pierce oscillator. The recommendations in the "Comments" columns alert the user to the design problems caused by the maximum conversion efficiency design.

Algorithm Figure 12.1 Pierce Oscillator, Collector Limiting

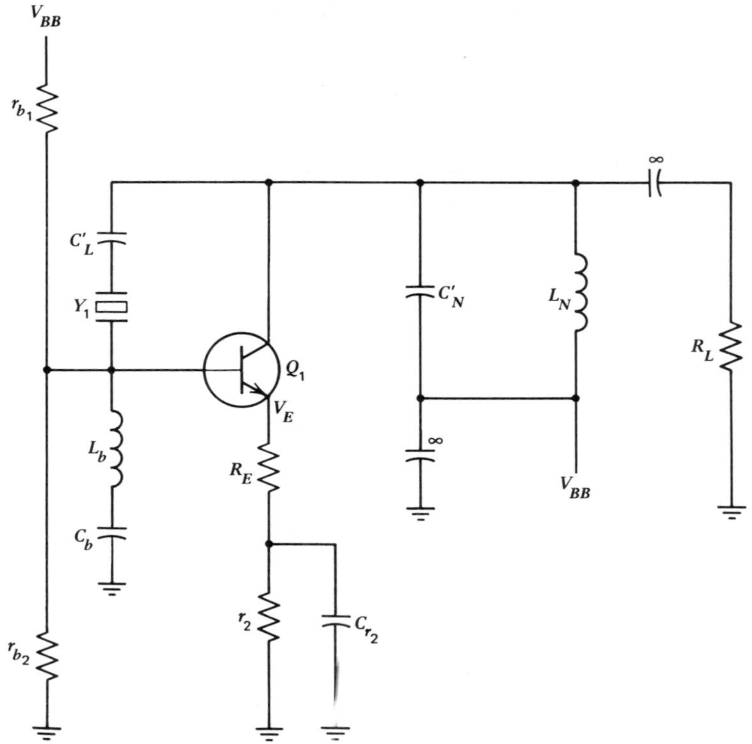

All units in....
℧ MHz Ω mW pF μH
mA, mV dc or rms

	Item	Value
Oscillator Performance	f	
	P_L	
	P_x	
	V_{BB}	
Principal Crystal Data	R_{df}	
	I_x	
	Cut	
	N	
Transistor Data	β_o	
	f_T	
	BV_{CE}	
	C_{cb}	
	C_{bet}	
	C_{ce}	
	P_{dis}	
	Type	
Circuit Parameters	A_{L_0}	
	η	
Calculated Data	I_{BB}	
	g_m	
	V_E	

Algorithm 12.1 Pierce Oscillator, Collector Limiting

Oscillator Performance
Crystal and Transistor Characteristics

Step No.	Item	Formula or Quantity	Units	Refer to Step Nos.	Comments	Text Equation No.
a1	f	Specified	MHz			
a2	P_L		mW		**Oscillator**	
a3	P_x		mW			
a4	V_{BB}		mV		performance	
b1	Cut	Specified				
b2	N				**Crystal**	
b3	df		Hz			
b4	R_L		Ω		**Characteristics**	
b5	C_L		pF		Notes:	

b6	C_1		pF		1. For production runs R_{df} = max	
b7	C_0		pF			
b8	C'_0		pF		2. P_x will decrease as R_{df} decreases	
b9	Q_x	$10^6/(2\pi C_1 fR1)$		b6, b11, a1	3. P_L will remain constant as R_{df} decreases	3.29a
b10	$\Delta X/$ $(\Delta f/f)$	$10^6/(\pi C_1 f)$	Ω	b6, a1		3.20a
b11	R_1	$[(C_L/(C_L + C_0)]^2 R_L$	Ω	b5, b7, b4		3.19
b12	R_{df}	$R_1/[1 - C'_0/(C_L + C_0) - 2C'_0 df/(C_1 f 10^6)]^2$	Ω	b11, b6, b7 a1, b5, b8		3.26a
b13	I_x	$(1000 P_x/R_{df})^{1/2}$	mA	a3, b12		3.27a
b14	C_{Ldf}	$\left[\dfrac{1}{C_L + C_0} + \dfrac{2(10)^{-6} df}{C_1 f}\right]^{-1} - C'_0$	pF	b5, b7, b3 b6, a1, b8		3.24a

Algorithm 12.1 (*Continued*)

Oscillator Performance
Crystal and Transistor Characteristics

Step No.	Item	Formula or Quantity	Units	Refer to Step Nos.	Comments	Text Equation No.
c1	Type No.	Specified				
c2	β_o				$\beta_o > 1200 g_m$	5.116
c3	f_T		MHz		$f_T > 600 f g_m$	5.113
c4	C_{bet}		pF			
c5	C_{cb}		pF		Transistor	
c6	C_{ce}		pF			
c7	BV_{CB}		mV		Characteristics	
c8	BV_{CE}		mV			
c9	BV_{BE}		mV			
c10	P_{dis}		mW		at maximum temperature	
c11	r_{bb}		Ω			

Circuit Parameters

		Specified			Recommend $A_{L_0} = 2$	
	d1	A_{L_0}				6.18
		Calculation of Component Values				
		Note: ⊕ *signifies physical component*				
	1	η	P_L/P_x			7.10
	2	V_L	the smaller of	mV	a2, a3	
	2a		$0.33 BV_{CE}$ or			see discussion of R_L and V_L
	2b		$\sqrt{10^7 P_L/\sqrt{f}}$		c8	7.23a
⊕	3	R_L	$V_L^2/(1000\,P_L)$	Ω	a2, a1	7.30
	4	R_2	$\eta R_{df} - R_{df}^2/R_L$	Ω	2, a2	7.9
	5	X_2	$\sqrt{R_L R_2}$	Ω	1, b12, 3	7.16
	6	R_T'	$R_2 + R_{df}$	Ω	3, 4	7.4
	7	$(g_m X_1)$	R_T'/X_2		4, b12	7.17
	8	I_e	$I_x(g_m X_1)$	mA	6, 5	7.1
	9	I_E	$\geqq 1.4 I_e$	mA	b13, 7	7.7
					8	6.17

Algorithm 12.1 (*Continued*)

Oscillator Performance
Crystal and Transistor Characteristics

	Step No.	Item	Formula or Quantity	Units	Refer to Step Nos.	Comments	Text Equation No.
⊕	10	R_E	$5[26/I_E + 1]$	Ω	9, c11, c2		6.20
	11	$g_{m_{L_0}}$	$0.83/R_E$	$\mho$	10		6.20a
	12	g_m	$g_{m_{L_0}}/A_{L_0}$	$\mho$	11, d1		6.18
	13	X_1	$g_m X_1 / g_m$	Ω	7, 12		SE
	14	V_b	$I_x X_1$	mV	b13, 13		7.5a
	15	V_E	$V_{BB} - 1.4(V_L + V_b)$	mV	a4, 2, 14	> 2000	7.27
⊕	16	r_2	$V_E/I_E - R_E$	Ω	15, 9, 10		2.99
	17	r_b	$\beta_o(r_2)/5$	Ω	c2, 16		2.100
⊕	18	r_{b_2}	$.83 r_b [V_{BB}/(V_E + 700)]$	Ω	17, a4, 15		2.103
⊕	19	r_{b_1}	$1/(1/r_b - 1/r_{b_2})$	Ω	17, 18		2.104
	20	R_{in}	$g_m X_1^2 / \beta_o$	Ω	12, 13, c2		6.21
	21	R_b	X_1^2 / r_b	Ω	13, 17		7.3

			Ω	b12, 4, 20, 21	R_T should $\simeq R_T'$	7.1
22	R_T	$R_{df} + R_2 + R_{in} + R_b$				
23	C_1	$159{,}000/X_1 f$	pF	13, a4		7.36
24	C_{bed}	$g_m(159{,}000/f_T)$	pF	12, c3		6.22
25	M_M	V_b/V_L		14, 2		5.70
26	C_{1M}	$C_{cb}(1 + 1/M_M)$	pF	c5, 25		5.75
27	C_1'	$C_1 - C_{bed} - C_{1M}$	pF	23, 24, 26		Fig. 7.2b
28		Check whether $N = 1$		b2		
28a		If yes, $s^2 = 0.2$				5.82a
28b		If no, $s = 1 - 1.5/N$				5.82b
29	X_{CN}	$(1 - s^2)X_2$	Ω	28, 5		5.83
30	C_N	$159{,}000/(X_N f)$	pF	29, a1		5.84
31	C_N'	$C_N - C_{cb}(1 + M_M) - C_{ce}$	pF	30, c5, 25, c4		5.73 and Fig. 7.
32	X_{LN}	X_{CN}/s^2	Ω	29, 28		5,85
33	L_N	$X_{LN}/6.28f$	μH	32, a1		5.85a
34		Check if the crystal cut is SC		b1		

Algorithm 12.1 (*Continued*)

Oscillator Performance
Crystal and Transistor Characteristics

	Step No.	Item	Formula or Quantity	Units	Refer to Step Nos.	Comments	Text Equation No.
⊕	35a	C_b	if no: C'_1	pF	27		SE
	35b	L_b	0			Continue at Step 37	SE
⊕	36a	C_b	if yes: $0.156\, C'_1$	pF	27		5.86
⊕	36b	L_b	$10^6 / [C_b (6.84 f)^2]$	μH	36a, a1		5.87
	37	X_{cr_2}	$0.06 / g_m$	Ω	12		7.38
⊕	38	C_{r_2}	$> 159{,}000 / (X_{cr_2} f)$	pF	37, a1		7.35
	39	X_{C_L}	$159{,}000 / (C_{L_{df}} f) - (X_1 + X_2)$	Ω	b14, a1, 13, 5		7.39
⊕	40	C'_L	$159{,}000 / (X'_{C_L} f)$	pF	39, a1		7.40
	41	Q_{op}	$< Q_X R_{df} / R_T$		b12, b9, 22		5.77a

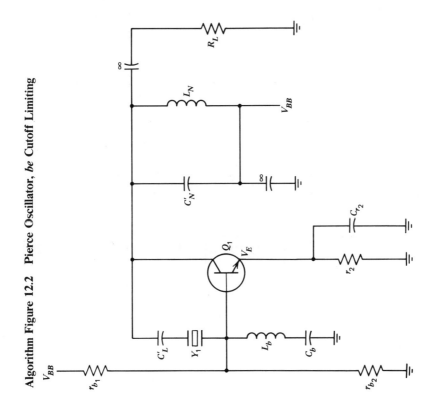

Algorithm Figure 12.2 Pierce Oscillator, be Cutoff Limiting

All units in ... fF MHz Ω mW pF μH mA, mV dc or rms	Item	Value
Oscillator Performance	f	
	P_L	
	P_x	
	V_{BB}	
Principal Crystal Data	R_{df}	
	I_x	
	Cut	
	N	
Transistor Data	β_o	
	f_T	
	BV_{CC}	
	C_{cb}	
	C_{bet}	
	C_{ce}	
	P_{dis}	
	Type	
Circuit Parameters	α	0.3
	γ_1	1.4
	V_{pe}	113
Calculated Data	I_{BB}	
	g_m	
	V_E	

Algorithm 12.2 Pierce Oscillator, *be* Cutoff Limiting

Oscillator Performance
Crystal and Transistor Characteristics

Step	Item	Formula or Quantity	Units	Refer to Step Nos.	Comments	Text Equation No.
a1	f	Specified	MHz			
a2	P_L		mW		Oscillator	
a3	P_x		mW			
a4	V_{BB}		mV		Performance	
b1	Cut	Specified				
b2	N		Hz		Crystal	
b3	df					
b4	R_L		Ω			
b5	C_L		pF		Characteristics	
					Notes:	

294

					1. For production runs	
b6	C_1			pF	R_{df} = max	
b7	C_0			pF		
b8	C_0'			pF	2. P_X will increase as R_{df} decreases	
b9	Q_x	$10^6/(2\pi C_1 f R_1)$			b6, b11, a1	3.29a
b10	$\Delta X/(\Delta f/f)$	$10^6/(\pi C_1 f)$		Ω	b6, a1	3.20a
					3. P_L will increase as R_{df} decreases	
b11	R_1	$[(C_L/(C_L + C_0)]^2 R_L$		Ω	b5, b7, b4	3.19
b12	R_{df}	$R_1/[1 - C_0'/(C_L + C_0) - 2C_0' df/(C_1 f \times 10^6)]^2$		Ω	b11, b5, b7	3.26a
					a1, b5, b8	
b13	I_x	$(1000 P_x/R_{df})^{1/2}$		mA	a3, b12	3.27a
b14	$C_{L_{df}}$	$\left[\dfrac{1}{C_1 + C_2} + \dfrac{2 \times 10^{-6} df}{C_1 f}\right]^{-1} - C_0'$		pF	b5, b7, b3	3.24a
					b6, a1, b8	

Algorithm 12.2 (*Continued*)

Oscillator Performance
Crystal and Transistor Characteristics

Step	Item	Formula or Quantity	Units	Refer to Step Nos.	Comments	Text Equation No.
c1	Type No.	Specified				
c2	β_o				$\beta_o > 1200\, g_m$	5.116
c3	f_m		MHz		$r_T > 600 f g_m$	5.113
c4	C_{bet}		pF			
c5	C_{cb}		pF		Transistor	
c6	C_{ce}		pF			
c7	BV_{CE}		mV		Characteristics	
c8	BV_{CE}		mV			
c9	BV_{BE}		mV			
c10	P_{dis}		mW		at maximum temperature	
c11	r_{bb}		Ω			

				Circuit Parameters		
	d1	α	0.3			See Section 7.6.1
	d2	γ_1	1.4			
	d3	V_{be}	113	mV		
			Calculation of Component Values			
	1	η	P_L/P_X		a2, a3	See discussion on η	7.10
	2	V_L	the smaller of	mV		See discussion on	
	2a		$0.22 BV_{CE}$ or		c8	R_L and V_L	7.26
	2b		$\sqrt{10^7 P_L/\sqrt{f}}$		a2, a1	in Section	7.30
⊕	3	R_L	$V_L^2/(1000\, P_L)$	Ω	2, a2	7.2.4	7.9
	4	V_E	$V_{BB} - [2.1(V_L + V_b)] - 1700$	mV	a4, 2, d3	> 2000 See Note A	7.28a
	5	R_2	$\eta R_{df} - R_{df}^2/R_L$	Ω	1, b12, 3	NOTE: If R_2 is $-$,	7.16
	6	X_2	$\sqrt{R_2 R_L}$	Ω	5, 3	the design is	7.4
	7	X_1	V_b/I_x	Ω	d3, b13	unsound, see	7.5a
	8	g_m	$((R_{df} + R_2)/[X_1(X_2 - X_1/\beta_o)]$	$\mho$	b12, 5, 7, c2	Section 7.2.3.1	7.34

Algorithm 12.2 *(Continued)*

Oscillator Performance
Crystal and Transistor Characteristics

Step	Item	Formula or Quantity	Units	Refer to Step Nos.	Comments	Text Equation No.
9	I_e	$g_m V_b$	mA	8, d3		7.7
10	I_E	I_e/γ_1	mA	9, d2		2.69
⊕ 11	r_2	V_E/I_E	Ω	4, 10		2.99
12	r_b	$\beta_o r_2/5$	Ω	c2, 11		2.100
⊕ 13	r_{b_2}	$0.83 r_b V_{BB}/(V_E + 700)$	Ω	12, a4, 4		2.103
⊕ 14	r_{b_1}	$1/(1/r_b - 1/r_{b_2})$	Ω	12, 13		2.104
15	R_{in}	$X_1^2 g_m/\beta_o$	Ω	7, 8, c2		7.2
16	R_b	X_1^2/r_b	Ω	7, 12		7.3
17		check that $(R_{in} + R_v) \ll (R_{df} + R_2)$		15, 16, b12, 5		
18	C_1	$159{,}000/(X_1 f)$	pF	7, a1		7.35
19	C_{bed}	$g_m\, 159{,}000/f_T$	pF	8, c3		2.75
20	M_M	$\simeq X_1/X_2$		7, 6		from 5.70

Note A: If V_E calculates $> V_{BB}/2$, make $V_E = V_{BB}/2$

	21	C_{1M}	$C_{cb}(1 + 1/M_M)$	pF	c5, 20	5.71
	22	C'_1	$C_1 - C_{bed} - C_{1M} - C_{bet}$	pF	18, 19, 21, c4	Fig. 7.2
	23		Check whether $N = 1$		b2	
	23a	s	If yes, $s^2 = 0.2$			5.82a
	23b	s	If no, $s = 1 - 1.5/N$		b2	5.82b
	24	X_{CN}	$(1 - s^2)X_2$	Ω	23, 6	5,83
	25	C_N	$159{,}000/(X_{CN}f)$	pF	24, a1	5.84
⊕	26	C'_N	$C_N - C_{ce} - C_{cb}(1 + M_M)$	pF	25, c6, c5, 20	5.73 and Fig. 7.2b
	27	X_{LN}	X_{CN}/s^2	Ω	24, 23	5.85
⊕	28	L_N	$X_{LN}/6.28f$	μH	27, a1	5.85a
	29		Check if the crystal		b1	
			cut is SC			
			if no			
⊕	29a	C_b	C'_1	pF	22	SE

Algorithm 12.2 *(Continued)*

Oscillator Performance
Crystal and Transistor Characteristics

	Step	Item	Formula or Quantity	Units	Refer to Step Nos.	Comments	Text Equation No.
⊕	29b	L_b	0			Continue at Step 31	SE
			if yes				
⊕	30a	C_b	$0.156 C_1'$	pF	22		5.86
⊕	30b	L_b	$10^6/[C_b(6.84f)^2]$	μH	30a, a1		5.87
	31	X_{cr_2}	$0.05/g_m$	Ω	8		7.38
⊕	32	C_{r_2}	$159{,}000/(X_{cr_2} f)$	pF	32, a1		7.35
	33	X_{C_L}	$159{,}000/(C_{L_{df}} f) - (X_1 + X_2)$		b14, a1, 7, 6		7.39
⊕	34	C_L'	$159{,}000/(X_{C_L} f)$	pF	33, a1		7.40
	35	Q_{op}	$Q_X R_{df}/(R_{df} + R_{in} + R_2 + R_b)$		b9, b12, 15, 5, 16		5.77a

All units in.... ℧ MHz Ω mW pF μH mA, mV dc or rms		Item	Value
Oscillator Performance		f	
		P_L	
		P_x	
		V_{BB}	
Principal Crystal Data		R_{df}	
		I_x	
		Cut	
		N	
Transistor Data		β_o	
		f_T	
		BV_{CE}	
		C_{cb}	
		C_{bet}	
		C_{ce}	
		P_{dis}	
		Type	
Circuit Parameters		A_{L_o}	
		n	
		η	
Calculated Data		I_{BB}	
		g_m	
		V_E	
		R_{df}/R_T	

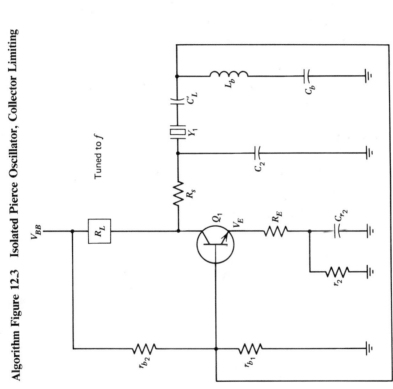

Algorithm Figure 12.3 Isolated Pierce Oscillator, Collector Limiting

Algorithm 12.3 Isolated Pierce Oscillator, Collector Limiting

Oscillator Performance
Crystal and Transistor Characteristics

Step No.	Item	Formula or Quantity	Units	Refer to Step no.	Comments	Text Equation No.
a1	f	Specified	MHz			
a2	P_L		mW		**Oscillator**	
a3	P_x		mW			
a4	V_{BB}		mV		**Performance**	
a5	η	P_L/P_x		a2, a3		
b1	Cut	Specified				
b2	N		Hz		**Crystal**	
b3	df					
b4	R_L		Ω		**Characteristics**	
b5	C_L		pF		Notes:	

b6	C_1		pF		1. For production runs	
b7	C_0		pF		R_{df} = max	
b8	C_0'		pF		2. P_x will increase as R_{df} decreases	
b9	Q_x	$10^6/\pi C_1 f R_1$		b6, b11, a1	3. P_L will remain constant as R_{df} decreases	3.29a
b10	$\Delta X/(\Delta f/f)$	$10^6/(\pi C_1 f)$	Ω	b6, a1		3.20a
b11	R_1	$[(C_L/(C_L + C_0)]^2 R_L$	Ω	b5, b7, b4		3.19
b12	R_{df}	$R_1/[1 - C_0'/(C_L + C_0) - 2C_0' df/(C_1 f \times 10^6)]^2$	Ω	b11, b6, b7 a1, b5, b8		3.26a
b13	I_x	$(1000\, P_x/R_{df})^{1/2}$	mA	a3, b12		3.27a
b14	$C_{L_{df}}$	$\left[\dfrac{1}{C_L + C_0} + \dfrac{2 \times 10^{-6} df}{C_1 f}\right]^{-1} - C_0'$	pF	b5, b7, b3 b6, a1, b8		3.24a

Algorithm 12.3 *(Continued)*

Oscillator Performance
Crystal and Transistor Characteristics

Step No.	Item Type No.	Formula or Quantity	Units	Refer to Step no.	Comments	Text Equation No.
c1		Specified				
c2	β_o				$\beta_o > 1200 g_m$	5.116
c3	f_T		MHz		$f_T > 600 f g_m$	5.113
c4	C_{bet}		pF		Transistor	
c5	C_{cb}		pF			
c6	C_{ce}		pF		Characteristics	
c7	BV_{CB}		mV			
c8	BV_{CE}		mV			
c9	BV_{BE}		mV			
c10	P_{dis}		mW		at maximum temperature	
c11	r_{bb}		Ω			

Circuit Parameters

		Specified				
d1	A_{L_0}				Recommend $A_{L_0} = 2$	6.18
		Calculation of Component Values				
		Note: ⊕ *signifies physical component*				
1	V_L	The smaller of	mV		See discussion	
1a		$0.33 BV_{CE}$ or		c8	on R_L and V_L	7.23
1b		$\sqrt{10^7 P_L / \sqrt{f}}$		a2, a1		7.30
2 ⊕	R_L	The smaller of	Ω			
2a		$V_L^2 / (1000 P_L)$ or		1, a2		7.9
2b		$(64{,}000)^2 / (f^2 n^2 R_{df} \eta)$		a1, b12, a5	Recommend $n = 5$	8.14
						8.28
1c	V_L'	$\sqrt{1000 P_L R_L}$	mV	a2, 2		7.9
3	I_L	V_L' / R_L	mA	1c, 2		SE
4 ⊕	R_s	$n \sqrt{\eta R_{df} R_L}$	Ω	2b, a5, b12, 2	$R_{s\,max} = 64 k / f$	8.19

Algorithm 12.3 (*Continued*)

Oscillator Performance
Crystal and Transistor Characteristics

Step No.	Item	Formula or Quantity	Units	Refer to Step no.	Comments	Text Equation No.
5	m_r	R_s/R_L		7, 2		8.7
6	I_c	$(1 + 1/m_r)I_L$	mA	8, 3		8.6a
7	I_E	$1.4 I_c$	mA	6		6.17
8	R_E	$5[(26/I_E) + 1]$	Ω	7		6.20
9	$g_{m_{L_0}}$	$0.83/R_E$	$\mho$	8		6.20a
10	g_m	$g_{m_{L_0}}/A_{L_0}$	$\mho$	9, d1		6.18
11	g_{me}	$g_m(1 + m_r)$	$\mho$	10, 5		8.10a
12	$(g_m X_1)$	I_c/I_x		6, b13		8.9a
13	X_1	$(g_m X_1)/g_m$	Ω	12, 10		SE
14	R_{in}	$g_m X_1^2/\beta_o$	Ω	10, 13, c2		7.2
15	R_t	$R_{\text{in}} + R_{df}$	Ω	14, b12		8.13
16	X_2	nR_t	Ω	2b, 15		8.14
17	R_T	$R_t + X_2^2/(R_s + R_L)$	Ω	15, 16, 4, 2		8.12
17a	X_2'	nR_T	Ω	2b, 17		8.23

18	V_b	$I_x X_1$		mV	b13, 13		8.8
19	V_E	$V_{BB} - 1.4(V'_L + V_b)$		mV	a4, 1c, 18		7.27
20	r_2	$V_E/I_E - R_E$	⊕	Ω	19, 7, 8		2.99
21	r_b	$\beta_o(r_2 + R_E)/5$		Ω	c2, 20, 8		2.100
22	r_{b_2}	$0.83 r_b V_{BB}/(V_E + 700)$	⊕	Ω	21, a4, 19		2.103
23	r_{b_1}	$1/(1/r_b - 1/r_{b_2})$	⊕	Ω	21, 22		2.104
24	C_1	$159{,}000/(X_1 f)$		pF	13, a1		7.36
25	C_{bed}	$g_m 159{,}000/f_T$		pF	10, c3		6.22
26	M_M	$\approx V'_L/V_b$			1c, 18		5.70
27	C_M	$(M_M + 1)C_{cb}$		pF	26, c5		5.75
28	C'_1	$C_1 - C_{bed} - C_M - c_{bet}$	⊕	pF	24, 25, 27, c4		Fig. 7.2b
29		Check if the crystal cut is SC			b1		

307

Algorithm 12.3 (*Continued*)

Oscillator Performance
Crystal and Transistor Characteristics

	Step No.	Item	Formula or Quantity	Units	Refer to Step no.	Comments	Text Equation No.
⊕	30a	C_b	if no C'_l	pF	28		SE
	30b	L_b	0			Continue at Step 33	SE
			if yes				
⊕	31a	C_b	$0.156 C'_l$	pF	28		5.86
⊕	31b	L_b	$10^6 / [C_b (6.84 f)^2]$	μH	31a, a1		5.87
⊕	32	C_2	$159{,}000 / (X_2 f)$	pF	17a, a1		8.25
	33	X_{cr_2}	$0.05 / g_m$	Ω	10		7.38
⊕	34	C_{r_2}	$159{,}000 / (X_{cr_2} f)$	pF	33, a1		7.35
	35	X_{C_L}	$159{,}000 / (C_{L_{df}} f) - (X_1 + X_2)$	Ω	b14, a1, 13, 18		7.39
⊕	36	C'_L	$159{,}000 / (X_{C_L} f)$	pF	35, a1		7.40
	37	Q_{op}	$< Q_x R_{df} / R_T$		b9, b12, 17		5.77a

308

	Item	Value	All units in … ℧ MHz Ω mW pF μH mA, mV dc or rms
Oscillator Performance	f		
	P_L		
	P_x		
	V_{BB}		
Principal Crystal Data	R_{df}		
	I_x		
	Cut		
	N		
Transistor Data	β_o		
	f_T		
	BV_{CE}		
	C_{cb}		
	C_{bet}		
	C_{ce}		
	P_{dis}		
	Type		
Circuit Parameters	A_{L_0}	A	
	η		
Calculated Data	I_{BB}		
	g_m		
	V_E		

Algorithm Figure 12.4 Colpitts Oscillator Collector Limiting

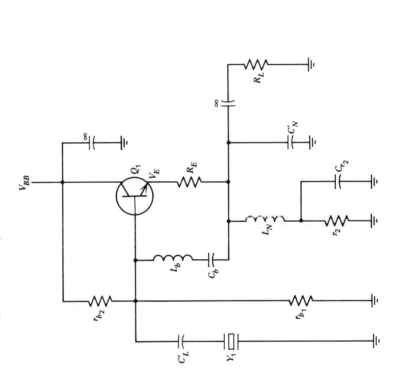

Algorithm 12.4 Colpitts Oscillator Collector Limiting

Oscillator Performance
Crystal and Transistor Characteristics

Step No.	Item	Formula or Quantity	Units	Refer to Step Nos.	Comments	Text Equation No.
a1	f	Specified	MHz		**Oscillator**	
a2	P_L		MW			
a3	P_x		mW		**Performance**	
a4	V_{BB}		mV			
b1	Cut	Specified			**Crystal**	
b2	N					
b3	df		Hz			
b4	R_L		Ω		**Characteristics**	
b5	C_L		pF		Notes:	

b6	C_1		pF		1. For production runs R_{df} = max	
b7	C_0		pF		2. P_x will decrease as R_{df} decreases	
b8	C_0'		pF		3. P_l will remain constant as R_{df} decreases	
b9	Q_x	$10^6/(2\pi C_1 f R_1)$		b6, b11, a1		3.29a
b10	$\Delta X/(\Delta f/f)$	$10^6/(\pi C_1 f)$	Ω	b6, a1		3.20a
b11	R_1	$[C_L/(C_L + C_0)]^2 R_L$	Ω	b5, b7, b4		3.19
b12	R_{df}	$R_1/[1 - C_0'/(C_L + C_0 -)]$	Ω	b11, b6, b7		3.26a
		$2C_0' df/(C_1 f \times 10^6)]^2$		a1, b5, b8		
b13	I_x	$(1000\, P_x/R_{df})^{1/2}$	mA	a3, b12		3.27a
b14	$C_{L df}$	$\left[\dfrac{1}{C_L + C_0} + \dfrac{2 \times 10^{-6} df}{C_1 f}\right]^{-1} - C_0'$	pF	b5, b7, b3 b6, a1, b8		3.24a

Algorithm 12.4 *(Continued)*

Oscillator Performance
Crystal and Transistor Characteristics

Step No.	Item	Formula or Quantity	Units	Refer to Step Nos.	Comments	Text Equation No.
c1	Type No.	Specified				
c2	β_o				$\beta_o > 1200\, g_m$	5.116
c3	f_T		MHz		$f_T > 600 f g_m$	5.113
c4	C_{bet}		pF		Transistor	
c5	C_{cb}		pF			
c6	C_{ce}		pF		Characteristics	
c7	BV_{CB}		mV			
c8	BV_{CE}		mV			
c9	BV_{BE}		mV			
c10	P_{dis}		mW		at maximum temperature	
c11	r_{bb}		Ω			

Circuit Parameters

		Specified		Recommend $A_{L_o} = 2$	
d1	A_{L_o}			Recommend $A_{L_o} = 2$	6.18
		Calculation of Component Values			
		Note: ⊕ *signifies physical component*			
1	η	P_L/P_x		a2, a3	See discussion	7.10
2	V_L	The smaller of	mV		of η	
2a		$0.33 BV_{CE}$ or		c8	See discussion of	9.3
2b		$\sqrt{10^7 P_L/\sqrt{f}}$		a2, a1	V_L and R_L	7.30
3	R_L	$V_L^2/(1000\, P_L)$	Ω	2, a2		7.9
4	R_2	$\eta R_{df} - R_{df}^2/R_L$	Ω	1, b12, 3		7.16
5	X_2	$\sqrt{R_L R_2}$	Ω	3, 4		7.4
6	R_T'	$R_2 + R_{df}$	Ω	4, b12		7.17
7	$(g_m' X_1)$	R_T'/X_2		6, 5		7.1
⊕ 8	I_e'	$I_x(g_m' X_1)$	mA	c13, 7		7.7

Algorithm 12.4 (*Continued*)

Oscillator Performance
Crystal and Transistor Characteristics

Step No.	Item	Formula or Quantity	Units	Refer to Step Nos.	Comments	Text Equation No.
9	I'_E	$1.4 I'_e$	mA	8		6.17
10	R'_E	$5(26/I'_E + 1)$	Ω	9, b11, b2		6.20
11	$g'_{m_{L_0}}$	$0.83/R_E$	$\mho$	10		6.20a
12	g'_m	$g'_{m_{L_0}}/A_{L_0}$	$\mho$	11, d1		6.18
13	X_1	$(g'_m X_1)/g'_m$	Ω	7, 12		SE
14	V_{be}	$I_x X_1$	mV	b13, 13		5.52a
15	V_E	$[V_{BB} - 1.4(V_L + V_{be})]$	mV	a4, 2, 14	> 2000	9.7
16	r'_2	$V_E/I'_E - R'_E$	Ω	15, 9, 10		2.99
17	r'_b	$\beta_o(r'_2)/5$	Ω	c2, 16		2.100
18	R_{in}	$g'_m X_1^2/\beta_o$	Ω	12, 13, c2		6.21
19	R_b	$(X_1 + X_2)^2/r'_b$	Ω	13, 5, 17		5.57
20	R_T	$R_{df} + R_2 + R_{in} + R_b$	Ω	b12, 4, 18, 19		5.40
21		Check whether $R_T < 1.1 R'_T$		20, 6		7.46
21a		If yes, go to Step 30				

314

Step	⊕	Variable	Formula	Units	If no, go to Step 22	Eq.
21b			If no, go to Step 22			
22		I_e	$I_x R_T / X_2$	mA	b13, 20, 5	5.58a
23		I_E	$1.4 I_e$	mA	22	6.17
24		g_m	$R_T/(X_1 X_2)$		20, 13, 5	5.40
25		$g_{m_{L_0}}$	$A_{L_0} g_m$		d1, 24	6.18
26	⊕	R_E	$1/g_{m_{L_0}} - 26/I_E - 1$	Ω	25, 23, c11, c2	6.19a
27	⊕	r_2	$(V_E/I_E) - R_E$	Ω	15, 23, 26	2.99
28		r_b	$\beta_o r_2/5$	Ω	27	2.100
29			Note: in the following Step, substitute (r_b') for (r_b) when Step 21a is yes.			
30	⊕	r_{b_2}	$0.83 r_b [V_{BB}/(V_E + 700)]$	Ω	28, a4, 15 or 17	2.103
31	⊕	r_{b_1}	$1/(1/r_b - 1/r_{b_2})$	Ω	28, 30 or 17	2.104
32		C_1	$159{,}000/(X_1 f)$	pF	13, a1	7.36
33		C_{bed}	$g_m 159{,}000/f_T$	pF	2, c3 or 1	6.22
34		M_M	V_{be}/V_L		14, 2	5.70
35		C_{1M}	$C_{cb}[1 + 1/M_M]$	pF	c5, 34	5.75

Algorithm 12.4 (*Continued*)

Oscillator Performance
Crystal and Transistor Characteristics

Step No.	Item	Formula or Quantity	Units	Refer to Step Nos.	Comments	Text Equation No.
36	C'_1	$C_1 - C_{bed} - C_{1M} - C_{bet}$	pF	32, 33, 35, c4		Fig. 5.9b
37		Check whether $N = 1$		b2		
27a		If yes, $s^2 = 0.2$				5.82a
37b		If no, $s = 1 - 1.5/N$				5.82b
38	X_{CN}	$(1 - s^2)X_2$	Ω	37, 5		5.83
39	C_N	$159{,}000/(X_{CN}f)$	pF	38, a1		5.84
40	C'_N	$C_N - C_{cb}(1 + M_M) - C_{ce}$	pF	39, c5, 34, c6		5.73
						Fig. 5.9b
41	X_{LN}	X_{CN}/s^2	Ω	38, 37		5.85
42	L_N	$X_{LN}/(6.28f)$	μH	41, a1		5.85a
43		Check if the crystal cut		b1		

316

⊕	44a	C_b	is SC		pF	SE	
	44b	L_b	If no, 0	Continue at Step 46		SE	
			If yes,				
⊕	45a	C_b	$0.156 C_1'$	36	pF	5.86	
⊕	45b	L_b	$10^6/[C_b(6.84f)^2]$	45a, a1	μH	5.87	
	46	X_{cr_2}	$\leqq 0.02 X_{LN}$	41	Ω	9.10	
⊕	47	C_{r_2}	$159{,}000/(X_{cr_2} f)$	46, a1	pF	7.35	
	48	$X_{C_L'}$	$[159{,}000/C_{L_{df}} f)] - (X_1 + X_2)$	b14, a1, 13, 5	Ω	7.39	
⊕	49	C_L'	$159{,}000/(X_{C_L'} f)$	48, a1	pF	7.40	
	50	Q_{op}	$< Q_x R_{df}/R_T$	b12, 20, b9		5.77a	

	Item	Value	All units in... $\mho$ MHz Ω mW pF μH mA, mV dc or rms
Oscillator Performance	f		MHz
	P_L		mW
	P_x		mW
	V_{BB}		mV
Principal Crystal Data	R_{df}		Ω
	I_x		mA
	Cut		
	N		
Transistor Data	β_o		
	f_T		MHz
	BV_{CE}		
	C_{cb}		pF
	C_{bet}		pF
	C_{ce}		pF
	P_{dis}		mW
	Type		
Circuit Parameters	α	0.3	
	Y_1	1.4	
	V_{be}	113	mV
	η		
Calculated Data	I_{BB}		mA
	g_m		$\mho$
	V_E		mV

Algorithm Figure 12.5   Colpitts Oscillator, *be* Cutoff Limiting

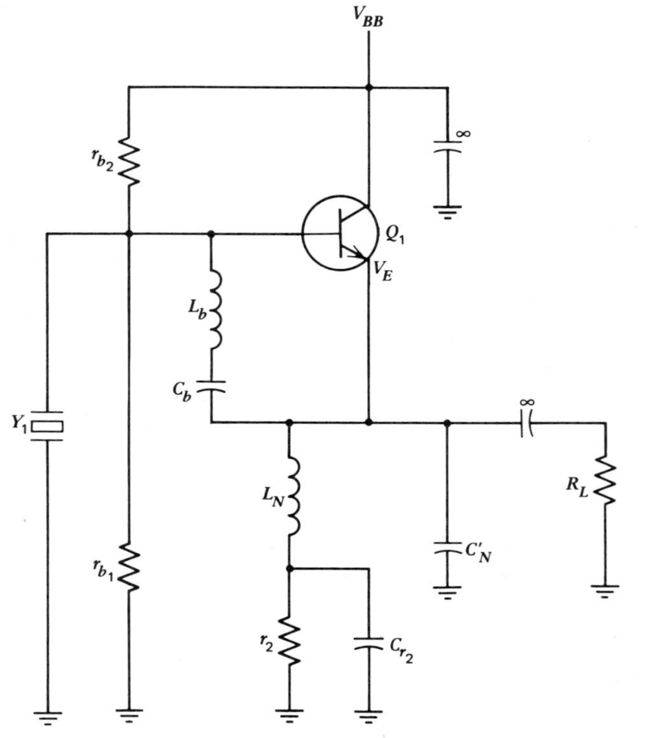

Algorithm 12.5 Colpitts Oscillator, *be* Cutoff Limiting

Oscillator Performance
Crystal and Transistor Characteristics

Step No.	Item	Formula or Quantity	Units	Refer to Step Nos.	Comments	Text Equation No.
a1	f	Specified	MHz		**Oscillator**	
a2	P_L		mW			
a3	P_x		mW		**Performance**	
a4	V_{BB}		mV			
b1	Cut	Specified			**Crystal**	
b2	N					
b3	dt		Hz			
b4	R_L		Ω		**Characteristics**	
b5	C_L		pF		Notes:	

Algorithm 12.5 (*Continued*)

Oscillator Performance
Crystal and Transistor Characteristics

Step No.	Item	Formula or Quantity	Units	Refer to Step Nos.	Comments	Text Equation No.
b6	C_1		pF		1. For production runs	
b7	C_0		pF		R_{df} = max	
b8	C_0'		pF		2. P_x will increase as R_{df} decreases	
b9	Q_x	$10^6/(2\pi C_1 f R_1)$		b6, b11, a1	3. P_L will	3.29a
b10	$\Delta X/(\Delta f/f)$	$10^6/(\pi C_1 f)$	Ω	b6, a1	increase as R_{df} decreases	3.20a
b11	R_1	$[(C_L/(C_L + C_0)]^2 R_L$	Ω	b5, b7, b4		3.19
b12	R_{df}	$R_1/[1C_0'/(C_L + C_0) - 2C_0' df/(C_1 f \times 10^6)]^2$	Ω	b11, b6, b7		3.26a
				a1, b5, b8		
b13	I_x	$(1000 P_x/R_{df})^{1/2}$	mA	a3, b12		3.27a

b14	$C_{L_{df}}$	$\left[\dfrac{1}{C_L+C_0}+\dfrac{2\times 10^{-6}\,df}{C_1 f}\right]^{-1}-C_0'$	pF				3.24a
c1	Type No.	Specified		b5, b7, b3 b6, a1, b8			
c2	β_o				$\beta_o > 1200\,g_m$		5.116
c3	f_T		MHz		$f_T > 600 f g_m$		5.113
c4	C_{bet}		pF			Transistor	
c5	C_{cb}		pF				
c6	C_{ce}		pF			Characteristics	
c7	BV_{CB}		mV				
c8	BV_{CE}		mV				
c9	BV_{BE}		mV			at maximum temperature	
c10	P_{dis}		mW				
c11	r_{bb}		Ω				

Algorithm 12.5 (*Continued*)

Oscillator Performance
Crystal and Transistor Characteristics

Step No.	Item	Formula or Quantity	Units	Refer to Step Nos.	Comments	Text Equation No.
		Circuit Parameters				
d1	α	0.3			See Section 7.6.1	
d2	γ_1	1.4				
d3	V_{be}	113	mV			
		Calculation of Component Values				
		Note: $\oplus$ *signifies physical component*				
1	η	P_L/P_x		a2, a3	See discussion on η	7.10
2	V_L	The smaller of	mV		See discussion of	
2a		$0.22 BV_{CE}$ or		c8	R_L and V_L	9.6
2b		$\sqrt{10^7 P_L/\sqrt{f}}$		a2, a1		7.30
3	R_L	$V_L^2/(1000 P_L)$	Ω	2, a2		7.9
⊕ 4	V_E	$V_{BB} - [> 2.1(V_L + V_{be}] - 1700$	mV	a4, 2, d3	> 2000, See Note A	9.8
5	R_2	$\eta R_{df} - R_{df}^2/R_L$	Ω	1, b12, 3	See Section 7.2.3.1	7.16

6	X_2	$\sqrt{R_2 R_L}$	Ω	5, 3		7.4
7	R'_T	$R_{df} + R_2$	Ω	b12, 5		5.40
8	X_1	V_{be}/I_x	Ω	d3, b13		5.52
9	g'_m	$R'_T/[X_1(X_2 - X_1/\beta_o)]$	$\mho$	7, 8, 6, c2		7.34
10	I'_e	$g'_m V_{be}$	mA	9, d3		7.7
11	I'_E	I'_e/γ_1	mA	10, d2		2.69
12	r'_2	V_E/I'_E	Ω	4, 11		2.99
13	r'_b	$\beta_0 r'_2/5$	Ω	c2, 12		2.100
14	R'_{in}	$g'_m X_1^2/\beta_0$	Ω	9, 8, c2		7.2
15	R'_b	$(X_1 + X'_2)^2/r'_b$	Ω	8, 6, 13		5.57
16	R_T	$R'_T + R'_{in} + R'_b$	Ω	7, 14, 15		5.40
17		Check that $R_T \leqq 1.1 R'_T$		16, 7		7.46
17a		If yes, go to Step 24				
17b		If no, go to Step 18				
		Note A: If V_E calculates $> V_{BB}/2$, make $V_E = V_{BB}/2$.				
18	I_e	$I_x R_T/X_2$	mA	b13, 16, 6		5.58a

Algorithm 12.5 (*Continued*)

Oscillator Performance
Crystal and Transistor Characteristics

	Step No.	Item	Formula or Quantity	Units	Refer to Step Nos.	Comments	Text Equation No.
	19	I_E	I_e/γ_1	mA	18, d2		2.69
	20	g_m	$R_T/(X_1 X_2)$	℧	16, 8, 6		5.40
⊕	21	r_2	V_E/I_E	Ω	4, 19		2.99
	22	r_b	$\beta_0 r_2/5$	Ω	c2, 21		2.100
	23				Note: In the following steps, substitute $(r_b')^0$ for (r_b) when Step 17 is yes.		
⊕	24	r_{b_2}	$0.83 r_b V_{BB}/(V_E + 700)$	Ω	22 or 13, a4, 4		2.103
⊕	25	r_{b_1}	$1/(1/r_b - 1/r_{b_2})$	Ω	22 or 13, 24		2.104
	26	C_1	$159{,}000/(X_1 f)$	pF	8, a1		7.35
	27	C_{bed}	$g_m 159{,}000/f_T$	pF	9 or 20, a1		2.75
	28	M_M	V_{be}/V_L		d3, 2		5.70
	29	C_{1M}	$C_{cb}(1 + 1/M_M)$	pF	c5, 28		5.71
	30	C_1'	$C_1 - C_{bed} - C_{1M} - C_{bet}$	pF	26, 27, 29, c4		Fig. 5.9b
	31		Check whether $N = 1$		b2		
	31a	s	If yes, $s^2 = 0.2$				5.82a
	31b	s	If no, $s = 1 - 1.5/N$				5.82b

	Step	Symbol	Formula	Unit	Ref	Note	Eq
	32	X_{CN}	$(1-s^2)X_2$	Ω	31, 6		5.83
	33	C_N	$159{,}000/(X_{CN}f)$	pF	32, a1		5.84
⊕	34	C'_N	$C_N - C_{cb}(1+M_M) - C_{ce}$	pF	33, c5, 28, c6		5.73, Fig. 5.9b
	35	X_{LN}	X_{CN}/s^2	Ω	32, 31		5.85
⊕	36	L_N	$X_{LN}/(6.28f)$	μH	35, a1		5.85a
	37		Check if the crystal cut is SC				
			If no,				
⊕	38a	C_b	C'_i	pF	30		SE
	38b	L_b	0			Continue at Step 40	SE
			If yes,				
⊕	39a	C_b	$0.156 C'_i$	pF	30		5.86

Algorithm 12.5 (*Continued*)

Oscillator Performance
Crystal and Transistor Characteristics

	Step No.	Item	Formula or Quantity	Units	Refer to Step Nos.	Comments	Text Equation No.
⊕	39b	L_b	$10^6/[C_b(6.84f)^2]$	µH	39a, a1		5.87
	40	X_{cr_2}	$0.02 X_{LN}$		35		9.10
⊕	41	C_{r_2}	$159{,}000/(X_{cr_2}f)$	pF	40, a1		7.35
	42	$X_{C_L'}$	$[159{,}000/(C_{Ld}f)] - (X_1 + X_2)$	Ω	b14, a1, 8, 6		7.39
⊕	43	C_L'	$159{,}000/(X_{C_L'}f)$	pF	42, a1		7.40
	44	Q_{op}	$Q_x R_{df}/R_T$		b9, b12, b16		5.77a

All units in... µS MHz Ω mW pF µH mA, mV dc or rms	Item	Value
Oscillator Performance	f	
	P_L	
	P_x	
	V_{BB}	
	M	
Principal Crystal Data	R_{df}	
	I_x	
	Cut	
	N	
Transistor Data	β_o	
	f_T	
	BV_{CE}	
	C_{cb}	
	C_{bet}	
	C_{ce}	
	P_{dis}	
	Type	
Circuit Parameters	α	
	Y_1	
	Y_M	
	V_{be}	
	I_{BB}	
Calculated Data	g_m	
	V_E	

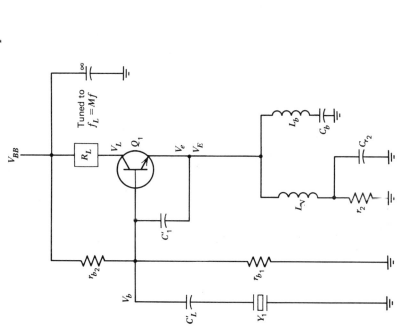

Algorithm Figure 12.6.1 Semi-isolated Colpitts Oscillator, Form 1 Mode Selector. *Note:* For $N = 1$, $L_N = C_{r_2} = 0$

327

Algorithm Figure 12.6.2 Semi-isolated Colpitts Oscillator Form 2 Mode Selector

Item	Value												
f													
P_L													
P_x													
V_{BB}													
M													
R_{df}													
I_x													
Cut													
N													
β_o													
f_T													
BV_{CE}													
C_{cb}													
C_{bet}													
C_{ce}													
P_{dis}													
Type													
α													
γ_1													
Y_M													
V_{be}													
I_{BB}													
g_m													
V_E													

All units in... ℧ MHz Ω mW pF μH mA, mV dc or rms

Groupings: Oscillator Performance; Principal Crystal Data; Transistor Data; Circuit Parameters; Calculated Data.

Algorithm 12.6 Semi-Isolated Colpitts Oscillator, be Cutoff Limiting

Oscillator Performance
Crystal and Transistor Characteristics

Step	Item	Formula or Quantity	Units	Refer to Step Nos.	Comments	Text Equation No.
a1	f	Specified	MHz		**Oscillator**	
a2	P_L		mW			
a3	P_x		mW		**Performance**	
a4	V_{BB}		mV			
a5	M					
b1	Cut	Specified			**Crystal**	
b2	N		Hz			
b3	df					
b4	R_L		Ω		**Characteristics**	
b5	C_L		pF		Notes:	

Algorithm 12.6 (*Continued*)

Oscillator Performance
Crystal and Transistor Characteristics

Step	Item	Formula or Quantity	Units	Refer to Step Nos.	Comments	Text Equation No.
b6	C_1		pF		1. For production runs R_{df} = max	
b7	C_0		pF			
b8	C_0'		pF		2. P_x will increase as R_{df} decrease	
b9	Q_x	$10^6/(2\pi C_1 f R_1)$		b6, b11, a1	3. P_I will remain constant as R_{df} decreases	3.29a
b10	$\Delta X/(\Delta f/f)$	$10^6/(\pi C_1 f)$	Ω	b6, a1		3.20a
b11	R_1	$[(C_L/(C_L + C_0)]^2 R_L$	Ω	b5, b7, b4		3.19
b12	R_{df}	$R_1/[1 - C_0'/(C_L + C_0) - 2C_0' df/(C_1 f \times 10^6)]^2$	Ω	b11, b6, b7 a1, b5, b8		3.26a
b13	I_x	$(1000 P_x/R_{df})^{1/2}$	mA	a3, b12		3.27a

b14	$C_{L_{vf}}$	$\left[\dfrac{1}{C_L+C_0}+\dfrac{2\times 10^{-6}\,df}{C_1 f}\right]^{-1}-C'_0$	pF	b5, b7, b3 b6, a1, b8			3.24a
c1	Type No.	Specified					
c2	β_o				$\beta_o > 1200\,g_m$		5.116
c3	f_T		MHz		$f_T > 600\,f g_m$		5.113
c4	C_{bet}		pF				
c5	C_{cb}		pF			Transistor	
c6	C_{ce}		pF			Characteristics	
c7	BV_{CB}		mV				
c8	BV_{CE}		mV				
c9	BV_{BE}		mV				
c10	P_{dis}		mW		at maximum temperature		
c11	r_{hb}		Ω				

Circuit Parameters

d1	α	0.3/M		a5	10.8

Algorithm 12.6 (*Continued*)

Oscillator Performance
Crystal and Transistor Characteristics

Step	Item	Formula or Quantity	Units	Refer to Step Nos.	Comments	Text Equation No.
d2	γ_1	1.4				10.9a
d3	γ_M	$1.6 - 0.2M$		a5		10.9
d4	V_{be}	$36/\alpha$	mV	d1		10.10
		Calculation of Component Values				
1	V_L	The smaller of	mV		See discussion of	
1a		$0.27 BV_{CE}$ or		c8	R_L and V_L	10.11
1b		$\sqrt{10^7 P_L / \sqrt{fM}}$		a2, a1		10.13
⊕ 2	R_L	$V_L^2/(1000 P_L)$	Ω	1, a2,		7.9
3	I_M	V_L/R_L	mA	1, 2		SE
4	I_E	I_M/γ_M	mA	3, d3		2.69
5	V_E	$V_{BB} - [1.7 V_L] - 1700$	mV	a4, 1	> 2000	10.11a
					If V_E calculates	10.11b
⊕ 6	r_2	V_E/I_E	Ω	5, 4	$> V_{BB}/2$,	2.99

7	r_b	$\beta_o(r_2)/5$	Ω	c2, 6		2.100
8	r_{b_2}	$0.83 r_b V_{BB}/(V_E + 700)$	Ω	7, a4, 5	make $V_E = V_{BB}/2$	2.103
9	r_{b_1}	$1/(1/r_b - 1/r_{b_2})$	Ω	7, 8		2.104
10		Check whether $M = 1$		a5		
10a	R_{L_1}	If yes, $R_{L_1} = R_L$	Ω	2		10.15
10b	R_{L_1}	If no, $R_{L_1} = 0$	Ω			10.14
11		Check whether $N = 1$		b2		
11a		If yes, $r_{2ac} = r_2$	Ω	6		10.16
11b		If no, $r_{2ac} \to \infty$	Ω			10.17
12	I_{e_1}	$I_E \gamma_1$	mA	4, d2		2.69
13	X_1	V_{be}/I_x	Ω	d4, b13		5.52
14	g_m	$\alpha/(26/I_E)$	℧	d1, 4		2.70
						2.72
15	R_{in}	$X_1^2 g_m/\beta_o$	Ω	13, 14, c2		7.2
16	R_{bp}	X_1^2/r_b	Ω	13, 7		5.60
17	R_p	$R_{df} + R_{in} + R_{bp}$	Ω	b12, 15, 16		5.60
18	a_i	I_x/I_{e_1}		b13, 12		5.58a

⊕ marks appear on rows 8 and 9.

Algorithm 12.6 (*Continued*)

Oscillator Performance
Crystal and Transistor Characteristics

Step	Item	Formula or Quantity	Units	Refer to Step Nos.	Comments	Text Equation No.
19	X_2	$a_i R_p (1 + 2a_i X_1/r_b)$	Ω	18, 17, 13, 7		5.63
20	V_e	$I_x \sqrt{\left[R_{df} + R_{in} + \dfrac{(X_1 + X_2)}{r_b}(X_1) \right]^2 + X_2^2}$	mV	b13, b12, 15,		5.66
				19, 13, 7		
21	V_b	$I_x \left\{ \sqrt{\left[(X_1 + X_2)^2 + R_{df}^2 \right]} \right\}$	mV	b13, 13, 19, b12		5.50a
22	V_{L_1}	$I_{e_1} R_{L_1}$	mV	12, 10		SE
23	C_1	$159000/(X_{1f})$	pF	13, a1		7.35
24	C_{bed}	$g_m 159000/f_T$	pF	14, c3		2.75
25	M_{cb}	$1 + V_{L_1}/V_b$		22, 21		10.4
26	C_M	$M_{cb} C_{cb}$	pF	25, c5		10.3
27	M_M	V_{be}/V_e		d4, 20		10.6
28	C_{1M}	$C_M(1 + 1/M_M)$	pF	26, 27		10.5
29 ⊕	C_1'	$C_1 - C_{bed} - C_{1M} - C_{bet}$	pF	23, 24, 28, c4		Fig. 5.91
30	C_{2M}	$C_M(1 + M_M)$	pF	26, 27		10.7

31	M_{ce}	$1 + V_{L_t}/V_e$		22, 20		10.2
32	C_{cem}	$C_{ce}M_{ce}$	pF	c6, 34		10.1
33		For nondoubly rotated cuts		b1		
33a		for $N = 1, L_N = 0$ $X_{CN} = X_2$		b2	See the note in the legend of Algorithm Figure 12.6.1. Go to step 35	
33b	s	for $N \neq 1, s = 1 - 1.5/N$		b2		5.82b
34	X_{CN}	$(1 - s^2)X_2$	Ω	30, 19	or as per 33a	5.83
35	C_N	$159{,}000/(X_{CN}f)$	pF	31, a1	$\rightarrow$	5.84
36	C'_N	$C_N - C_{2M} - C_{cem}$	pF	35, 30, 32		SE
37	X_{LN}	X_{C_N}/s^2	Ω	34, 33	(For $N = 1$ see note in the legend of Algorithm Figure 12.6.1	5.85
38	L_N	$X_{L_N}/(6.28f)$	μH	37, a1		5.85a
$\oplus$						
39a	C_b	C'_N	pF	36		SE
$\oplus$						

Algorithm 12.6 (*Continued*)

	Step	Item	Formula or Quantity	Units	Refer to Step Nos.	Comments	Text Equation No.
			Oscillator Performance Crystal and Transistor Characteristics				
	39b	L_b	0			Continue at Step 45	SE
	40		For SC-cut		b1		
	40a	X_2'	$1/\left(\frac{1}{X_2} - \frac{f(C_{2M} + C_{cem})}{159{,}000}\right)$	Ω	19, a1, 30, 32	See Fig. 59b	SE
	41		For SC-cut, $N = 1$		b1, b2		
⊕	41a	C_b	$0.156\frac{159{,}000}{X_2^1 f}$	pF	40a, a1		5.86
⊕	41b	L_b	$10^6/[C_b(6.84f)^2]$	μH	41a, a1	Continue at Step 46.	5.87
	42		For SC-cut, $N = 3$ or 5		b1, b2	See Section 5.6.5.4	
⊕	43a	C_b	$159{,}000/(5.37fX_2')$	pF	a1, 40a	Form 1 network	5.92a
							5.93
⊕	43b	L_N	$5.01X_2'/(2\pi f)$	μH	40a, a1	Algorithm Figure 12.6.1	5.88a
							5.99
⊕	43c	L_b	$0.905L_N$	μH	43b	Continue at Step 46.	5.91a
⊕	44a	C_b	$159{,}000/(1.48fX_2')$	pF	a1, 40a	Form 2 network	5.97a

⊕	44b	L_N	$2.62 X_2'/(2\pi f)$	μH	40a, a1	Algorithm Figure 12.6.2	5.98
							5.94a
⊕	44c	L_b	$0.905 L_N$	μH	44b	Continue at Step 46	5.95
							5.96a
	45		Not used				
	46	$X_{C_{r2}}$	$0.02 X_{LN}$	Ω	37 or X_{LN} of 43 b or 44b		9.9
⊕	47	C_{r2}	$159{,}000/(X_{c_{r2}} f)$	pF	46, a1	for $N=1$ $C_{r_2} = 0$	9.10
	48	X_{C_L}	$159{,}000/(C_{L_d} f) - (X_1 + X_2)$	Ω	b14, a1, 13, 19		7.39
⊕	49	C_L'	$159{,}000/(X_{C_L} f)$	pF	48, a1		7.40
	50	Q_{op}	$Q_x R_{df} / \left(R_{df} + R_{in} + \dfrac{X_2^2}{r_{2ac}} + \dfrac{(X_1 + X_2)^2}{r_b} \right)$		b9, b12, 15, 13, 19 11, 7		5.77a

All units in.... U MHz Ω mW pF μH mA, mV dc or rms	Item	Value
Oscillator Performance	f	
	P_L	
	P_x	
	V_{BB}	
Principal Crystal Data	R_{df}	
	I_x	
	Cut	
	N	
Transistor Data	β_o	
	f_T	
	BV_{CE}	
	C_{cb}	
	C_{bet}	
	C_{ce}	
	P_{dis}	
	Type	
Circuit Parameters	A_{L_0}	
	Q_{L_3}	
	R_A/X_2	
Calculated Data	I_{BB}	
	$|Z_c|$	
	R_N	
	X_A	
	X_L	

Algorithm Figure 12.7 Butler Oscillator, $X_A = 0$ Design.

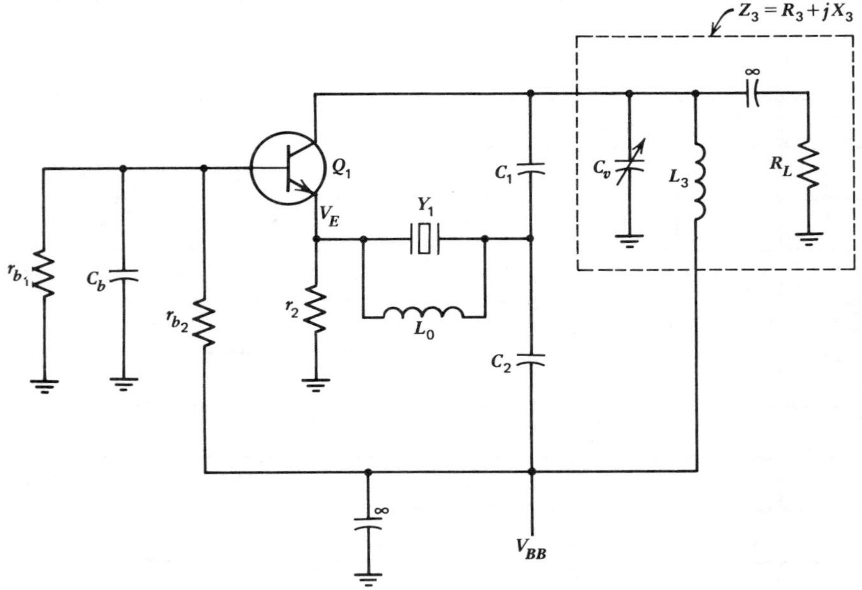

Algorithm 12.7 $X_A = 0$ Butler Oscillator, R_{IN} Limiting

Oscillator Performance
Crystal and Transistor Characteristics

Step No.	Item	Formula or Quantity	Units	Refer to Step No.	Comments	Text Equation No.
a1	f	Specified	MHz			
a2	P_L		mW		**Oscillator**	
a3	P_x		mW			
a4	V_{BB}		mV		**Performance**	
b1	Cut	Specified				
b2	N		Hz		**Crystal**	
b3	df					
b4	R_L		Ω		**Characteristics**	
b5	C_L		pF		Notes:	

Algorithm 12.7 (*Continued*)

Oscillator Performance
Crystal and Transistor Characteristics

Step No.	Item	Formula or Quantity	Units	Refer to Step No.	Comments	Text Equation No.
b6	C_1		pF		1. For production runs R_{df} = max	
b7	C_0		pF			
b8	C_0'		pF		2. P_x will decrease as R_{df} decreases.	
b9	Q_x	$10^6/(2\pi C_1 f R_1)$		b6, b11, a1	3. P_L will remain constant	3.29a
b10	$\Delta X/(\Delta f/f)$	$10^6/(\pi C_1 f)$	Ω	b6, a1	as R_{df} decreases.	3.20a
b11	R_1	$[1 C_L/(C_L + C_0)]^2 R_L$	Ω	b5, b7, b4		3.19
b12	R_{df}	$R_1/[1 - C_0'/(C_L + C_0) - 2C_0'\,df/(C_1 f \times 10^6)]^2$	Ω	b11, b6, b7 a1, b5, b8		3.26a
b13	I_x	$(1000 P_x/R_{df})^{1/2}$	mA	a3, b12		3.27a

b14	$C_{L_{df}}$	$\left[\dfrac{1}{C_L+C_0}+\dfrac{2\times 10^{-6}\,df}{C_1 f}\right]^{-1} - C_0'$	pF	b5, b7, b3 b6, a1, b8	3.24a
c1	Type No.	Specified			
c2	β_o			$\beta_o > 1200/R_{\text{IN}}$	5.116
c3	f_T		MHz	$f_T > 4fI_E$	11.1a
c4	C_{ber}		pF		
c5	C_{cb}		pF	Transistor	
c6	C_{ce}		pF		
c7	BV_{CB}		mV	Characteristics	
c8	BV_{CE}		mV		
c9	BV_{BE}		mV		
c10	P_{dis}		mW	at maximum temperature	
c11	r_{bb}		Ω		
				Circuit Parameters	
d1	A_{L_0}	Specified		Recommend $A_{L_0} = 2$	11.45

Algorithm 12.7 (*Continued*)

Oscillator Performance
Crystal and Transistor Characteristics

	Step No.	Item	Formula or Quantity	Units	Refer to Step No.	Comments	Text Equation No.
	d2	Q_{L_3}	Specified				
			Calculation of Component Values				
			Note: ⊕ *signifies physical component*				
	1	I_E	$1.4 I_x$	mA	b13		6.23
	2	R_{IN_0}	$26/I_E$	Ω	1		2.87
	3	R_A	$A_L(R_{IN_0} + R_{df})$	Ω	d1, b12, 2		11.45a
	4	P_A	$I_x^2 R_A / 1000$	mW	b13, 3		11.25
	5	n	P_L / P_A		a2, 4		11.40
⊕	6	R_L	$[(n+1)I_x R_A]^2 / (10^3 P_L)$	Ω	5, b13, 3	$< 10{,}000/\sqrt{f}$	11.42
	7	X_2	$R_A / 5$	Ω	3		11.35
	8	$X_1 + X_2$	$(n+1) X_2$	Ω	5, 7		11.37
	9	X_1	$n X_2$	Ω	5, 7		11.37
	10	X_L	$(X_1 + X_2)\left(1 + \left(\dfrac{X_1 + X_2}{R_L}\right)^2\right)$	Ω	8, 6		11.27

	#	Symbol	Formula	Unit	Refs	Condition	Section
⊕	11	C_2	$159{,}000/(X_2 f)$	pF	7, a1		5.84
⊕	12	C_1	$159{,}000/(X_1 f)$	pF	9, a1		5.84
⊕	13	C_V	0 to $C_1/2$	pF	12		SE
	14	X_{L_3}	$1/(1/X_L + C_{V_{\text{mean}}} f/159{,}000)$	Ω	10, a1, 13		11.16
⊕	15	L_3	$X_{L_3}/(6.28 f)$	μH	14, a1		5.85a
⊕	16	C_b	$> 1.59 \times 10^6/(f(R_A - R_{df}))$	pF	a1, 3, b12		11.48
	17	P'_L	$P_L[1 + R_L/(Q_{L_3} X_{L_3})]$	mW	a2, 6, d2, 14	See Note A	11.52
	18		Check that $P'_L < 1.1 P_L$			If not, see Note A	
	19	V_L	$\sqrt{1000 P_L R_L}$	mV	a2, 6	$< 0.25 BV_{CE}$	7.9
	20		Not used				
	21	V_E	$V_{BB} - 1.7 V_L - 1700$	mV	a4, 19	> 2000	11.46
						If V_E calculates	11.46a
⊕	22	r_2	V_E/I_E	Ω	21, 1	$> V_{BB}/2$,	2.99
						make $V_E = V_{BB}/2$	
	23	r_b	$\beta_o r_2/5$	Ω	c2, 22		2.100

Algorithm 12.7 (*Continued*)

Oscillator Performance
Crystal and Transistor Characteristics

	Step No.	Item	Formula or Quantity	Units	Refer to Step No.	Comments	Text Equation No.
⊕	24	r_{b_2}	$0.83 r_b V_{BB}/(V_E + 700)$	Ω	23, a4, 21		2.103
⊕	25	r_{b_1}	$1/(1/r_b - 1/r_{b_2})$	Ω	23, 24		2.104
⊕	26	L_0	$10^6/[C_0'(6.28f)^2]$	μH	b8, a1	Omit for $f < 50$ MHz	11.49
	27	Q_{op}	$< Q_x R_{df}/R_A$		b9, b12, 3		11.55
			Note A: If necessary and desirable, redo Steps 5 to 27 substituting P_L' for P_L in Step 5.				

All units in.... ℧ MHz Ω mW pF μH mA, mV dc or rms	Item	Value																			
Oscillator Performance	f																				
	P_L																				
	P_x																				
	V_{BB}																				
Principal Crystal Data	R_{df}																				
	I_x																				
	Cut																				
	N																				
Transistor Data	β_o																				
	f_T																				
	BV_{CE}																				
	C_{cb}																				
	C_{bet}																				
	C_{ce}																				
	P_{dis}																				
	Type																				
Circuit Parameters	A_{L_0}																				
	Q_{L_3}																				
	R_A/X_2																				
Calculated Data	I_{BB}																				
	$|Z_c|$																				
	R_N																				
	X_A																				
	X_L																				

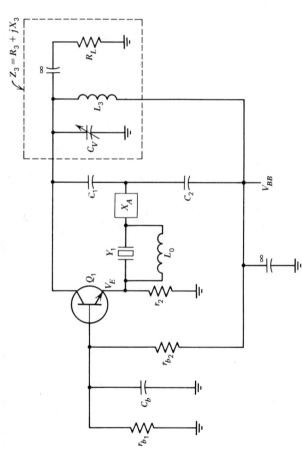

Algorithm Figure 12.8   Stable Butler Oscillator

Algorithm 12.8 Stable Butler Oscillator, R_{IN} Limiting

Oscillator Performance
Crystal and Transistor Characteristics

Step No.	Item	Formula or Quantity	Units	Refer to Step Nos.	Comments	Text Equation No.
a1	f	Specified	MHz			
a2	P_L		mW		**Oscillator**	
a3	P_x		mW			
a4	V_{BB}		mV		**Performance**	
b1	Cut	Specified			**Crystal**	
b2	N					
b3	df		Hz			
b4	R_L		Ω		**Characteristics**	
b5	C_L		pF		Notes:	

b6	C_1		pF		
b7	C_0		pF		1. For production runs R_{Df} = max
b8	C_0'		pF		2. P_x will decrease as R_{df} decreases
b9	Q_x	$10^6/(2\pi C_1 f R_1)$		b6, b11, a1	3.29a 3. P_L will remain constant
b10	$\Delta X/(\Delta f/f)$	$10^6/(\pi C_1 f)$		b6, a1	3.20a as R_{df} decreases
b11	R_1	$[(C_L/(C_L + C_0)]^2 R_L$	Ω	b5, b7, b4	3.19
b12	R_{df}	$R_1/[1 - C_0'/(C_L + C_0)] -$ $2C_0' df/(C_1 f \times 10^6)]^2$	Ω a1, b5, b8	b11, b6, b7	3.26a
b13	I_x	$(1000 P_x/R_{df})^{1/2}$	mA	a3, b12	3.27a
b14	$C_{L_{df}}$	$\left[\dfrac{1}{C_L + C_0} + \dfrac{2 \times 10^{-6} df}{C_L f}\right]^{-1} - C_0'$		b6, a1, b8	3.24a
c1	Type No.	Specified			

Algorithm 12.8 *(Continued)*

Oscillator Performance
Crystal and Transistor Characteristics

Step No.	Item	Formula or Quantity	Units	Refer to Step Nos.	Comments	Text Equation No.
c2	β_o				$\beta_o > 1200/R_{IN}$	5.116
c3	f_T		MHz		$f_T > 4fI_E$	11.1a
c4	C_{bet}		pF		Transistor	
c5	C_{cb}		pF			
c6	C_{ce}		pF		Characteristics	
c7	BV_{CB}		mV			
c8	BV_{CE}		mV			
c9	BV_{BE}		mV			
c10	P_{dis}		mV		at maximum temperature	
c11	r_{bb}		Ω			

Circuit Parameters

		Specified					
d1	A_{L_0}					Recommend $A_{L_0} = 2$	11.45
d2	Q_{L_3}						
		Calculation of Component Values					
		Note: ⊕ signifies physical component					
1	I_E	$1.4 I_x$	mA	b13			6.23
2	R_{IN_0}	$26/I_E$	Ω	1			2.87
3	R_{A_0}	$R_{IN_0} + R_{df}$	Ω	2, b12			11.43
4	R_A	$A_{L_0} R_{A_0}$	Ω	d1, 3			11.45
⊕ 5	R_L	$2000 P_L / I_x^2$	Ω	a2, b13		$< 10{,}000/\sqrt{f}$	11.65
6	n	$0.5 R_L / R_A - 1$		5, 4			11.73
7	R_A / X_2	Choose a value				Recommend 2.5 to 5	
8	X_{L+}	$R_L / (2 R_A / X_2 + 1)$		5, 7		See Section 11.4.6.	11.74
9	X_2	$R_A / (R_A / X_2)$	Ω	4, 7			SE
10	X_1	$n X_2$	Ω	6, 9			11.71
11	$X_1 + X_2$	$(n+1) X_2$	Ω	6, 9			11.37

Algorithm 12.8 (*Continued*)

Oscillator Performance
Crystal and Transistor Characteristics

Step No.	Item	Formula or Quantity	Units	Refer to Step Nos.	Comments	Text Equation No.
12	$-R_N$	$\dfrac{X_1 X_2 \dfrac{R_L}{1+(R_L/X_{L+})^2}}{\left(\dfrac{R_L}{1+(R_L/X_{L+})^2}\right)^2 + \left(\dfrac{X_{L+}}{1+(X_{L+}/R_L)^2} - (X_1+X_2)\right)^2}$	Ω	10, 9, 5		11.56
		$-X_1 X_2 \left(\dfrac{X_{L+}}{1+(X_{L+}/R_L)^2} - (X_1+X_2)\right)$ Denominator		8		
13	X_A		Ω	10, 9, 5		11.57
				8		
14		Check that $-R_N \approx R_A$		12, 4		
15	C_2	$159{,}000/(X_2 f)$	pF	9, a1		5.84

⊕	16	C_1	C_2/n	pF	15, 6		11.37
⊕	17	C_o	$C_1/2$	pF	16		SE
	18	X_{L_3}	$1/(1/X_{L+} + fC_{v_{\text{mean}}}/159{,}000)$	Ω	8, a1, 17		11.16
⊕	19	L_3	$X_{L_3}/(6.28f)$	μH	18, a1		5.85a
⊕	20	C_b	$1.59 \times 10^6/(f(R_A - R_{df}))$	pF	a1, 3, b12		11.48
⊕	21	L_A	$X_A/(6.28f)$	μH	13, a1		5.85a
	22	P'_L	$P_L(1 + R_L/(Q_{L_3}X_{L_3}))$	mW	a2, 5, d2, 18	Note A	11.52
	23		Check that $P'_L < 1.1P_L$		a2, 22		
	24	V_L	$\sqrt{1000 P_L R_L}$	mV	a2, 5		7.9
	25	V_E	$V_{BB} - 1.7V_L - 1700$	mV	a4, 24	> 2000	11.46
						If V_E calculates	11.46a
⊕	26	r_2	V_E/I_E	Ω	25, 1	> $V_{BB}/2$	2.99
						make $V_E = V_{BB}/2$	
	27	r_b	$\beta_o r_2/5$	Ω	c2, 26		2.100
⊕	28	r_{b_0}	$0.83 r_b V_{BB}/(V_E + 700)$	Ω	27, a4, 25		2.103
⊕	29	r_{b_1}	$1/(1/r_b - 1/r_{b_2})$	Ω	27, 28		2.104
⊕	30	L_0	$10^6/[C_0(6.28f)^2]$	μH	b8, a1	omit for $f < 50$	11.49

Algorithm 12.8 *(Continued)*

Oscillator Performance
Crystal and Transistor Characteristics

Step No.	Item	Formula or Quantity	Units	Refer to Step Nos.	Comments	Text Equation No.
31	Q_{op}	$< Q_x R_{df}/ - R_N$		b9, b12, 12		11.81
		Note A: If necessary and desirable, redo Steps 5 to 31 substituting P'_L for P_L in Step 5.				
		For the X_{L-} design.				
	X_{L-}	$R_L/(2R_A/X_2 - 1)$	Ω	5, 7		11.75
		Continue with Steps 9 to 31. Note that L_A, in Step 21, 11.75 is a capacitor.				

13

Other Oscillator Configurations

13.1 INTRODUCTION

This chapter treats oscillator configurations not considered in the previous chapters. The treatment, of necessity, will be qualitative rather than quantitative because space limitations do not permit a full exposition of this extremely extensive subject. Also, because of the enormous number of available circuits, only a sampling can be made.[13.7]

The chapter is concerned with the following:

1. Single transistor oscillators not covered in the previous chapters.
2. Multiple transistor circuits in which a transistor is added to the circuits of Chapters 5 to 11 to improve their performance.
3. Multiple transistor circuits which are substantially different from the basic circuits of Chapters 5 to 11. ALC type oscillators are discussed in Chapter 14.

13.2 SINGLE TRANSISTOR OSCILLATORS

13.2.1 Low-Frequency Crystal Oscillators

As the frequency decreases, the crystal resistance increases markedly, and the circuits previously described are either unsuitable or must be modified to accommodate the higher resistance. The FET is particularly useful because of its very high input impedance. Figures 13.1, 13.2, and 13.3 show oscillator circuits using FETs as the active device. These circuits are suitable for

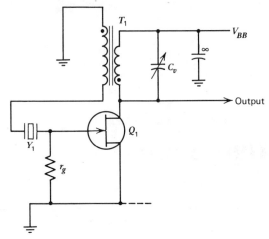

Figure 13.1 Series resonance low-frequency oscillator.

low-frequency crystals, from 16 kHz to above 1 MHz. The amplitude limiting process is the clamp biased limiting described in Sections 6.4 and 6.2.3.

13.2.1.1 Oscillator of Fig. 13.1

The crystal operates near series resonance, so that the operating Q, $Q_{op} \approx Q_x R_1/(R_1 + r_g)$. It is therefore seen that for small deteriorations of Q, R_1

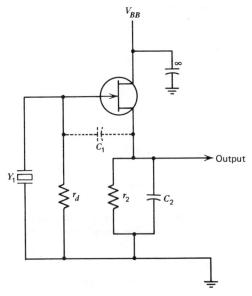

Figure 13.2 Colpitts low-frequency oscillator.

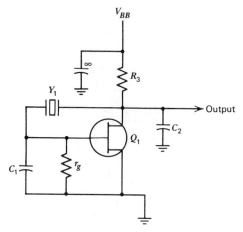

Figure 13.3 Pierce low-frequency oscillator.

must be comparable to or larger than r_g and the circuit is therefore suitable for crystals having very large R_1.

The circuit has the obvious disadvantage that it employs transformer T_1 to provide the 180° phase shift necessary for oscillation. The transformer may be a bulky and relatively expensive device and is therefore undesirable.

If the crystal is short-circuited, the circuit will oscillate at the resonant frequency of T_1 and C_V. When the crystal is used, T_1 and C_V serve to choose the correct response of the crystal.

13.2.1.2 Oscillator of Fig. 13.2

Figure 13.2 shows the Colpitts oscillator described in Chapter 9, adapted for a FET. This circuit is suitable for frequencies from 100 kHz up. C_1, shown in dashed lines, is often omitted and the transistor gate drain capacitance then supplies the functions of C_1. However, this is very poor practice as the transistor capacitance varies widely from transistor to transistor. Also, the circuit performance depends excessively on the transistor characteristics.

13.2.1.3 Oscillator of Fig. 13.3

This is the Pierce oscillator equivalent of the Colpitts oscillator of Fig. 13.2, and its performance is quite similar.

13.2.1.4 Oscillator of Fig. 13.4

This oscillator is basically the isolated Pierce oscillator of Chapter 8. It is suitable for frequencies from 50 kHz up. Resistor R_s is set for the proper crystal drive level and for dependable oscillator starting.

356   **Other Oscillator Configurations**

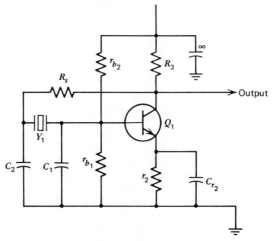

Figure 13.4   Low-frequency version of isolated Pierce oscillator.

13.2.2   Other Oscillator Variations

13.2.2.1   Oscillator of Fig. 13.5

This oscillator is basically the Colpitts oscillator of Chapter 9. The difference is the point of extracting the power output. Because C_L is in series with the crystal, the output is harmonic-free and the noise floor is lower than that of the normal Colpitts oscillator. However, the power output is made very small, in order not to excessively deteriorate the oscillator operating Q.

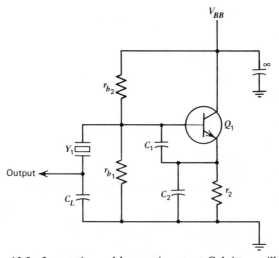

Figure 13.5   Low-noise and harmonic output Colpitts oscillator.

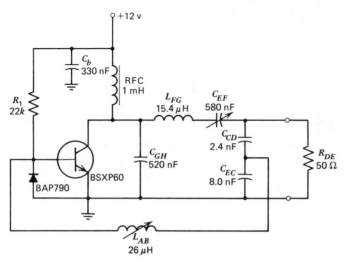

Figure 13.6 Class E high-efficiency tuned power oscillator (from Ebert, J. and Kazimierczuk, M., *IEEE J. Solid-State Circuits* **SC-16**, No. 2, 65 (April 1981).

13.2.2.2 Oscillator of Fig. 13.6 [13.1]

This oscillator is basically a Pierce oscillator but it is unique in that the transistor operates in Class E. By Class E operation is meant that the transistor operates as a switch with a duty cycle of 50%. This results in extremely high conversion efficiency. For example, the subject oscillator provides 3 W output at 2 MHz at a collector efficiency of over 95%. The output wave form is relatively sinusoidal.

13.2.2.3 Oscillator of Fig. 13.7

This oscillator may be considered as a variation of the Butler oscillator discussed in Chapter 11. However, in this case it is more convenient to consider it as a Pierce oscillator with the crystal providing degenerative emitter feedback. To emphasize the similarity between this circuit and the Pierce oscillator (see Fig. 5.1a), the like components have been assigned the same symbol number.

The oscillator is another member of the family of oscillators in which the emitter current is also the crystal current. As a result, the output has good wave form and a low noise floor, which is considerably below that of the normal Pierce oscillator. A further advantage is that considerable power output can be obtained without the serious deterioration of the operating Q that exists in the normal Pierce oscillator. However, at low power output, the normal Pierce oscillator has a better operating Q.

The circuit as shown is not popular because it requires more components than the Pierce or Butler oscillator and it does not have superior performance.

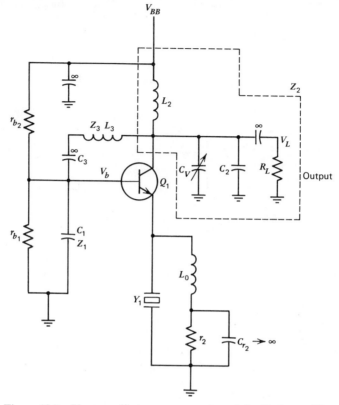

Figure 13.7 Pierce oscillator representation of the Butler oscillator.

However, it can be modified, as explained in Section 13.5.1, to have some very desirable characteristics.

A major problem in this circuit at high P_L/P_x is that V_L/V_b can be large and, therefore, Miller effects play an important role.

13.3 THE DARLINGTON TRANSISTOR PAIR[2.4, 2.5]

Figure 13.8a shows two transistors, Q_1 and Q_2, connected in cascade, called the Darlington pair. In many cases the pair can be replaced by an equivalent single transistor Q_{eq}. The approximate relationships between the properties of the pair and the equivalent transistor are now developed.

Q_1 has the parameters listed below. Each of these parameters is defined and calculated in Chapter 2.

$$\beta_{o_1}$$

13.3 The Darlington Transistor Pair

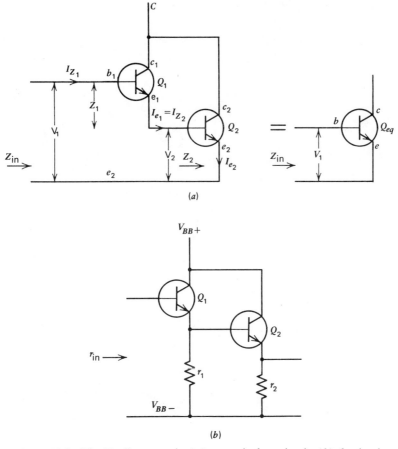

Figure 13.8 The Darlington pair. (a) ac equivalent circuit. (b) dc circuit.

The impedance between b_1 and $e_1 = Z_1$
The capacitance $C_{c_1 b_1}$
The capacitance $C_{c_1 e_1}$

$$g_{m_1} = \frac{I_{e_1}}{V_{b_1}} \quad \text{assuming } i_C \approx i_E$$

Similarly, Q_2 has the parameters

$$\beta_{o_2}, \quad Z_2, \quad C_{c_2 b_2}, \quad C_{c_2 e_2}, \quad g_{m_2}$$

and Q_{eq} has the parameters

$$\beta_o, \quad Z_{in}, \quad C_{cb}, \quad C_{ce}, \quad g_m$$

13.3.1 Calculation of the Equivalent ac Parameters

From inspection of Fig. 13.8a,

$$V_1 = Z_1 I_{Z_1} + Z_2 I_{Z_2} \tag{13.1}$$

$$Z_{in} = \frac{V_1}{I_{Z_1}} = Z_1 + Z_2 \frac{I_{Z_2}}{I_{Z_1}} \approx Z_1 + \beta_1 Z_2 \tag{13.2}$$

At low frequencies

$$Z_2 \approx \frac{\beta_{o_2}}{g_{m_2}} \tag{13.3}$$

$$Z_{in} \approx \frac{\beta_{o_1} \beta_{o_2}}{g_{m_2}} \tag{13.4}$$

Also, from Eq. (13.1)

$$g_m \approx \frac{I_{e_2}}{V_1} = \frac{I_{e_2}}{V_2} \cdot \frac{V_2}{V_1} = g_{m_2} \frac{V_2}{V_1} \tag{13.5}$$

It is thus seen that, since $V_1 > V_2$,

$$g_m < g_{m_2} \tag{13.6}$$

Usually,

$$V_2 \to V_1$$

Therefore,

$$g_m \to g_{m_2} \tag{13.7}$$

Also from inspection,

$$C_{cb} \approx C_{c_1 b_1} \tag{13.8}$$

$$C_{ce} \approx C_{c_2 e_2} \tag{13.9}$$

13.3.2 Calculation of the Equivalent dc Parameters

Figure 13.8b shows the equivalent dc circuit which includes bias emitter resistors r_1 and r_2. Bias resistor r_1 is not required for the dc function, but is supplied to increase I_{E_1} and f_{T_1}.

$$r_{in} \approx \beta_{o_1}\left[r_1 \| (\beta_{o_2} r_2)\right] \tag{13.10}$$

If $r_1 \to \infty$,

$$r_{in} \approx \beta_{o_1}\beta_{o_2}r_2 \tag{13.11}$$

r_{in} is always much larger than the r_{in} of the single transistor circuit with the same I_E.

13.3.3 Darlington Pair Colpitts Oscillator

Figure 13.9 is the schematic of a Colpitts oscillator using two transistors in the Darlington connection. Because of this connection, r_{b_1} and r_{b_2} are very high-value resistors, and therefore do not deteriorate the circuit operating Q, even for crystals having low values of R_{df}.

R_E is provided to reduce the $1/f$ noise of Q_2.

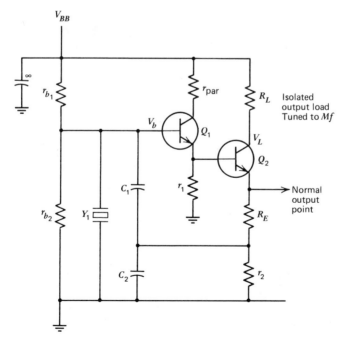

Figure 13.9 Darlington pair Colpitts oscillator.

In this circuit, the normal power output point is at the emitter of Q_2. The power, available at this point without excessive deterioration of the circuit operating Q, is quite small. However, power can be extracted from the collector circuit of Q_2 at fairly high levels.

Because of the excellent isolation provided by the Darlington pair, as stated in Eq. (13.8), V_L can be large compared to V_b without significant Miller effects. Also, the output circuit can be tuned to a multiple, M, of the oscillator frequency. It will be noted that this circuit has performance superior to that of the semi-isolated Colpitts oscillator of Chapter 10, but at the expense of one additional transistor and two resistors.

13.4 THE CASCODE AMPLIFIER[2.4, 2.5]

A circuit configuration which has proven to be useful in the field of oscillators is the well-known Cascode amplifier shown in Fig. 13.10. Its most important

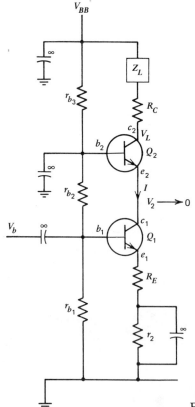

Figure 13.10 Cascode amplifier circuit.

13.4 The Cascode Amplifier

property is that V_L, which may be high, is isolated from the base of Q_1. Instead, C_{cb} of Q_2 is lumped into the load Z_L.

The voltage at the emitter of Q_2, V_2, is equal to Ir_e where r_e is the dynamic emitter resistance of Q_2 and is very small. Therefore, the voltage at that point is very small and the C_{ce} of Q_2 is therefore also effectively lumped into Z_L. For the same reason, C_{be} of Q_2 is also rendered ineffective.

Because V_2 is very small, C_{cb} of Q_1 is effectively lumped into the source impedance.

It is thus seen that all the transistor capacitances which may give rise to the harmful Miller effects, which seriously limit the performance of the $M = 1$ semi-isolated Colpitts oscillator and the Butler oscillator, are rendered ineffective.

R_c and R_E are provided to reduce the phase noise and to minimize the tendency for spurious oscillations.

An example of the use of the cascode circuit is the semi-isolated Colpitts oscillator of Figure 13.11, which is the Colpitts oscillator of Fig. 5.7b to which has been added a cascode type load isolator. As a result, this circuit will work well for all values of M. It is interesting to note that the Q_1 oscillator circuit operates identically to the normal Colpitts oscillator without external load.

The algorithm for the semi-isolated Colpitts oscillator can be easily modified to include this type of oscillator.

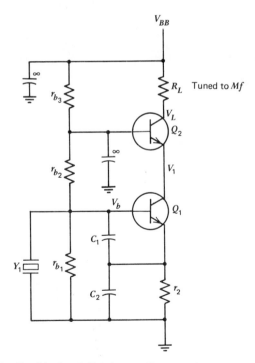

Figure 13.11 Semi-isolated Colpitts oscillator with cascode output circuit.

364 Other Oscillator Configurations

13.5 OTHER OSCILLATOR CIRCUITS

This section discusses miscellaneous oscillator circuits, other than ALC types, which are substantially unlike those treated in Chapters 7 to 11. Some of the circuits are quite similar except for the limiting means.

13.5.1 The Two-Transistor Oscillator and Limiter Configuration[13.2]

Figure 13.12 shows the oscillator of Fig. 13.7 combined with the cascode limiting stage of Fig. 13.11. This circuit has the advantage that the oscillator

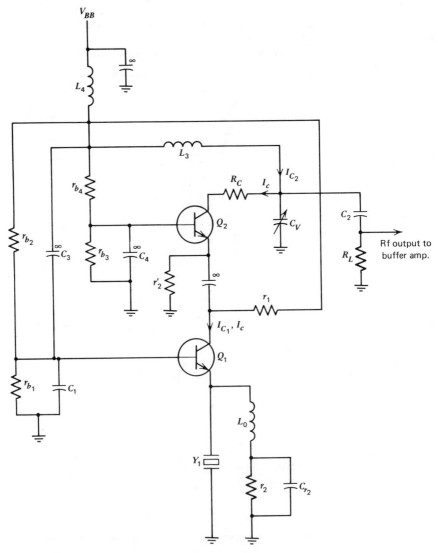

Figure 13.12 The two-transistor oscillator and limiter configuration.

function is separated from the limiting function. Q_1 functions as the oscillator and operates in a linear mode. Q_2 functions as the limiter and its operation is highly nonlinear. I_{C_1} is large to make R_{IN_1} small so as to produce a larger operating Q. I_{C_2} is made $\approx 1.4 I_c$ to fix the current I_c in accordance with Eq. (6.23). It should be noted that the isolation provided by Q_2, as described in Section 13.4, is partially, but not seriously, negated by the nonlinear operation of Q_2.

13.5.2 Two-Transistor Diode Limiting Oscillator[13.3]

Figure 13.13a shows the oscillator of Fig. 13.7 combined with the cascode isolating stage of Fig. 13.11. Both stages operate in essentially linear mode. Limiting is provided by the diodes CR_1 and CR_2 which function as described in Section 6.5, except that the limiting level is set by the values of $-V_C$ and $+V_C$. Two diodes are provided to produce symmetrical limiting action, which results in slightly less noise, but a single diode limiter is often more convenient as it does not require a $-V_C$ supply and does not markedly increase the noise. The single diode arrangement is shown in Fig. 13.13b.

13.5.3 Two-Transistor Oscillator with Separate Variable Gain Type Limiter[13.4]

Figure 13.14 shows the ac circuitry of the oscillator of Fig. 13.7 combined with the cascode isolating stage of Fig. 13.11. This oscillator is designed for very low-noise output and is capable of operation at very high frequencies (100 MHz or more).

Both Q_1 and Q_2 operate in essentially linear mode. The limiting function is provided by IC_1 which acts as a differential limiter and is connected in the feedback path between the output circuitry of Q_2 and the input to Q_1. The differential limiter operates as a variable gain amplifier, the gain of which decreases as the amplitude increases. The differential configuration has excellent phase noise properties as also pointed out by Baugh.[13.5] The amplitude may be conveniently set by changing the dc current in CCG_2.

IC_2 functions as a buffer output amplifier. IC_1 and IC_2 are low-noise integrated circuits suitable for operation at very high frequencies, and include their respective constant current generators CCG_2 and CCG_3.

The remaining circuitry is composed of discrete components.

The function of the constant current generators is to maintain the total I_C of each stage at a closely regulated fixed value. It should be noted that their ac output impedance is also very high to prevent loading of Y_1 and to ensure highly balanced operation of the limiting and output amplifiers for best noise operation.

13.5.4 Two-Transistor Emitter Coupled Oscillator

Figure 13.15 shows an oscillator circuit suitable for a wide frequency range, provided that the crystal resistance is not too large or too small. It has been popular for many years and was formerly known as the Butler oscillator.

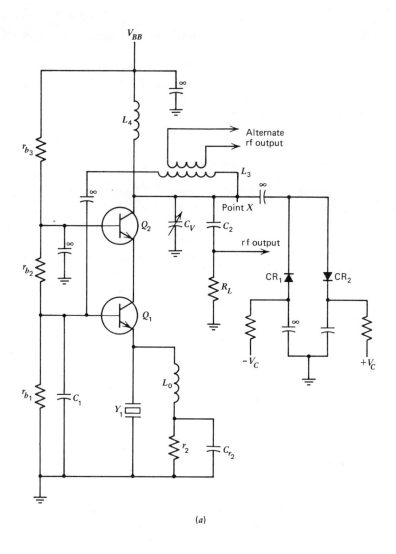

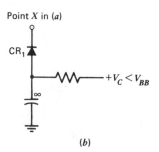

Figure 13.13 Two-transistor oscillator and limiter circuit. (*a*) With two diode limiter. (*b*) Single diode arrangement.

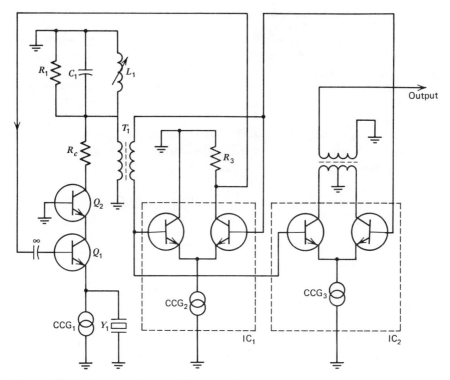

Figure 13.14 Two-transistor oscillator with separate variable gain limiter. (From Rhode, V. L., *Proceedings of the 32nd Annual Symposium on Frequency Control*, p. 418, 1978.)

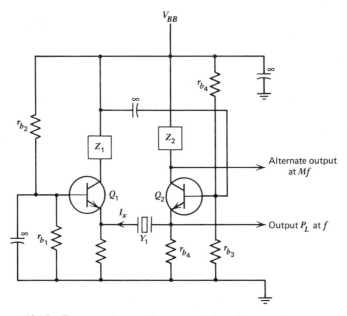

Figure 13.15 Two transistor emitter coupled oscillator (old Butler circuit).

The crystal operates near series resonance. Z_1 is a network for selecting the desired crystal response or overtone. In the simplest case, Z_1 can be a resistor. The limiting takes place in Q_1 and may be either the R_{IN} variation or the collector base voltage type. The operating Q is thus somewhat deteriorated. However, it has excellent noise performance at Fourier frequencies not too near the carrier.

For fundamental output $Z_2 = 0$. For harmonic output Z_2 is a circuit tuned to Mf and the output taken out at the "Alternate Output" point.

Groslambert et al.[13.6] describe an oscillator using this circuit.

14

ALC Oscillators

14.1 INTRODUCTION

As pointed out in Section 3.10, the frequency is a function of the crystal current, I_x (and power). Therefore, a major cause of instability and aging in high-performance oscillators is the variation in crystal current. All the oscillators considered thus far have the common characteristic that the crystal current is determined by the relatively noncontrolled nonlinear action of semiconductors and associated components. The action changes with time, temperature, and many other factors. It therefore follows that the most stable oscillators are most likely to be of the type wherein the crystal current is monitored and constrained to have a high degree of constancy as measured by $\partial I_x / I_x$. The type of oscillator performing the latter process is the Automatic Level Control oscillator. The word *level*, in this case, is synonymous with *crystal current*.

The single transistor oscillator performs the frequency generation and limiting in the same circuit. As a result, the frequency generation function cannot be optimized without destroying the limiting function. The same is true, but to a smaller extent, of the two-transistor oscillators of Chapter 13. However, it is extremely desirable to be able to design circuits in which the frequency generation and limiting are performed by separate relatively independent portions of the complete oscillator so that the operation of each can be optimized or modified without grossly affecting the other function. Again, the oscillator circuit, having this feature, is the ALC type.

The ALC oscillator is used only where the *highest* long- and short-term frequency stability is required. Usually, the oscillator is placed in a closely temperature controlled oven to minimize the instability due to varying temperature.

The price paid for the superior performance of the ALC oscillator is the complexity of the circuitry and the consequent poorer reliability and higher cost.

14.2 GENERAL DESCRIPTION OF THE ALC OSCILLATOR

This oscillator may be treated as a system composed of many components or subsystems, each of which can be specified and designed as necessary for the desired overall system performance.

Figure 14.1a shows the block diagram of the typical ALC oscillator. Figure 14.1b is Fig. 14.1a redrawn to demonstrate that the block diagram of the regulator type of control system is also applicable to the ALC oscillator. The tremendous amount of theory[14.1–14.3] developed to analyze and design this control system is therefore completely usable for the ALC oscillator.

14.2.1 The Steady-State g_m Loop and the Calculation of V_f

Figure 14.1b is somewhat incomplete in that it does not take into account the value of V_C requird by the g_m versus V_C characteristic of the oscillator. This is analogous to the voltage required to make the average frequency of the controlled oscillator equal to that of the reference oscillator in a phase-lock

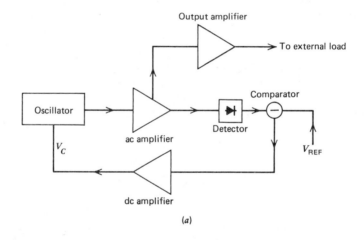

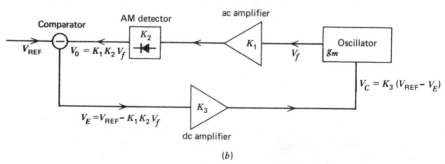

Figure 14.1 Block diagram of the ALC oscillator. (a) Basic oscillator. (b) (a) redrawn to show similarity to a regulator system.

14.2 General Description of the ALC Oscillator

loop. In the phase-lock system the steady-state control voltage is usually not important as its major effect is to create a constant phase difference more or less proportional to the difference between the actual operating frequency and the frequency of the controlled oscillator at $V_C = 0$. However, it does greatly affect the acquisition process if the required V_C is in the neighborhood of the maximum voltage available from the phase detector and may even make the acquisition process impossible, which is also analogously true in the ALC oscillator.

As seen from Chapters 1 and 5, every oscillator has a g_m or equivalent parameter which must have a unique value, g_{me}, to satisfy the conditions for equilibrium oscillation. In an ALC oscillator where all the frequency generating components are designed to operate in a completely linear fashion, the output amplitude *is independent of* g_{me}.

g_{me} is determined by the oscillator components as described in Chapter 5. This, in turn, fixes the value of V_C because of the relationship

$$g_m = F(V_C) \tag{14.1}$$

where

$$\frac{\partial g_m}{\partial V_C} \text{ is always positive} \tag{14.1a}$$

From Fig. 14.1*b*,

$$V_C = K_3(V_{REF} - K_1 K_2 V_f) \tag{14.2}$$

or

$$V_f = \frac{1}{K_1 K_2}\left(V_{REF} - \frac{V_C}{K_3}\right) \tag{14.2a}$$

Also, let

$$V_f = r_{osc} I_x \tag{14.3}$$

It is interesting to observe the oscillator amplitude behavior in accordance with Eqs. (14.1) to (14.2a).

When the oscillator is first started, $V_f = 0$ and $V_C = K_3 V_{REF}$. Therefore, $g_m \gg g_{me}$ and V_f increases. As V_f increases, V_C decreases in accordance with Eq. (14.2) and g_m decreases in accordance with Eq. (14.1). Eventually g_m reaches g_{me} and the oscillator assumes the equilibrium state. To ensure good oscillator starting performance g_m at $V_f = 0$ should be at least $4 g_{me}$.

If V_{REF} is varied, V_f also varies in accordance with Eq. (14.2a) since V_C remains constant.

372   ALC Oscillators

From the above it is seen that it is very important that K_1, K_2, and K_3, and r_{osc} remain constant in order to maintain a constant I_x.

14.2.2 Requirements for K_3 and V_C

The values of K_3 and V_C required to achieve a specified contribution, $\partial(\Delta f/f)$, by the variation of I_x, to the frequency instability, is now computed. From Section 3.10.2

$$\frac{\Delta f}{f} = aI_x^2 \tag{14.4}$$

where a is a constant depending upon the crystal cut, from which,

$$\partial\left(\frac{\Delta f}{f}\right) = \frac{2a\,\partial I_x}{I_x}I_x^2 \tag{14.5}$$

From Eqs. (14.2a) and (14.3),

$$I_x = \frac{1}{r_{osc}K_1K_2}\left(V_{REF} - \frac{V_C}{K_3}\right) \tag{14.6}$$

from which

$$\partial I_x = -\frac{\partial V_C}{K_1K_2K_3 r_{osc}} \tag{14.7}$$

Assume that due to instability of components or environment, V_C shifts by an amount ∂V_C; then from Eqs. (14.6) and (14.7)

$$\frac{\partial I_x}{I_x} = -\frac{\partial V_C}{K_3 V_{REF} - V_C} \tag{14.8}$$

From Eqs. (14.5) and (14.8)

$$\partial\left(\frac{\Delta f}{f}\right) = -\frac{2aI_x^2\,\partial V_C}{K_3 V_{REF} - V_C} \tag{14.9}$$

For the case where $K_3 = 1$

$$\partial\left(\frac{\Delta f}{f}\right) = -\frac{2aI_x^2\,\partial V_C}{V_{REF} - V_C} \tag{14.10}$$

14.2 General Description of the ALC Oscillator

For the case where $K_3 \gg 1$ and $V_{REF} > V_C$,

$$\partial\left(\frac{\Delta f}{f}\right) = -\frac{2aI_x^2 \partial V_C}{K_3 V_{REF}} \qquad (14.11)$$

Consider the following example:
From Ref. 3.50, $a = 2 \times 10^{-8}$ for the fifth overtone best SC-cut crystal at 5 MHz. Let

$$I_x = 0.1 \text{ mA}$$

then Eq. (14.9) becomes

$$\partial\left(\frac{\Delta f}{f}\right) = -\frac{4 \times 10^{-10} \partial V_C}{K_3 V_{REF} - V_C}$$

make $K_3 = 1$ (No K_3 amplifier)

$$V_{REF} = 3 \text{ V}$$

$$V_C = 1 \text{ V}$$

then for $|\Delta(\Delta f/f)| = 10^{-12}$, ΔV_C cannot exceed 5 mV which is a very small quantity. The advantage of having a large value for K_3 becomes obvious.

Since ΔV_C is caused by the necessity for a change in $g_m = \Delta g_m$, it follows that

$$\Delta V_C = \left(\frac{\partial V_C}{\partial g_m}\right) \Delta g_m \qquad (14.12)$$

So that for any Δg_m, ΔV_C is smallest when $(\partial g_m / \partial V_C)$ is largest.

In connection with the above, it should be noted that V_C is an excellent monitoring point. If V_C changes, g_m changes in accordance with Eq. (14.1). The reason for the change of g_m should be investigated, as it may be accompanied by relatively significant frequency shifts. For example, frequency aging of the crystal will not change the g_m. However, a change in the crystal resistance and changes in the other circuit parameters will change the g_m.

14.2.3 The Approximate Noise Performance of the ALC Loop

From Fig. 14.1b, one may intuitively surmise that since V_{REF} is assumed noise-free, then V_0 is also noise-free. If K_2 and K_1 are also noise-free, then V_f must also be noise-free. Furthermore, if r_{osc} is noise-free, then I_x must also be noise-free.

It is implied in the above that since the detector is responsive only to AM signals, V_f will be only AM noise-free. However, Kulagin[14.4] has shown that the control loop also tends to reduce the FM noise because of the interdependence of AM and FM noise.

14.3 DESCRIPTION OF THE COMPONENTS OF THE ALC OSCILLATOR

This section will consider those characteristics of the components of the ALC oscillator which must be optimized for their proper functioning in the ALC circuit.

14.3.1 The Oscillator Circuit Requirements

The oscillator circuit may be any of those discussed previously. It must have the following properties:

1. It must have an input terminal to which a dc signal is fed for varying the g_m.
2. Equation (14.3) must be satisfied, and r_{osc} must be as noise-free as possible.
3. The value of crystal current is set at the compromise value most suitable for the desired short-term and long-term frequency performance.
4. The electronic circuitry is designed for lowest noise and maximum stability.

14.3.2 The Pierce Oscillator

A circuit very often used in ALC oscillators is the Pierce oscillator discussed in Chapters 5 and 7; the schematic diagram is shown in Fig. 14.2.

It is seen that it is exactly the same as that previously described except that the transistor has been replaced by an amplifying system, the g_m of which is controlled by V_C. Since the limiting function is mainly provided by circuitry external to the Pierce oscillator, this oscillator can be designed for lowest noise output and highest stability. Some of the measures that should be taken to achieve the desired performance are as follows:

1. Make X_1 very small compared to the input impedance at point 1. This is tantamount to minimizing the source resistance seen at point 1.
2. Similarly, make X_2 small compared to the output impedance at point 2.
3. Make the g_m versus V_C characteristic of Y_A as stable as possible.
4. Make the active device operate in the linear mode to minimize phase noise.

14.3 Description of the Components of the ALC Oscillator

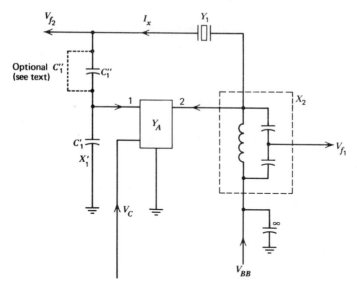

Figure 14.2 The general Pierce oscillator.

Figure 14.2 shows the output as V_{f_1} which is in the output circuit of Y_A and which is the output normally used. However, V_{f_2} has the advantage that at high Fourier frequencies, its noise content can be considerably less than that of V_{f_1}. This is due to the filtering action of the crystal at Fourier frequencies outside of the crystal bandwidth.

Figure 14.3a shows the theoretical phase noise characteristic of V_{f_1} and V_{f_2} in the frequency domain (see Section 14.4). The figure also shows the phase noise of the circuitry less the crystal $S_{\phi_\delta}(f)$. f_{δ_1} is the Fourier frequency where the flicker noise of the active device is equal to the white noise.

Figure 14.3b is the same as Fig. 14.3a except that the operating Q is larger. Both Figs. 14.3a and 14.3b are somewhat unrealistic in that they do not include the contribution in the buffer and final amplifiers additive noise, $S_{\phi_{\delta_a}}(f)$, to the overall noise. Figure 14.3c is Fig. 14.3b modified to show the effect of the amplifier noise.

To increase the V_{f_2} output, a relatively small-value capacitor, C_1'', is placed in series with C_1' and the voltage drop across it, produced by the crystal current, supplies the V_{f_2} output. This becomes a problem because as this capacitor approaches the value of the tuning capacitance it severely restricts the tuning range. This problem is discussed and solved by Burgoon and Wilson in Ref. 14.6.

Output V_{f_2} also has the advantage that the tuning in the Y_A output circuit does not affect the value of r_{osc}. Very often the output circuit tuning is used as a means of fine frequency tuning the oscillator and will incidentally influence the value of r_{osc}.

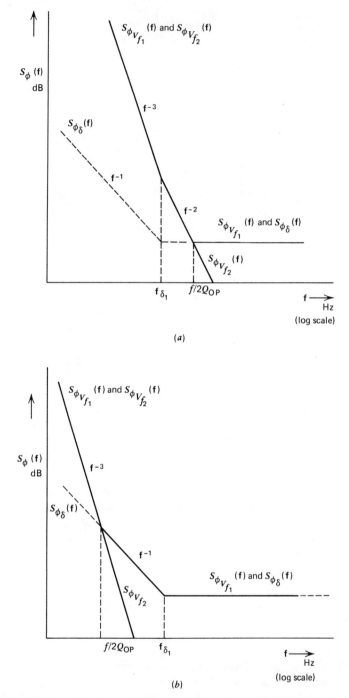

Figure 14.3 Phase noise characteristics of V_{f_1} and V_{f_2}. *Note*: All scales log. (*a*) Medium Q_{op}. (*b*) High Q_{op}. (*c*) (*b*) modified to include additive amplifier noise.

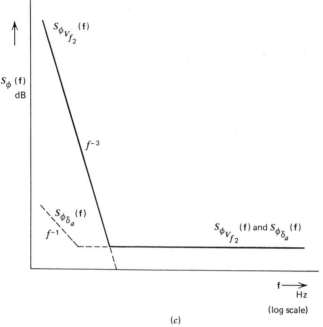

(c)

Figure 14.3 (*Continued*).

14.3.2.1 Y_A as a Voltage Variable g_m Transistor

Figure 14.4a shows the case where Y_A in Fig. 14.2 is made to be a single voltage variable g_m transistor. This is a very frequent realization of Y_A. R_E is a low-value resistor which reduces the transistor flicker phase noise.[14.10] r_2 serves to stabilize the g_m versus V_C characteristics but at the cost of a considerable reduction in the value of dg_m/dV_C.

It is interesting to compute the value of dg_m/dV_C.

$$I_E = \frac{V_C - 700}{r_2 + R_E}$$

From Eqs. (2.44) and (2.45a),

$$g_{m_0} \approx 0.038 I_E$$

where g_{m_0} is the small-signal g_m at $R_E = 0$,

$$g_{m_0} \approx \frac{0.038(V_C - 700)}{R_E + r_2} \; \mho \qquad (14.13)$$

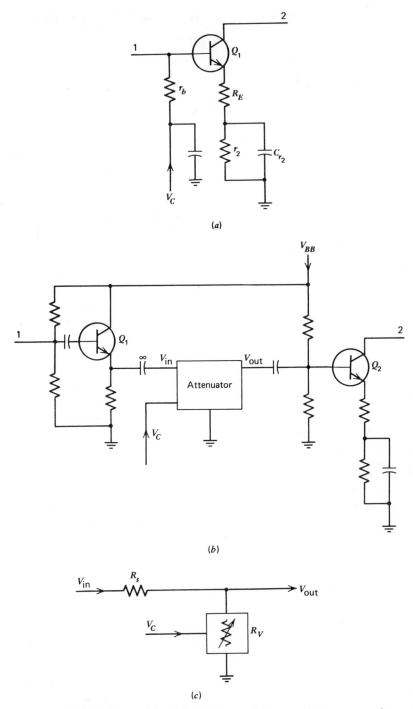

Figure 14.4 Realization of Y_A in Fig. 14.2. (a) Voltage variable g_m transistor. (b) Constant g_m transistors with auxiliary voltage variable attenuator. (c) Attenuator diagram.

14.3 Description of the Components of the ALC Oscillator

V_C is in mV, I_E is in mA and;

$$g_m = \frac{g_{m_0}}{1 + g_{m_0} R_E}$$

$$\approx \frac{0.038(V_C - 700)}{R_E[1 + 0.038(V_C - 700)] + r_2} \tag{14.14}$$

also

$$\frac{dg_m}{dV_C} \approx \frac{0.038}{R_E[1 + 0.038(V_C - 700)] + r_2} - \frac{(0.038)^2(V_C - 700)R_E}{\left[R_E[1 + 0.038(V_C - 700)] + r_2\right]^2} \tag{14.15}$$

$$\approx \frac{0.038}{R_E[1 + 0.038(V_C - 700)] + r_2} \tag{14.15a}$$

for most circuits.

Equations (14.14) and (14.15) are useful for circuit analysis and troubleshooting.

Section 14.3.2, item 4 states that the transistor should operate in the linear mode. Figure 2.12a shows that when $R_E = 0$, $V_b \leq 10$ mV. Figure 2.13b shows that higher values of V_b are permissible for the linear mode depending upon the value of R_E and I_E. However, if the signal is increased beyond these limits by increasing the crystal drive, the white phase noise will decrease but the noise at low Fourier frequencies may increase. Thus if low phase noise at Fourier frequencies greater than 10 Hz is desired, high crystal drive should be employed.

Evidence exists that for minimum noise the β_o and the f_T of the transistor should be as high as possible (see Healey, Ref. 14.5). This also has the effect that the input impedance of the transistor is increased so that the requirement of Section 14.3.2, item 1, should be easier to satisfy. The main disadvantage of using extremely high f_T transistors is that they are more prone to spurious oscillations, which may require suppression and consequently greater circuit complexity.

FETs also have desirable noise properties and consideration should be given to their use. FETs, however, have the disadvantage of having low values of dg_m/dV_C, which is not important when the FET is used as a fixed gain device.

When Q_1 is a bipolar transistor, consideration should be given to making it a PNP type as there is some evidence that *PNP* transistors have less flicker phase noise than *NPN* transistors.

14.3.2.2 Y_A as Constant g_m Transistors with Auxiliary Voltage Variable Attenuator

Figure 14.4b shows another realization of Y_A. It consist of an emitter follower feeding a voltage variable attenuator. Examples of suitable attenuators are an FET or PIN diode used as a voltage variable resistor, R_v, in series with a fixed resistor, R_s, as shown in Fig. 14.4c.

The output of the attenuator feeds a constant gain amplifying network, Q_2. This network can be designed for maximum performance since it is operating in a fixed mode.

If the voltage gain of Q_1 is assumed to be 1, then

$$g_m = \frac{V_{\text{out}}}{V_{\text{in}}} g_{m_{Q2}} \qquad (14.16)$$

An example of an oscillator using the attenuator type of Y_A is discussed by Babbitt in Ref. 14.7.

14.3.3 The ac Amplifier K_1 in Fig. 14.1b

This amplifier has the dual function of amplifying the oscillator level to a magnitude sufficient to drive the rectifier in its linear region and to supply a relatively large signal to the output amplifier.

The necessary characteristics of this amplifier are very high gain stability and noise freedom. A large amount of both local and overall negative feedback will achieve both of these characteristics. If the operating linearity of this amplifier is extremely good, it need not be tuned, provided the frequency is low enough to permit the necessary gain without tuned circuits.

The magnitude of $K_1 K_2$ is determined by Eqs. (14.2) and (14.6). Since, as shown in Section 14.3.4, K_2 has a relatively small range of values, Eq. (14.2) also determines the value of K_1.

Pustafari describes a typical amplifier circuit in Ref. 14.8.

14.3.4 The Detector K_2 in Fig. 14.1b

The detector is a rectifier which converts the ac output of K_1 to a dc signal. Its necessary characteristics are that it must be noise-free and its conversion ratio K_1 must be highly stable. To achieve both these characteristics, the circuit is usually a voltage doubler operating in its linear region, thus necessitating an input voltage of 1 V rms, minimum. The Schottky diode is recommended for best noise performance. The output must be well filtered to prevent the ac input from supplying an undesirable and unintentional coupling to the oscillator circuitry.

14.3.5 The dc amplifier K_3 in Fig. 14.1b

The gain K_3 of this amplifier is calculated as described in Section 14.2.2. Its necessary characteristics are noise freedom, gain constancy, and very low drift. It is usually a high-performance operational amplifier with massive feedback. Its output saturation voltage must be sufficiently large to ensure good oscillator starting.

14.3.6 The Reference Voltage V_{REF} in Fig. 14.1b

As seen in Section 14.2, this voltage plays a very important role. In many ALC oscillators, V_{REF} is not explicitly identifiable, but it must exist and very often it is not constant; for example, the turn-on voltage (or contact potential) of a transistor which is relatively ill-defined. It is therefore extremely desirable that V_{REF} be explicit and capable of being intentionally varied to produce the desired crystal current. Obviously, V_{REF} must be noise-free and extremely stable as already pointed out.

Felch and Israel, in Ref. 14.9, provide an analysis of the ALC system wherein V_{REF} is not explicit. They also describe a precision oscillator using this type of ALC system.

14.3.7 The Comparator

The comparator in Fig. 14.1b is a simple subtraction circuit. The subtraction function is usually obtained by making the polarity of the diodes in the detector such that its output polarity is opposite to that of V_{REF} and the sum of the two voltages is then obtained.

14.3.8 The Output Amplifier in Fig. 14.1a

The functions of the output amplifier are to

1. Provide power gain.
2. Not add discernable noise.
3. Provide isolation so that variation of the external load circuit does not produce a frequency change.

All the above functions are admirably performed by the Cascode Amplifier Circuit discussed in Section 13.3.

14.3.9 Noise Due to Components

The reader is alerted that where the best stability is required, due consideration should be given to the noise present in all components, including resistors, capacitors, inductors, transistors, and diodes. It is very difficult to make generalizations about the magnitude of the noises contributed by the components, but they may be significant. The noise magnitude depends upon the type of component, the individual component in a given lot, and upon the manufacturer. As the manufacturing technology changes, the noise also changes. For the best noise performance it is necessary to resort to selection. (See also Ref. 14.10.)

14.3.10 Output Crystal Filters

Where the noise, at Fourier frequencies relatively distant from the carrier, is considered excessive, it may be reduced by suitable crystal filters. However, care should be taken that the filter does not introduce noise due to excessive drive in the crystals in the filter.

14.4 THE PREDICTION OF THE APPROXIMATE PHASE NOISE PERFORMANCE OF THE OSCILLATOR

Figures 14.3 show the approximate phase noise performance in the frequency domain of the Pierce oscillator of Section 14.3.2. This section presents the theory on which the figures are based.

Leeson[14.11] has proposed a heuristically derived oscillator phase noise model. The validity of this model has been theoretically confirmed by Sauvage.[14.12]

According to this model,

$$S_{\phi V_{f_1}}(f) = S_{\phi_\delta}(f)\left(1 + \left(\frac{f}{2Q_{op}f}\right)^2\right) \quad (14.17)$$

Obviously,

$$S_{\phi V_{f_2}}(f) = S_{\phi V_{f_1}}(f) G_{Y_1}(f) \quad (14.18)$$

where $G_{Y_1}(f)$ is the transfer function of the Y_1, X_1 network. But

$$G_{Y_1}(f) \approx \frac{1}{1 + (2Q_{op}f/f)^2} \quad (14.19)$$

therefore

$$S_{\phi V_{f_2}}(f) \approx S_{\phi_\delta}(f)\left(\frac{f}{2Q_{op}f}\right)^2 \quad (14.20)$$

which is an extension of Leeson's model.

15

Gate Oscillators

15.1 INTRODUCTION

This and Chapter 16 treat the use of integrated circuits in oscillator design. Integrated circuits cannot be modified at will and therefore the performance characteristics of the integrated circuit oscillators cannot be tailored for the specific application as is possible in oscillators composed of discrete components. However, there are many applications wherein lesser and a wide range of performance can be tolerated, and for these applications the integrated circuit offers the advantage of economy because of the smaller part count as compared with that of the equivalent discrete element circuits.

The foregoing is not meant to imply that integrated circuits play no role in the design of high-performance oscillators. For example, Section 13.5.3 discusses a high-performance oscillator, wherein linear integrated circuits are combined with discrete active and passive elements. Furthermore, Ref. 15.1 shows a special purpose high-performance oscillator, wherein a digital integrated circuit supplies all the active elements.

This chapter discusses the use of a class of digital integrated circuits, called gates, in oscillators. The gates are primarily intended for digital use; that is, in logic, data processing, and computing systems, and are therefore rated, specified, and characterized for these applications. Digital integrated circuits, characterized for oscillator use, are discussed in Chapter 16.

Oscillators, using these gates for all the active elements, are of low to moderate quality and generally have outputs, such as TTL, suitable for driving other digital circuits. Often, the gates are spares forming part of a larger digital system, and therefore no additional cost is incurred for the oscillator active elements and many of the passive elements. Provided the performance is adequate, the cost of a gate oscillator is considerably less than that of the equivalent discrete element oscillator.

Because the gates are characterized only for digital use, there are many problems in using them for oscillators. These problems are admirably discussed by Holmbeck in Ref. 15.2. The problems are primarily due to two device characteristics: the spread of triggering levels and the propagation delay, t_{PR}, through the device. It is important to note that both of these characteristics vary with environmental conditions, power supply voltage, and with individual units. Due to the triggering characteristics of digital elements within an integrated circuit device, each individual circuit may require its individual adjustment of bias in the oscillator circuit to insure that the oscillator will always start. Also, when the digital element is biased for linear operation, the oscillator will oscillate readily at frequencies determined by the propagation delay of the element and time constants external to the element. For correct oscillator design it is necessary that the unwanted modes of operation be suppressed, a task difficult to accomplish reliably because of the variation between individual units.

15.2 CLASSIFICATION OF GATE ELEMENTS

There are many groups in which the elements may be classified. Among the important groups are as follows:

1 By type and input and output voltage levels

CMOS
TTL (bipolar)
ECL (bipolar)

Note: Under certain conditions, the CMOS and TTL gates are compatible.

2 By magnitude of t_{PR}

Approx. t_{PR} (ns)	
1	ECL III
2	ECL 10,000
3	STTL (Schottky)
5	LSTTL (low-power Schottky)
10	Standard TTL
70	CMOS

3 By magnitude of power consumption

Approx. power consumption
per gate (mW)
- 1.3/MHz at 5 V   CMOS
- 2   LSTTL (low-power Schottky)
- 10   Standard TTL
- 22   STTL (Schottky)
- 25   ECL 10,000
- 60   ECL III

4 By function

BUFFER (noninverting)
BUFFER (inverting)
AND gate (noninverting)
AND gate (inverting)
OR gate (noninverting)
NOR gate (inverting)
EXCLUSIVE OR gate (inverting and noninverting)
EXCLUSIVE NOR gate (inverting and noninverting)
LINE RECEIVER (inverting and noninverting)

Notes:

1 Above 2 MHz, the CMOS power consumption is greater than that of the LSTTL.
2 The standard TTL above is the 5400/7400 series. There are other series which may have a lower or higher power consumption and smaller or larger t_{PR}.
3 Not all functions are available in all element types.
4 Two inverting elements in cascade are equivalent to a noninverting element.
5 Manufacturers' catalogues should be consulted for the latest performance data, nomenclature, and symbols.

15.3 APPLICATION OF THE VARIOUS GATE TYPES

The CMOS gate is used below about 2 MHz, the TTL up to about 50 MHz, and the ECL above 50 MHz. The LSTTL gate is now used in many applications where STTL was formerly used.

386 Gate Oscillators

Because of the greater ease of biasing (see Section 15.4) and because of the effects of propagation delay, circuits using inverting gates have been found to be more practical than circuits using noninverting gates and are therefore in much greater use.

15.4 BIASING

As previously repeatedly stated, the oscillator small signal or the starting loop gain should be much greater than the equilibrium loop gain. Because of the noise immunity properties of the gates, the initial small-signal gain is zero, and therefore the oscillator will never start. It must therefore be biased into the quasi-linear region of operation.

15.4.1 Biasing TTL gates

Figure 15.1 shows the input output voltage characteristics for the typical TTL inverting gate. It will be noted that the desired operating point is where $V_{OUT} \approx 2$ V. This point may be obtained by connecting r_b as shown in Fig. 15.2a. Because of the variation between gates from unit to unit, r_b may have to be selected for each unit.

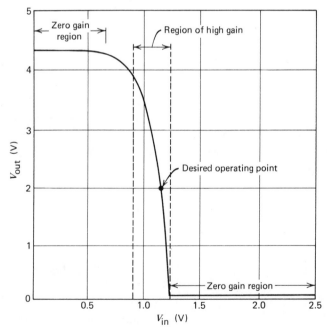

Figure 15.1 Typical TTL gate output versus input voltage characteristic.

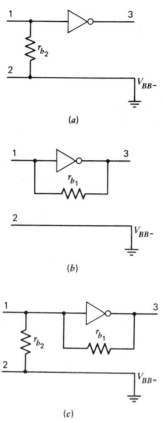

Figure 15.2 Methods of biasing TTL gates. (*a*) Method whereby the bias component does not load the resonator. (*b*) Preferred method for gate stabilization. (*c*) Combination of (*a*) and (*b*).

A more satisfactory biasing method is shown in Fig. 15.2*b*, wherein r_b is connected to provide negative feedback and will also provide a measure of stabilization for temperature variations.

Figure 15.2*c* shows the combination of Figs. 15.2*a* and 15.2*b*. In this case r_{b_1} supplies most of the bias and r_{b_2} is used to adjust the output voltage symmetry.

Often, in oscillator circuits, a resonator may be connected to points 1 and 3. r_{b_1} may then cause a serious deterioration of the resonator performance. This may be corrected by either of the two methods shown in Fig. 15.3. In Fig. 15.3*a*, L_b is inserted to make Z_b large compared to the resonator impedance. In Fig. 15.3*b*, r_{b_1} is split into r'_{b_1} and r''_{b_1} and the connecting point ac grounded by the capacitor. The result is to connect r'_{b_1} across points 3 and 2 and r''_{b_1} across points 1 and 2 which presumably already have low impedances, so that the effect of r''_{b_1} and r'_{b_1} become negligible.

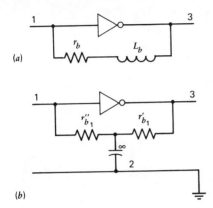

Figure 15.3 Means of eliminating the loading effect of r_{b_1}. (a) By inductor in series with r_{b_1}. (b) By partition and ac grounding.

15.4.2 Biasing CMOS Gates

Figure 15.2a is not applicable because of the internal configuration. However, Fig. 15.2b is applicable.

15.4.3 Biasing ECL Gates

Figure 15.4 shows that the proper bias is -1.3 V.

The ECL III gate is provided with a -1.3 V bias generating circuit. All that is necessary is to connect this circuit to point 1 in Fig. 15.2 via the proper ac isolating means.

The ECL 10,000 series gates must be provided with the power supply voltage divider necessary to generate the -1.3 V which is then applied to point 1 in Fig. 15.2 via the required ac isolating means.

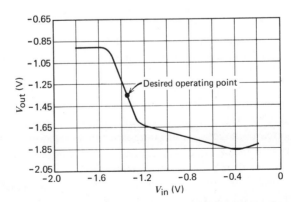

Figure 15.4 Typical ECL gate output versus input voltage characteristic.

15.5 CONVERSION OF GATES HAVING DIFFERENT FUNCTIONS INTO EQUIVALENT INVERTERS

Usually the simple inverter function is all that is required for an oscillator design. Often, other types of spare gates are available. This section demonstrates how to convert these other types into inverters.

Figure 15.5 shows how to convert NANDs, NORs, EXCLUSIVE ORs, and NORs into the equivalent inverter shown in Fig. 15.2. The inverter of Fig. 15.5a has a higher input impedance and is therefore suitable for an antiresonant oscillator circuit. The inverter of Fig. 15.5b has a lower input impedance and is therefore more suitable for series resonant oscillator circuits.

After having been converted, they are biased and otherwise treated as the inverters of Figs. 15.2 and 15.3.

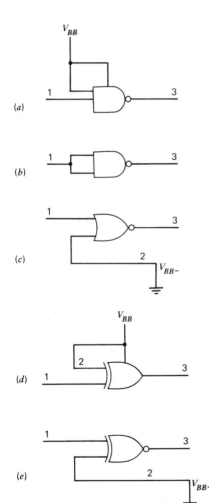

Figure 15.5 Connections for converting NAND and NOR gates into inverters. (a) NAND gate. (b) NAND gate for lower input impedance. (c) NOR gate. (d) Exclusive OR gate. (e) Exclusive NOR gate.

15.6 THE INVERTER APPROXIMATE EQUIVALENT CIRCUIT

Figure 15.6a shows the equivalent circuit of the inverter. In this circuit A is the gain between points 4 and 1 and is a function of frequency. It is seen from Figs. 15.1 and 15.4 that at low frequencies

$$A \approx 16 \quad \text{for the TTL gate}$$

$$A \approx 4 \quad \text{for the ECL gate}$$

In general

$$A = |a|\angle\theta \tag{15.1}$$

where both $|A|$ and θ are functions of f, and

$$\theta \approx -360°t_{PR} f \tag{15.2}$$

t_{PR} is in μs and f is in MHz.

From Eq. (15.2) it is seen that the gate becomes noninverting when

$$f_{NI} \approx \frac{1}{2t_{PR}} \tag{15.3}$$

No information is given in the normal gate description to evaluate the variation of A with f, but it can be stated that A decreases as f increases.

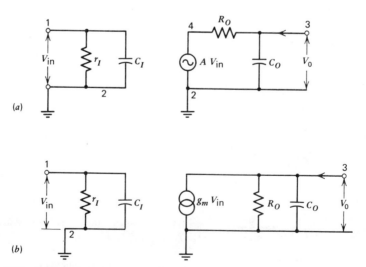

Figure 15.6 Inverter equivalent circuits. (a) Basic circuit. (b) Norton's equivalent of (a).

R_O has a low value and is approximately 5 Ω for the MEC III gate and 30 Ω for the standard TTL gate.

These values are at low frequencies and will increase and become complex as the frequency increases.

C_O is approximately 30 pF
C_I is approximately 7 pF
r_I is approximately 1200 Ω
All at low frequencies for standard TTL gates.

It should be noted that the input portion of the equivalent circuit is only valid where the effect of r_b is rendered negligible as in Fig. 15.3a. Otherwise,

1. r_I is in parallel with r_{b_2} in Fig. 5.2a.
2. r_I is in parallel with r_{b_1}/A in Fig. 15.2b by virtue of the Miller effect.
3. r_I is in parallel with r_b'' in Fig. 15.3b.

Upon applying Norton's theorem to the circuit of Fig. 15.6a, the circuit of Fig. 15.6b results. It will be noted that it is quite similar to the equivalent circuit of the transistor of Figs. 2.6a and 2.11. In Fig. 15.6b,

$$g_m = \frac{A}{R_O} \qquad (15.4)$$

Therefore much of the theory developed in Chapters 5 through 10 can be applied to the gate oscillator, at least qualitatively.

It is important to be aware of the important differences between the transistor and gate equivalent circuits:

1. The gate has a much larger g_m.
2. The gate θ_{g_m} is much larger.
3. R_O is much smaller than the equivalent r_{ce} of the transistor.
4. C_O is much larger than the equivalent C_{ce} of the transistor.

These differences cause major changes in the respective oscillator design procedures.

15.7 TYPE OF LIMITING USED IN GATE OSCILLATORS

The limiting means is analogous to the collector base voltage limiter described in Section 6.2.4. As a result, the output is relatively independent of the resonator resistance, and the frequency is relatively voltage sensitive.

15.8 GATE OSCILLATORS WITH CRYSTALS OPERATING IN THE INDUCTIVE REGION

This type of oscillator is more useful than that in which the crystal operates at series resonance. This is because the circuit tuning elements tend to suppress the spurious oscillations described in Section 15.1.1. However, it requires more external components than the series resonant oscillator.

Figure 15.7a shows the basic model for this class of oscillators and Fig. 15.7b shows a typical realization. It will be recognized as the Pierce oscillator described in Chapters 5 and 7. The major difference is that the load resistance R_L is quite small, which produces a slightly smaller operating Q.

It is interesting to compute the approximate frequency shift due to the gate propagation delay.

Consider the following example:

$$f = 10 \text{ MHz}$$

$$Q_{op} \approx 25{,}000$$

the gate is the standard TTL type, and

$$t_{PR} = 10 \text{ ns}$$

Therefore

$$\theta_{g_m} = -36°$$

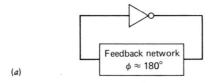

(a)

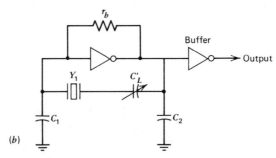

(b)

Figure 15.7 Oscillators with the crystal operating in the inductive region. (a) Basic model. (b) Typical realization of (a).

15.8 Gate Oscillators with Crystals Operating in the Inductive Region

from Eq. (15.2) and

$$\frac{\Delta f_{\theta_{g_m}}}{f} \approx \frac{\tan(-36°)}{2(25,000)} = -15 \times 10^{-6}$$

from Eq. (5.18).

It should be remembered that t_{PR} is variable for many reasons; therefore its contribution to the frequency becomes a cause of frequency instability. Usually C'_L can be adjusted to compensate for the mean value of $\Delta f_{\theta_{g_m}}/f$.

An equivalent circuit can be substituted for the actual circuit. In the equivalent circuit, $\theta_{g_m} = 0$ and an inductor L_{eq} is connected in series with the crystal (see Section 19.8 and Ref. 15.2, Fig. 8) where

$$L_{eq} \approx \frac{2 \times 10^6 \Delta f_{\theta_{g_m}}/f}{(2\pi f)^2 C_1} \quad (\mu\text{H}) \tag{15.5}$$

where C_1 is the crystal motional capacitance. L_{eq} is often considered a figure of merit of the circuit although it obviously is also a strong function of the crystal characteristics.

If the crystal is an overtone, then C_1 or C_2 can be replaced by one of the networks described in Section 5.6.4 to select the desired overtone.

Figure 15.8 shows the gate oscillator version of the isolated Pierce oscillator described in Chapter 8. R_s has the following functions.

1. It provides a means of setting the crystal drive.
2. It serves to isolate the gate properties from the oscillator, so that much of the theory developed in Chapter 7 is applicable. However, the θ_{g_m} of the gate must be taken into consideration in the calculations.

At high frequencies, R_s may be replaced by a capacitor C_s to compensate for the gate propagation delay.

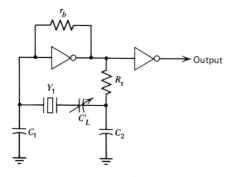

Figure 15.8 Gate oscillator version of the isolated Pierce oscillator.

15.9 GATE OSCILLATORS WITH CRYSTALS OPERATING NEAR SERIES RESONANCE

Figure 15.9a shows the basic model for this class of oscillators, and Fig. 15.9b shows a practical realization.

These oscillators, under the most favorable conditions, have the least part count, excluding the active elements, and are therefore considered desirable. However, in general their long-term frequency performance is inferior to those of Section 15.8 in which C_1 and C_2 tend to swamp the contribution of the gate to the frequency. Furthermore, they are even more prone to spurious oscillation than the oscillators of Section 15.8 because of the absence of C_1 and C_2, which limit the frequency regions of possible oscillation, and because of the higher gain due to the two gates in cascade.

In this circuit the gate inputs are connected in parallel to reduce the input impedance to better match the crystal.

Because of the two gates in cascade

$$\theta_{g_m} = 2(0.36 ft_{PR}) \tag{15.6}$$

and, therefore, for the same example in Section 15.8

$$\theta_{g_m} = -72°$$

and

$$\frac{\Delta f}{f} \theta_{g_m} \frac{\tan(-72°)}{2(25,000)} = -62 \times 10^{-6}$$

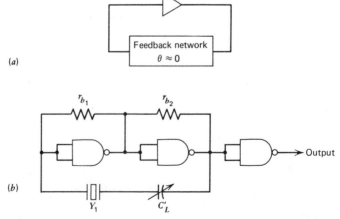

Figure 15.9 Gate oscillator with the crystal operating near series resonance. (a) Basic model. (b) Typical realization of (a).

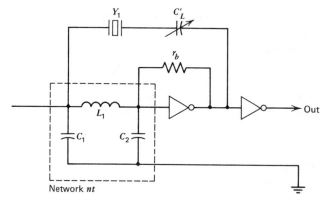

Figure 15.10 Typical high-frequency oscillator for crystal network operating near series resonance.

which is over four times that in the example in Section 15.8. Also, L_{eq} will be four times as large and C'_L will therefore become much smaller.

The oscillator of Fig. 15.9 is useful only up to moderately high frequencies as at higher frequencies the gate phase shift becomes excessive.

The oscillator of Fig. 15.10 is more suitable for high-frequency operation. In that oscillator, the required phase shift, θ, is made up of $180° + \theta_{g_m} + \theta_{nt}$.

The network, nt, performs the following functions:

1. Provides the phase shift $180° - \theta_{g_m}$.
2. Matches the gate input impedance to the crystal.
3. Selects the desired crystal overtone.

15.10 CLOSING REMARKS

The material has been presented to provide the reader with some basic theory dealing with gate oscillators. The theory, in general, will yield only qualitative results and will offer little help toward solving the problems of spurious oscillators which can only be accomplished experimentally. Accordingly, when designing gate oscillators for production it is necessary to conduct extensive experimentation to ensure the absence of spurious effects. It is quite difficult to completely eliminate the tendency for these spurious effects and a great deal of component modification may be necessary, as well as layout changes, addition of components for the sole purpose of spurious elimination, and improvements in the supply voltage bypassing. Even the choice of which gate on the chip is used may have an impact in the performance. In many cases, when spare gates are not available, it is found that the component count, using more gates for the active elements, may exceed the parts count for the more reliable discrete element circuitry or equivalent.

396 Gate Oscillators

Those interested in the measurement of the approximate circuit equivalent parameters are referred to Chapter 19, particularly Section 19.8.

In view of the rapid development of integrated circuitry, it is possible and highly probable that much of the material on specific devices in this chapter, as well as in Chapter 16, may be rendered obsolete with the passage of time, and much improved performance may become obtainable. However, the general theory and comments will always be applicable.

16
Integrated Circuit Oscillators

16.1 INTRODUCTION (SEE ALSO SECTION 15.1)

This chapter covers the integrated circuits which are explicitly rated and described for use in oscillators, in contrast to the gate oscillators of Chapter 15. These integrated circuits require only the addition of a crystal and perhaps some capacitors and overtone/mode selector to make up a complete *oscillator* which provides a rated output into a rated load. Sometimes the complete *oscillator* is included within the chip enclosure. Sometimes the oscillator function is only part of the total chip functions which may include dividers, phase detectors, and so on, but in that case specific information is given for the oscillator function.

The information available for using these chips ranges from extremely poor to extremely good. The performance of the chips tends to improve as the quality of the information improves. The minimum information supplied are the pin connections, power supply requirements, and the recommended operating frequency range. The latter information is insufficient to produce a satisfactory oscillator design. The minimum required, in addition, are descriptions of the recommended crystal characteristics, the total oscillator performance with these crystals as a function of power supply, and environmental conditions, assuming the crystal performance to be independent of environmental conditions.

Because of the large number of circuits introduced and/or available, a detailed treatment of all the circuits is impractical. A number of general principles and comments are stated, which should be helpful in the design and application of the circuits. Examples of some popular readily available circuits are given and references are cited which discuss their measured performance, which is often at variance with the circuit manufacturer's specifications. In view of the large discrepancies that often exist between the measured and specification performances, it is highly recommended that each lot of circuits be carefully screened to ensure performance adequate for the application.

The chapter discusses the following types of oscillators:

1. Gate oscillators not considered in Chapter 15.
2. Linear integrated circuit oscillators.
3. Miniature oscillators which include the crystal in the same package that contains the circuitry. The circuitry may be gate type, linear integrated circuit type, and/or hybrid types only, which may be composed solely of thick-film discrete elements or discrete components combined with integrated circuits.

16.2 GATE OSCILLATORS WITH DIGITAL OUTPUT

The operating principles of this class of oscillator is covered in Chapter 15. Several examples of commercially available oscillators (in the year 1981) are now discussed.

16.2.1 Clock Oscillator 8224 of the 8080 Microprocessor System

This oscillator normally operates at 4 MHz. The circuit is the series resonant type. Extensive tests were made at various frequencies and are reported by Holmbeck in Ref. 15.2. Curves of L_{eq} versus f from 4 to 12 MHz are included. L_{eq} is shown to increase as f increases. However, these curves do not state the crystal current or the characteristics of the crystals used for obtaining the data. Repeatability between different units is reported as being poor, particularly at the higher frequencies.

16.2.2 The 74124 and 74324 and Their Schottky and Low-Power Schottky Version Oscillators

These are basically voltage-controlled oscillators. When a crystal is provided instead of the frequency-determining capacitor, the oscillator is intended to operate as a crystal-controlled oscillator. As VCOs, they are rated for operation from several Hz to 30 MHz. As crystal oscillators, they will operate up to 20 MHz, depending upon the gates or chips, but up to several thousand ppm below the crystal frequency. No data is given for the required crystal characteristics.

Reference 15.2 includes curves of L_{eq} versus f from 10 to 20 MHz for the Schottky version. The repeatability between units also is reported to be poor.

In general these circuits do not appear very attractive for crystal oscillator use.

16.2.3 The Plessy SP 705 Oscillator

This is a series resonant crystal oscillator provided with $\div 2$ and $\div 4$ output circuits. These units are rated to operate from 1 to 10 MHz. The oscillator is discussed in Ref. 13.7 and is reported to be unreliable.

16.2.4 "Wristwatch" Oscillators

These oscillators are examples of the successful application of gates to oscillators. The oscillators use the antiresonant circuits of Fig. 15.7 and require two capacitors. One of the capacitors may be contained within the chip. The remaining capacitors and the crystal are external to the chip. The oscilllators operate in the frequency range from 30 kHz to 1 MHz and possess moderate and adequately repeatable performance. The descriptive data includes information on the required crystal characteristics.

The oscillators form part of the complete chip which also includes dividers, drivers, and so on. The chips generally are CMOS to conserve power, which increases as the oscillator frequency increases.

Examples of these "wristwatch" chips are as follows:

1 Motorola MC 14450 Oscillator/2^{16} Divider/Buffer for a 32-kHz crystal.
2 Motorola MC14451 Oscillator/2^{11} to 2^{19} Divider/Buffered Duty Cycle Control for 30 kHz to 1 MHz.

16.3 LINEAR INTEGRATED CIRCUIT OSCILLATORS

The operating principles of these oscillators are covered in Chapters 5 to 13. The oscillators may consist of

1 Discrete passive and active components combined with integrated circuit active elements. The design is identical and the performance is equal to or only slightly inferior to the oscillators described in those chapters. Reference 16.1 describes the application of some general purpose linear integrated circuits to oscillator design.

2 Integrated circuit chips which require only the addition of some capacitor and/or overtone/mode selector to make up a complete *oscillator* which provides a rated output into a rated load. The output signal may be sine wave or digital. Also, the oscillator may be part of a chip containing a system. Meyer and Soo present, in Ref. 16.8, excellent detailed treatments of the design of this type of oscillator, using NMOS active elements and the NMOS large-signal characteristics.

Oscillators of the first type have already been completely described in the previous chapters. Several examples of commercially available oscillators of the second type are now discussed.

16.3.1 National DS 8907 Digital Phase-Locked Loop Frequency Synthesizer

This chip contains the greater part of the frequency synthesizer for FM/AM receivers. The internal Colpitts oscillator operates at 4 MHz. The crystal and the oscillator capacitors are external to the chip. The performance is moderate and repeatable. Many of these chips are in successful use.

16.3.2 The MC-12060, MC 12560, 12061, and the 12561 Oscillator Groups

These are chips which only require an external crystal and bypass capacitors to produce a complete *oscillator*. The outputs are sine wave, TTL, and ECL. The −60 series is intended for frequency ranges of 100 kHz to 2 MHz, and the −61 series for 2 to 20 MHz. The data sheets include considerable information, including performance and crystal requirements.

The oscillator circuit is a modification of the series resonant circuit described in Section 13.5.4, but is provided with ALC. The performance of these oscillators is discussed in Refs. 13.7, 15.2, and 16.2, wherein they are reported as being better than most IC oscillators but not in compliance with the ratings. Also, considerable care must be taken in their application.

16.3.3 The Plessy SL 680A Oscillators

This unit is rated to operate from 100 kHz to 100 MHz. The crystal operates near series resonance. The circuit includes ALC and is unique in that it maintains the crystal current constant at about 0.2 mA. The output is sine wave. The performance of this oscillator is discussed in Ref. 13.7 wherein it is reported that the unit operated reliably at the fundamental overtone, up to about 18 MHz, but at frequencies considerably lower than the crystal f_s.

Reference 16.3 describes another application of this unit in a 10-MHz temperature-controlled *oscillator*.

16.4 MINIATURE PACKAGED CRYSTAL *OSCILLATORS*

For many years considerable effort has been expended toward the development of miniature packaged oscillators in which the crystal is within the *oscillator* enclosure. This section presents a sampling of the *oscillators* which have resulted from these efforts.

16.4.1 Microminiature *Oscillator* in TO-5 Enclosure

Reference 16.4, which was published 16 years ago, reports some early efforts in microminiaturization. The *oscillator* operated at 10 MHz used the Colpitts circuit and had moderate performance.

The paper contains a general discussion of the relative merits of monolithic, compatible, and hybrid integrated circuitry. The *oscillator* described used hybrid circuitry. The crystal was unpackaged and used the same TO-5 enclosure for hermetic sealing.

16.4.2 Tactical Miniature Crystal Oscillator

Reference 16.5 describes a high-performance ovenized 5.115-MHz miniature *oscillator*. It uses hybrid circuitry and the crystal is sealed in a flat pack ceramic enclosure adding little to the overall crystal size. The oscillator uses the Colpitts circuit with ALC.

The paper includes much valuable information on component performance and techniques for setting the oscillator frequency to the rated frequency at the crystal temperature turnover point.

16.4.3 Miniature Packaged Crystal *Oscillators* of Ref. 16.6

Reference 16.6 describes a series of miniature *oscillators* for the frequency range of 0.5 to 22 MHz. The performance is moderate, being ± 100 ppm over a temperature range of -40 to $95°C$. The output is TTL. The oscillators having outputs below 8 MHz use crystals in the 8 to 22-MHz region combined with frequency dividers. The crystals operate near series resonance.

The circuitry is hybrid, which includes a monolithic silicon integrated circuit. The *oscillators* are housed in TO-8, TO-5, and DIP enclosures, which also provide the crystal hermetic sealing.

The paper is excellent for describing the present state of development of miniature *oscillators*. It also includes much information on the theoretical and practical aspects of oscillator and TTL output circuit design.

16.4.4 Military Specification Miniature *Oscillators*

Reference 16.7 describes a series of miniature *oscillators*, covering the wide frequency range of 1 kHz to 50 MHz. The lower frequencies are obtained by means of frequency dividers. The *oscillators* are housed in a hermetically sealed DIP can. The output is TTL. The rated performance is moderate, being

1. Initial accuracy— ± 15 ppm at $25°C$.
2. Frequency stability versus temperature— ± 25 ppm from 0 to $70°C$; ± 50 ppm from -55 to $125°C$.
3. Frequency stability versus supply voltage— ± 5 ppm for a 0.25-V change in supply voltage which is nominally 5 V.

17

Circuitry Frequency Stability Requirements in Crystal Oscillators, Excluding the Crystal

17.1 INTRODUCTION[17.1]

This chapter considers the crystal oscillator as a system composed of several subsystems. The subsystems must be completely specifiable as regards inputs, outputs, interfaces, and performance.

Preferably, the subsystem stability performance should be describable in the same terms as those of the system. Frequently this is impossible, undesirable, or inconvenient. In that case, relationships, preferably simple, should be established between the terms of the subsystem performance and the terms of the system performance.

This chapter identifies the minimum subsystems present in all crystal oscillators, and, indeed, in all oscillators, and relates the stability of each subsystem to the total oscillator stability.

All crystal oscillators are made up of at least two distinct subsystems.

1. The crystal resonator hereafter also called the osci (see Section 1.3.1, item 1).
2. The remaining circuitry, hereafter called the llator.

It would be very desirable to evaluate the stability of each one independently. In the present state-of-the-art, it is not feasible to simply and completely evaluate the crystal by itself, although extensive work has been carved out by Walls, Wainwright, and others to independently determine the crystal short-term performance.[3.44, 3.59] However, it turns out that the llator is readily amenable to such evaluation and that quantitative limits can be set for the performance of the llator, relative to the oscillator performance, under certain operating conditions of the llator and the crystal described later.

This chapter develops the approximate relations between the oscillator and llator performance for a given crystal and describes the experimental technique for measuring the llator performance. The derivation is relatively simple but it yields considerable information concerning the design of oscillators and the measurement of the llator performance.

It should be noted that the material in this chapter is particularly applicable to the medium and long-term frequency stability but does include some treatment of the short-term stability.

The llator evaluation is useful for the three following broad categories of applications:

1 *In new designs*, it permits the budgeting of the llator for an overall oscillator stability performance and the experimental confirmation of whether the budget is being met.

2 *In old designs*, or in manufacture of oscillators based on already available designs, the situation often arises where the oscillator performance is unsatisfactory and the crystal designer is convinced that the fault lies in the circuitry, while the circuit designer is equally convinced that it is a crystal problem. At present, the solution for this dilemma has been to replace the crystal and/or circuitry until satisfactory performance has been achieved, usually at great cost of time and money. The llator evaluation procedure would markedly facilitate the resolution of this problem.

3 It provides information on the crystal stability performance.

17.2 THE OSCI AND LLATOR CONCEPTS (*See Section 1.3.1, Item 1*)

Figure 17.1 shows the block diagram of a typical oscillator using the llator concept. It is seen that it consists of only two subsystems:

1 The osci.
2 The llator.

In a crystal oscillator, the osci would be a crystal. In a noncrystal harmonic oscillator, the osci would be a resonator or part of a resonator. The llator is the

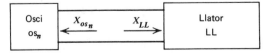

Figure 17.1 Oscillator block diagram.

rest of the oscillator including the oven and any other means for maintaining the electrical, mechanical, and thermal environments.

A very important and necessary condition for the llator evaluation is that the osci is considered "perfect," meaning that the osci is completely stable, that is, independent of the environment and time; this condition can usually be practically satisfied during the time necessary for measuring the llator performance.

Obviously the total instability of the actual oscillator is the sum of the llator instability and the actual osci instability. Conversely, the instability of the osci is the difference between the actual oscillator instability and instability of the llator referred to the perfect osci.

It is noted from Fig. 17.1 that the osci has associated with it a reactance X_{os} which is a function of only the operating frequency, f. Also the llator has associated with it a reactance, X_{LL}, which is a function of not only the physical reactances in the llator but also includes the contribution of the active circuitry and the environmental conditions.

17.3 OSCILLATOR RELATIONSHIPS USING THE OSCI LLATOR CONCEPT

For a given osci, os_n, and a llator, LL, in Fig. 17.1

$$\frac{\Delta f_{os_n}}{f} = \frac{\Delta f_{os_n}}{\Delta X_{os_n}} \cdot \frac{\Delta X_{os_n}}{f} \qquad (17.1)$$

and from Eq. (1.48)

$$X_{os_n} = -X_{LL} \qquad (17.2)$$

from which

$$\Delta X_{os_n} = -\Delta X_{LL} \qquad (17.3)$$

for

$$\Delta f \ll f \qquad (17.4)$$

f can be assumed constant

Equation (17.3) states that any ΔX_{LL} produced by instability of the llator will be counterbalanced by an equal $-\Delta X_{os_n}$ which thus produces a $\Delta f_{os_n}/f$ in accordance with Eq. (17.1).

If the osci is made to be a crystal, then in accordance with Section 3.2

$$\Delta X_{xtal} \approx 2 Q_{xtal} R_{xtal} \frac{\Delta f_{xtal}}{f} \qquad (17.5)$$

so that from Eqs. (17.3) and (17.5),

$$\frac{\Delta f_{xtal}}{f} \approx \frac{\Delta X_{LL}}{2Q_{xtal} R_{xtal}} \qquad (17.6)$$

17.4 CRYSTAL OSCILLATORS WHEREIN THE CRYSTAL NETWORK OPERATES IN THE INDUCTIVE REGION

Consider an oscillator wherein the crystal network operates within its inductive region, such as the Pierce oscillator and investigate the effect of replacing the crystal with the perfect os_2 shown in Fig. 17.2.

L in os_2 is chosen so that the frequency of oscillation is approximately that of the crystal oscillator.

Obviously,

$$X_{os_2} = 2\pi L f \qquad (17.7)$$

and

$$\Delta X_{os_2} = 2\pi L \Delta f_{os_2} \qquad (17.8)$$

so that from Eqs. (17.2), (17.3), (17.7), and (17.8)

$$\frac{\Delta f_{os_2}}{f} \approx \frac{\Delta X_{LL}}{X_{LL}} \qquad (17.9)$$

Combining Eq. (17.5) and (17.9),

$$\frac{\Delta f_{os_2}}{f} = \frac{-\Delta f_{xtal}}{f} \cdot \frac{2Q_{xtal} R_{xtal}}{X_{LL}} \qquad (17.10)$$

or the converse

$$\frac{\Delta f_{xtal}}{f} = \frac{-\Delta f_{os_2}}{f} \frac{X_{LL}}{2Q_{xtal} R_{xtal}} \qquad (17.10a)$$

The significance of Eq. (17.10) is illustrated by the following examples:

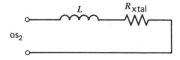

Figure 17.2 Osci os_2.

Example 1

$Q_{xtal} = 100{,}000$, $R_{xtal} = 20\ \Omega$, $X_{LL} = -100\ \Omega$ and $\Delta f_{xtal}/f$ desired $= 10^{-9}$, then

$$\frac{\Delta f_{os_2}}{f} = 4 \times 10^{-5}$$

Example 2

$Q_{xtal} = 2 \times 10^{6}$, $R_{xtal} = 100\ \Omega$, $X_A = -400\ \Omega$ and $\Delta f_{xtal}/f$ desired $= 10^{-12}$, then

$$\frac{\Delta f_{os_2}}{f} = 10^{-6}$$

It is interesting to note from Eq. (17.10) that for maximum permissible llator instability, Q_{xtal} and R_{xtal} should be maximized and X_{LL} should be minimized. It follows that, for equal Q_{xtal}, the crystal with greater R_{xtal} (which is equivalent to smaller motional capacitance C_1) will permit greater llator instability. Example 2 demonstrates the severe requirements imposed on the llator. This is emphasized by noting that the Q of the oscillator with os_2 is less than 4 for a stability of 10^{-6}, an achievement which offhand appears impossible but direct experiment has proven to be feasible, a fact corroborated by the existence of crystal oscillators having stabilities better than 10^{-12}.

17.5 EXPERIMENTAL PROCEDURE FOR THE OSCILLATOR OF SECTION 17.4

An oscillator is available with known crystal parameters and a required stability. The problem is to determine whether the llator is adequate for this required stability.

The procedure is as follows:

1 Remove the crystal.

2 Replace the crystal by os_2 making R the same as R_{xtal} and making L a value which will yield approximately the same operating frequency. This is facilitated by determining the value of X_{LL} with the H.P. 4815 Vector Impedance Meter and calculating and trimming L until the right frequency is obtained. When the oscillator is energized, the current in os_2 should be approximately the same as in the crystal. os_2 is realized by a stable inductor in series with a resistor and this normally is a relatively easy procedure, provided

17.6 Crystal Oscillators Wherein the Crystal Network Operates Near Series Resonance

that care is taken to maintain the environmental conditions for os_2 constant. The oscillator output frequency is then monitored by a frequency counter and the stability $\Delta f_{os_2}/f$ determined. The resolution of the counter need not be very high since a relatively unstable oscillator frequency is being measured.

3 The value of $\Delta f_{os_2}/f$ determined above is compared against that called for by Eq. (17.10). It if is less, then the llator is satisfactory. Otherwise efforts should be exerted to improve it as necessary.

17.6 CRYSTAL OSCILLATORS WHEREIN THE CRYSTAL NETWORK OPERATES NEAR SERIES RESONANCE

In this oscillator, the osci should take the form shown in Fig. 17.3. L_1 and C_1 are tuned to the approximate operating frequency.

For this osci in the region of series resonance,

$$\Delta X_{os_3} = 2Q_{os_3} R_{xtal} \frac{\Delta f}{f} \qquad (17.11)$$

where

$$Q_{os_3} = \frac{X_{C_1}}{R_{xtal}} \qquad (17.12)$$

and Eq. (17.10a) becomes, for this osci,

$$\frac{\Delta f_{xtal}}{f} = \frac{\Delta f_{os_3}}{f} \frac{Q_{os_3}}{Q_{xtal}}$$

The rest of the analysis and the measurement procedures are the same as for os_2.

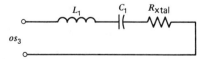

Figure 17.3 Osci os_3.

17.7 EXPERIMENTAL DETERMINATION OF THE OSCILLLATING POINT IN THE OSCILLATOR USING THE OSCI AND LLATOR CONCEPTS

Following the procedure described in Section 19.8, the llator impedance, $Z_{LL} = R_{LL} + jX_{LL}$, is measured at the nominal I_x and operating frequency. When $-R_{LL}$ is plotted against $-X_{LL}$ a curve, similar to that shown in Fig. 17.4a, is obtained. An approximation for this curve can be calculated from Eq. (5.38) wherein $-R_{LL}$ and $-X_{LL}$ are functions of the parameter g_m.

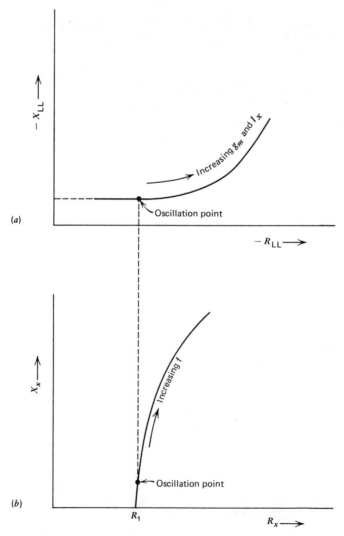

Figure 17.4 Determination of the point of oscillation. (a) X_{LL} versus R_{LL}. (b) X_x versus R_x.

Another curve of R_x versus X_x can be obtained for the crystal by measurement or by calculation from the material in Chapter 3. A typical curve is shown in Fig. 17.4b.

When the two curves are superimposed, it is seen that oscillation occurs where R_x coincides with $-R_{LL}$, in the flat region of $-X_{LL}$. The corresponding X_x is then read off Fig. 17.4b and the operating frequency computed from the crystal characteristics.

17.8 CONDITIONS FOR THE APPLICABILITY OF SECTIONS 17.1 TO 17.7

The material presented thus far is only applicable to the medium- and long-term frequency stability since the analysis does not include the crystal filtering action and the additive phase noises such as those described in Section 14.3.2 and Fig. 14.3.

Also, some practical problems are created during the analysis and measurement procedures when the crystal is replaced by another osci which markedly decreases the oscillator operating Q. Among these problems are the following:

1 The analysis does not take into consideration the C_0 of the crystal.

2 During the measurement procedure, some spurious oscillation, because of the new osci, may be caused and must be suppressed. A greater problem is that the oscillator may squegg because the shorter time constant of the new osci may be smaller or the same order of magnitude of the other time constants in the oscillator, thus causing squegging. This squegging will not exist in the actual oscillator because of the very high Q of the crystal which produces a much larger time constant. The squegging may be corrected by changing some of the time constants in the oscillator biasing and filtering circuits (see Section 18.4).

17.9 SHORT-TERM PERFORMANCE

The long-term performance has been described, up to now, in the time domain. However, the short-term performance is more conveniently analyzed in the frequency domain. After the analysis has been completed the results may be translated into the time domain. Also, it may be more convenient to obtain the raw experimental data in the time domain; the data will then require translation into the frequency domain.

When analyzing the short-term performance the noise contributions of many more subsystems require consideration. Among these subsystems are the ALC amplifiers (if present), the output amplifiers, the oven (if any), and the power supply.

An approximate evaluation of the circuit noise $S_{\phi_s}(f)$ can be obtained by replacing the crystal with a type os_3 osci (see Fig. 17.3). The operating Q, Q_{os_3}, and the $S_\phi(f)$ of the oscillator with the os_3 osci are measured.

$S_{\phi_s}(f)$ is then computed by means of the converse of Eq. (14.17) or (14.20), as applicable.

Knowing the Q_{os_3}, the oscillator operating Q with the crystal can be computed from the crystal parameters. The $S_\phi(f)$ of the oscillator with the crystal is then computed from Eq. (14.17) or (14.20) as applicable.

It should be remembered that the above procedure will yield only the $S_\phi(f)$ of the oscillator for the case where the crystal is considered perfect; that is, the crystal itself does not generate significant noise and does not exhibit frequency changes due to change in time or environment.

18

Special Problems in Oscillators

18.1 INTRODUCTION

This chapter considers some problems encountered in oscillators. Some of these problems exist in all high-frequency circuitry. Others are peculiar to oscillators. Many of these problems are due to transistor parasitic elements and to the oscillator physical layout and could, therefore, not be included in the design procedures. Others could have been anticipated by the designer and the design should therefore have been performed so that these problems are not encountered.

The chapter discusses the probable cause of these problems and suggests some possible cures.

18.2 SPURIOUS OSCILLATIONS

Spurious oscillations are oscillations other than those due to the crystal responses. Most spurious oscillations are caused by the parasitic inductance and capacitances in the transistors, by the parasitic inductances and capacitances created by poor design and circuit layout, and by ground loop couplings caused by poor design and circuit layout.

Oscillations are also caused by inductors used for adjusting the oscillator frequency and for neutralizing the C_0 of the crystal.

18.2.1 Transistor Parasitic Elements

Spurious oscillations due to transistor parasitic elements can be eliminated by resistors placed in series with the transistor collector and base circuits, preferably located adjacent to the transistor. At very high frequencies, lossy ferrous beads are effective.

412 Special Problems in Oscillators

18.2.2 Layout Parasitic Elements and Couplings

One of the most important steps in designing the completed oscillator is making the physical layout of the components for the oscillator. Unfortunately, this task is usually relegated to a draftsman who makes the layout on the basis of fitting the components into the available space with little or no regard to, and understanding of, the lead inductances and capacitances and undesirable couplings between the various circuits. This causes spurious oscillations as well as poor isolation. These spurious oscillations an be eliminated by additional bypassing and inserting lossy elements in the circuits, but at the cost of deteriorating the basic oscillator performance. It is not a rare occurrence for the breadboard model to operate well but for the final layout model to be plagued by spurious oscillations. This indicates that more attention should have been paid to the layout.

18.2.3 Spurious Oscillations Due to Poor Design

Poor design includes insufficient or incorrect decoupling and the creation of unnecessarily large circulating currents in the ground plane. Figure 18.1 shows an example of a tuned circuit having a Q of about 10. Figure 18.1a shows the correct method of bypassing since the bypass capacitor, C_{bp}, and the ground plane carry only 1 mA. Figure 18.1b shows the incorrect method since C_{bp} and the ground plane carry 9 mA, which can cause much larger undesirable coupling to adjacent circuits. In addition, the latter method also requires a larger value and better C_{bp} since it is a larger part of the tuned circuit.

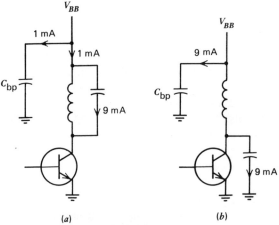

Figure 18.1 Examples of bypassing. (a) Correct. (b) Incorrect.

18.2.4 Oscillation Caused by the Frequency Adjusting and C_0 Neutralizing Inductors

Frequently, because of the crystal calibration error or because it is desired to pull the oscillator frequency a relatively large amount as in VCOs, an inductor is placed in series with the crystal. Also, at very high frequencies, where the crystal operates near series resonance, a C_0 neutralizing capacitor is placed in parallel with the crystal. Both of these inductors produce an additional resonance which may cause spurious oscillations. These oscillations may be suppressed by "de-Qing" the inductor, which consists of placing a resistor in parallel with the inductor. A recommended value for the resistor is

$$R \geqq 10 X_L \tag{18.1}$$

where X_L is the reactance of the inductor at the operating frequency.

The price of the resistor is that the oscillator performance is deteriorated by decreasing the operating Q.

18.3 SPURIOUS SIGNALS DUE TO CRYSTAL RESPONSES

Figure 3.16 shows that the true crystal equivalent circuit is not that of Fig. 3.5 but is much more complex. Each branch of the circuit, called a mode, can produce a signal near the mode resonant frequency, provided that the oscillator circuit conditions are such as to maintain the signal. The strength of the mode is measured by the mode resistance, being inversely proportional to the resistance.

Typical modes are:

1. Fundamental, third overtone, and so on, which are predictable.
2. a, b, and c modes in SC-cut crystals, which are also predictable. Each overtone has a set of a, b, and c modes.
3. Spurious modes which are generally unpredictable as they depend upon the crystal manufacturing process. These modes are normally weak compared to modes 1 and 2.

18.3.1 Spurious Overtone Signals

These can be easily suppressed using the techniques described in Chapter 5.

18.3.2 Strong Spurious Mode Signals[18.1]

These, too, can be suppressed by means of techniques described in Chapter 5, but with greater difficulty. If the techniques are not properly applied, it is possible for a strong spurious signal to exist at the b mode frequency, particularly in self-limiting oscillator circuits. To minimize the possibility of generating the spurious signal, the mode selector circuitry should be checked for adequate safety margin as described in Section 19.6.3.2.

Reference 18.2 discusses a circuit configuration wherein the desired modes can be selected by adjusting the transistor-bias conditions. However, it is not considered practical.

There are cases where it is desired that both the b and c modes exist simultaneously. In those cases, it is usually required that the oscillator output signal magnitudes be almost equal, but the crystal b mode drive be controllably less than the c mode drive. Oscillators with these characteristics have been constructed and operate as desired.

18.3.3 Weak Spurious Signals

Often when the oscillator output spectrum is examined, very weak signal components corresponding to the weak modes are found. Their amplitude may be 60 dB below the principal signal or smaller. These weak signals are not true oscillations but noise enhanced by the crystal spurious modes. The mechanism, by which the noise is enhanced, is not well understood. Qualitatively, it seems reasonable that such enhancement could result from a recirculation of the noises, at the weak mode frequencies, at those discrete frequencies where the corresponding mode motional impedance can provide essentially 360° loop phase closure even though the gain condition for the oscillation cannot be satisfied.

The above type of spurious oscillations can only be effectively reduced by additional filtering.

18.4 SQUEGGING

By squegging (sometimes called motorboating) is meant the self-produced amplitude modulation of the high-frequency oscillation. This is caused by interaction between the time constants of the bias and coupling circuits and the time constants of the high-frequency tuned circuits of the oscillator loop. The low-frequency variations of the envelope of the high-frequency signal may be sinusoidal, exponential, or trapezoidal.

This phenomenon is more likely to be present in self-limiting oscillators having relatively low operating Q's. Crystal oscillators have high operating Q's and are therefore free from squegging. Where squegging does appear to be present in a crystal oscillator, it is more likely to be a spurious oscillation at a frequency close to the desired frequency and should be eliminated in the same manner as any other spurious oscillation.

Squegging can be stopped by

1. Reducing the bias and coupling time constants (the coupling time constant has greater effect).

2   Raising the operating Q.
3   Reducing the active element driving voltage.

For a more detailed treatment of squegging see Ref. 2.3.

18.5   CRYSTAL PHYSICAL LOCATION AND CONNECTIONS

Another oscillator problem, associated with poor layout, is the location of the crystal and its connections to the rest of the oscillator circuitry. It is important that the crystal leads be made as short as possible so as not to increase the effective C_0. This is particularly true for ovenized oscillators where the crystal is operated in its inductive region and, for thermal reasons, it is located relatively distant from the circuitry. The result of increasing the effective C_0 is to make it harder to set the oscillator to its specified operating frequency and to magnify the effective crystal resistance, R_{df}, which in turn decreases the operating Q and may even cause the cessation of oscillation.

If the leads to the crystal must be long, then the means of adjusting the oscillator frequency, such as the trimming capacitor, inductor, and varactors should be located very close to the crystal.

18.6   OSCILLATOR STARTING

After the steady state performance of the oscillator has been evaluated as being satisfactory, the oscillator should be checked for good starting characteristics. A rule of thumb is that the starting time does not exceed $\frac{1}{2}$ s unless the specifications require a shorter starting time. In general, higher Q circuits tend to have larger starting times.

If the starting time is excessive or it does not start, the small signal loop gain should be increased sufficiently to insure satisfactory starting. As described in Chapter 3, the crystal resistance at zero current, which is the current at the moment of starting, is considerably larger than the resistance at the operating current. Therefore, the actual small signal loop gain is smaller than that calculated with the normal resistance and due allowance should be made for this increase of crystal resistance.

The larger zero current resistance phenomenon also goes far toward explaining many of the hystereris effects such as that illustrated in Fig. 18.2.

The starting problems are more pronounced in the self-limiting and gate type oscillators. The ALC oscillator has the advantage that the small-signal gain can be made very large since the frequency generating function is independent of the limiting function and therefore starting can be readily facilitated. Some oscillators are so difficult to start that they are provided with auxiliary starting circuits. A contrary situation, wherein the oscillator does not start because of excessive small-signal gain is described in Ref. 18.3.

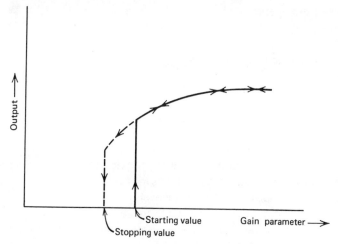

Figure 18.2 Hysteresis effects in oscillator stopping and starting.

18.7 ISOLATION

By isolation is meant the effect of changes in the load impedance upon the oscillator frequency. As stated in Table 12.1, the isolation of the basic oscillators is relatively poor. To improve the isolation, the cascode amplifier of Fig. 13.10 and/or the neutralized C_{cb} amplifier of Fig. 18.3 are interposed between the oscillator and the load. Using these types of amplifiers, almost any degree of isolation required can be obtained. However, where extremely good

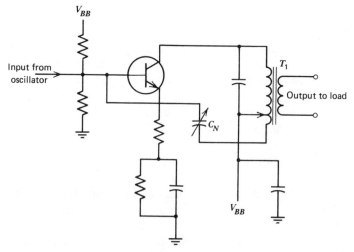

Figure 18.3 Neutralized C_{cb} isolating amplifier.

isolation is necessary, care must be taken that the power supplies are adequately decoupled and that coupling via ground currents is minimized.

18.8 AGING OF COMPONENTS

In addition to the noise generated by the various components, as described in Chapter 14, the components also have long-term effects which can be classified as aging. Over long periods of time the components gradually change value which, in turn, cause oscillator frequency changes which cannot be distinguished from the aging of the crystal. The aging of the components set limits to the minimum possible aging of any oscillator. On the other hand, if the aging of the components and the crystal can be controlled, it can be used to cancel the crystal aging and thus a morstable oscillator may result.

The total aging of the components can be evaluated using the llator evaluation technique described in Chapter 17.

Typical aging rates per day are:

2-μH toroid coil wound on phenolic form with O.D. of 0.12 in.	$- 2$ ppm
Porcelain capacitor	± 3 ppm
Silver mica capacitor	$- 30$ ppm
Glass capacitor	$- 70$ ppm
Ceramic capacitor	± 10 ppm

18.9 OSCILLATOR TESTING

To ensure that none of the problems described in this chapter are present in the prototype models or final units of the oscillator, the prototypes and sample quantities of the final units should be tested for the presence of these problems.

The overall performance of the *oscillators* should be checked for compliance with the applicable items of Section 4.8 of Ref. 13.8 and the sample specifications in Section 1.4.3.

19

Component and Circuitry Measurements

19.1 INTRODUCTION

To facilitate oscillator circuit design, troubleshooting, and trimming, it is absolutely necessary to have full knowledge of the characteristics of the components and circuitry at the operating frequency.

Too often, component characteristics are only approximately known from sketchy catalogue information and guesswork. In particular, guessing is intolerable in oscillator design and, for that matter, in any type of design.

In addition to component characteristics, it is also required to determine the magnitudes of the voltages and currents at the different circuit points. Knowledge of these magnitudes is essential for the following purposes:

1. Troubleshooting.
2. Circuit trimming.
3. Adjusting circuit networks to precalculated values.
4. Gathering data for recordkeeping which will be useful for later design, maintenance, and troubleshooting.

It can be categorically stated that the industry has been remiss in performing adequate recorded measurements during the design and production phases in the mistaken belief that such measurements consume unnecessary time and expense. This is extremely false economy and results in

1. Marginal and often more expensive designs.
2. Inability to duplicate performance in another but supposedly identical unit.

3 Excessive production problems.
4 Difficulty in later repair because of the lack of reference data.

The types of measurements considered in this chapter are:

1 dc voltage, current, and resistance.
2 ac current and voltage at the oscillator operating frequency.
3 Linear component and small-signal immittances at the oscillator operating frequency.
4 Large-signal immittance and phase at the oscillator operating frequency.

19.2 GENERAL CHARACTERISTICS OF THE MEASUREMENT PROCEDURES

1 The measurements should be performed at the oscillator operating frequency.
2 The measuring devices should not excessively load the circuit being measured; that is, the magnitude does not change markedly upon connecting the measuring device to the measured point.
3 The accuracy should be adequate for the intended purpose. In general, $\pm 5\%$ is sufficient for most measurements but it should be checked that greater accuracy is not required.

19.3 dc MEASUREMENTS

19.3.1 Voltage

dc voltages are usually measured with a voltmeter having a dc input resistance of at least 10 MΩ. Therefore the dc loading is normally negligible. However, the ac loading due to the capacitance of the voltmeter and its leads may be large and will markedly change the ac voltages existing at the measuring points. This, in oscillators, may cause a significant change in the dc voltage and result in a serious error in the voltage measurement. As shown in Fig. 19.1, R_{IS} are provided to minimize the effective dc loading due to the ac loading.

The resistors, R_{IS}, are low-capacitance ($< \frac{1}{2}$ pF) isolating elements inserted in series with the voltmeter leads *at the measuring points*. Their resistance may be 100 kΩ which will reduce the 10 MΩ input resistance voltmeter reading by about 1% for each isolating resistor for which a correction may be made in the final voltmeter reading. If one lead is connected to a point having no ac signal present, its R_{IS} may be omitted.

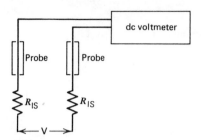

Figure 19.1 Setup for dc voltage measurements.

19.3.2 Current

In most cases, the current can be determined by measuring the voltage drop across a known resistance as described above and then computing the current. In those cases where this is not feasible the clip on dc milliammeter is recommended.

19.3.3 Resistance

The most convenient insrument for measuring dc resistance at reasonable accuracy is the digital volt-ma-ohmmeter. Of course, the oscillator must be deenergized when the measurements are being made. Also, in many cases, one of the leads of the device being measured must be disconnected for meaningful readings.

19.4 ac VOLTAGE AND CURRENT MEASUREMENTS

Often these measurements are difficult to make without significantly loading the circuit being measured, particularly at very high frequencies. Therefore the discussion of these measurements includes means of minimizing the loading.

19.4.1 ac Voltage Measurements

The most suitable instruments for these measurements are the high-impedance RF millivoltmeters with an input capacitance of less than 2 pF. When higher voltages are to be measured, the same instrument should be used with suitable multipliers.

In many cases, a meter input capacitance of even only 2 pF will cause considerable mistuning. If, in those cases, a tapped-down point is available, the ratio, n, between the readings of the two points should be determined by using two similar meters simultaneously, as shown in Fig. 19.2. Obviously, the V_2 reading will be more accurate because of the lower source impedance level. Once n is determined, only V_2 is measured and V_1 is computed using the relationship

$$V_1 = nV_2 \tag{19.1}$$

In making the measurements, the location of the voltmeter ground terminal on the device being measured is very critical, particularly for low voltages at high frequency because of ground signal couplings.

The above measurements are based upon the premise that one of the device measuring points is ground. In those cases where the voltage between two nonground points is required, the only practical method is to take measurements between each point and ground, using a vector voltmeter (see Section 19.6) and then to compute the voltage between the two points.

In interpreting the voltmeter readings, it should be kept in mind that the response of the meter is strongly dependent upon the wave form being measured, and different voltmeter designs may have different wave-form responses which may also be functions of the magnitude of the signals. The vector voltmeter is unique in that it responds only to the fundamental component of the signal.

Another point to consider is that the voltmeter input impedance is often a function of the signal magnitude.

19.4.2 ac Current Measurements

This is a relatively difficult measurement to make but fortunately there are few cases where ac current measurements are required.

In many cases, the current can be determined by measuring the voltage drop across a known impedance and then computing the current (see, e.g., Section 7.7.1).

If the current cannot be measured as described above, a current probe in combination with a millivoltmeter or oscilloscope must be used. Commercial current probes are available for the entire useful frequency range. However, a common problem is that it is difficult to insert the current probe because of the size of the clip-on jaws of the probe. It is, therefore, recommended that permanent half-loops of sufficient size be provided at those points where the current is likely to be measured.

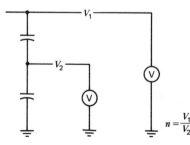

$n = \dfrac{V_1}{V_2}$ Figure 19.2 Setup for ac voltage measurements where a tapped point is available.

19.5 THE OSCILLOSCOPE AS A MEASUREMENT TOOL

One of the most useful instruments for electronics measurement is the oscilloscope. This section discusses the precautions that must be taken for its use in oscillators. These precautions are the same as when measuring other devices at high frequency and at high impedance points.

The oscilloscope should be used principally for wave-form observations and timing information. It should not be used for accurate measurements of amplitudes. The direct reading meter is more accurate, generally has higher input impedance, and is easier to use.

Extreme care should be taken that the oscilloscope probe does not load the circuit excessively. For that reason, the probe input impedance should be consistent with the type of measurement. In practice, many measurements made with an oscilloscope are excessively loaded by its probe impedance. The probe input impedance should be measured to ensure it is sufficiently high for the intended application. The manufacturer's specifications are usually optimistic.

19.6 LINEAR AND SMALL-SIGNAL IMMITTANCE MEASUREMENTS

This is a very important type of measurement as it provides the oscillator designer with much needed information on component and circuit behavior at the actual operating frequency and eliminates considerable guesswork and conjecture. It also facilitates the adjustment of the critical networks prior to activating the oscillator.

19.6.1 Necessary Characteristics of the Instrumentation

1 It should be easy to operate and be direct reading.

2 It should be capable of measuring in-circuit components and networks as well as out of circuit. To that end the measuring terminals or probe should be as small as possible and extremely portable to be able to gain access to the points being measured.

3 The instrumental signal level should be small to ensure small-signal operation.

4 It should have a wide frequency operating range. In addition, it is desirable that the frequency setting resolution be capable of being made extremely fine.

5 It should be capable of measuring a wide range of impedances in all four quadrants, that is, negative and positive reactance and resistance in all combinations.

6 It should be capable of measuring active circuitry for determining transistor characteristics.

7 It should have an accuracy of at least $\pm 5\%$.

The instruments having most of the above characteristics are the Hewlett Packard 4193A Vector Impedance Meter for frequencies from 0.4 to 110 MHz and the Hewlett Packard 4815A Vector Impedance Meter for frequencies from 0.5 to 108 MHz.

The instruments are quite similar except that the 4193A output data is presented in digital form, while the 4815A data is in analogue form. The digital presentation affords greater impedance resolution but not necessarily greater accuracy. The 4193A has considerably greater built-in frequency settability, resolution, and accuracy, but the 4815A frequency performance can be improved as desired by the addition of an external frequency source.

The instruments operate on the principle of injecting a constant current into the impedance being measured and displaying magnitude and phase of the resultant voltage drop across the impedance.

The instruments present the impedance data as $|Z|$, θ, and have the disadvantages that one terminal of the impedance being measured must be at ground potential and that the accuracy for the R component of high Q impedances is quite poor.

19.6.2 The Measurement of Passive Linear Components and the Small-Signal Characteristics of Nonlinear Components

These instruments are extremely useful for measuring the small-signal reactance and resistance characteristics of such components as resistors, capacitors, inductors, and voltage or current variable components such as varactors, pin diodes, Zener diodes, and other semiconductors, all at the frequency of operations.

19.6.3 Examples of Measurement of Active and Passive Networks in Oscillators

The impedance measurements which can be made with these instruments are very useful in both the design and production phases of oscillators.

Some examples are now given in the following subsections.

19.6.3.1 Measurement of the Small-Signal Llator Impedance, Z_{LL} (See Fig. 5.9)

The crystal or other osci, Z_3, is disconnected and the $|Z|$-θ meter connected in its place. The llator is energized, Z and θ are read, and the corresponding X_{LL}

424 Component and Circuitry Measurements

and R_{LL} calculated. θ should be between $-90°$ and $-180°$ so that R_{LL} is negative. The small-signal $|R_{LL}| \geq 2R_3$ for oscillation and proper limiting.

19.6.3.2 Adjustment and Checking of the Overtone and c Mode Selector (See Fig. 10.3)

With the oscillator deenergized, connect the $|Z|$-θ meter to the Q_1 emitter and to ground. The meter frequency is scanned across the oscillator frequency range and X_2 as read on the $|Z|$-θ meter should be as shown in Fig. 19.3.

19.6.3.3 Determination of Crystal Parameters

In this example the parameters of Fig. 19.3 are determined for a high Q crystal.

As the internal frequency generator is not sufficiently stable for this measurement, the $|Z|$-θ meter will be driven by a high-resolution frequency synthesizer controlled by a very stable oscillator, as shown in Fig. 19.4.

The measurements are performed as follows:

1 The frequency is set for $|Z| = $ min and $\theta = 0°$. The $|Z|$ reading is then R_1 and the frequency is f_s.

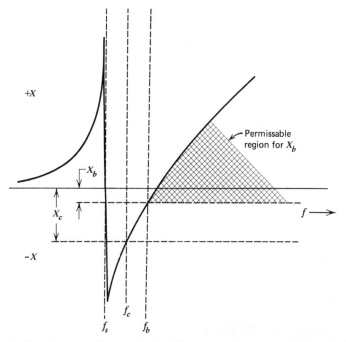

Figure 19.3 Relations for X_2 network for SC-cut crystal as measured by the $|Z| - \theta$ meter. *Notes:* (1) f_s is designed to be $0.75f_c$. (2) f_s should be $\geq 0.70f_c$: critical for $N = 5$, noncritical for $N = 3$. (3) $X_b \leq \frac{1}{4}X_c$.

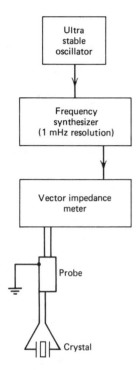

Figure 19.4 Test equipment setup for determining high Q crystal parameters.

2. The frequency is then changed slightly until $\theta = +45°$ and $f_{45°}$ noted.
3. The frequency is then adjusted so that $\theta = -45°$ and $f_{-45°}$ noted.
4. $$Q_x = \frac{f_s}{f_{45°} - f_{-45°}} \tag{19.2}$$
5. Knowing $Q_x, f_s,$ and $R_1, C_1,$ and L_1 are easily computed.
6. C_0 is obtained by shifting the frequency to $0.99 f_s$, reading X_0, and then computing C_0.

It should be noted that the crystal must be protected against meaningful temperature changes during the measurement period.

Limitations of the above procedure are that the crystal drive cannot be varied and the measurements are performed at room temperature.

19.6.3.4 Measurement of Transistor Small-Signal Parameters

As pointed out in Chapters 2 and 5, the transistor small-signal parameters are extremely useful in determining the starting conditions for oscillation and also are the basis for the large-signal characteristics. It is therefore desirable to be able to simply obtain the parameters at the actual operating frequency, dc current, and dc voltage. This section demonstrates the procedures for determining many of the parameters using the vector impedance meter and accessory equipment.

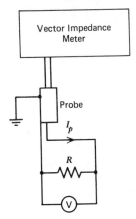

Figure 19.5 Test setup for measuring the probe current in the vector impedance meter.

Before starting the measurements it is necessary to know the actual probe current, I_p. This current is stated in the equipment specification but it is often incorrect, so it should be checked.

Figure 19.5 shows the equipment setup for determining I_p. The $|Z|$-θ meter is set to the desired frequency and a convenient resistor connected to the $|Z|$-θ meter probe in parallel with voltmeter V. $|Z|$ is read on the $|Z|$-θ meter and V on the voltmeter; then

$$I_p = \frac{V}{|Z|} \tag{19.3}$$

It will be found that I_p is fairly independent of the frequency but may be a function of the $|Z|$ range. (See the specifications for the applicable instrument.)

Figure 19.6 shows the equipment arrangement for measuring the small-signal parameters of transistor Q_1. In that figure, the 50-Ω resistor is provided for suppressing spurious oscillations and to enable the determination of the ac current I_c, with voltmeter V. r_{b_1} and r_{b_2} are selected so that their resistance $\ggg Z$ or its components.

It is obvious that

$$|I_c| = \frac{|V|}{50} \tag{19.4}$$

$$|V_b| = |Z||I_p| \tag{19.5}$$

therefore

$$|\beta| = \frac{|I_e|}{|I_p|} \approx \frac{|I_c|}{|I_p|} \tag{19.6}$$

$$|g_m| = |y_{21}| \approx \frac{I_c}{V_b} = \frac{|I_c|}{|I_p||Z|} \tag{19.7}$$

19.6 Linear and Small-Signal Immittance Measurements

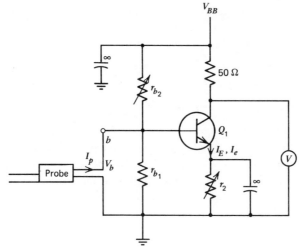

Figure 19.6 Test setup for measuring transistor parameters.

Z can be separated into two parallel components as shown in Fig. 19.7, where

$$r_{be} = \frac{|Z|}{\cos \theta} \tag{19.8}$$

$$X_{be} = \frac{|Z|}{\sin \theta} \tag{19.9}$$

$$C_{be} = \frac{159{,}000}{fX_{be}} \tag{19.10}$$

If

$$r_{bb'} \text{ is neglected,} \tag{19.11a}$$

and

$$\theta_{g_m} \text{ assumed } 0, \tag{19.11b}$$

then

$$g_{m_0} \approx |g_m| \tag{19.12}$$

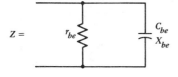

Figure 19.7 Separation of Z into r_{be} and X_{be} components.

and

$$\beta_o \approx g_{m_0} r_{be} \qquad (19.13)$$

From Eq. (2.47),

$$f_T \approx \frac{159,000}{C_{be}} g_{m_0} \qquad (19.14)$$

assuming $C_{bet} \ll C_{be}$. It should be noted that if V in Fig. 19.6 is made a vector voltmeter, then the assumptions in Eqs. (19.11) are not necessary and more accurate results may be obtained, but with more complicated equations.

It is obvious that additional parameters of the transistor can be obtained using procedures similar to those described above.

19.7 LARGE-SIGNAL IMMITTANCE AND PHASE MEASUREMENTS

The type of instrument most useful in the above class of measurements is the vector voltmeter such as the Hewlett Packard 8405A, combined with a suitable signal generator. Its frequency range of operation is 1 to 1000 MHz.

The voltmeter measures two voltages and the phase between them. One voltage is called Channel A and the other Channel B. There is a probe for each voltage. The instrument has two meters. One meter indicates the voltage of the channel selected by a manually operated switch on the meter front panel. The other meter reads the phase difference between the two. Obviously the two voltages must have the same basic frequency, otherwise their steady-state phase relationship cannot be displayed. The A channel is called the reference channel and should be connected to the signal which has the cleanest wave form. The meters respond only to the fundamental component of the signals. However, a jack is provided for viewing the signals translated to a 20-kHz carrier which has the same harmonic wave form (but not the noise) of the signals. The Channel A signal should have a relatively high signal-to-noise ratio.

The probe input capacitance is relatively high about 2.5 pF and therefore caution should be exerted that the probes do not excessively load the circuit under test.

Figure 19.8 shows the equipment setup for measuring the input impedance of the device under test (DUT), such as an amplifier. Transformer T1 and low value resistance R are used for obtaining the current flowing into the DUT. These two elements can be replaced by a suitable current probe feeding probe B. When using a current probe, due allowance should be made for the phase shift in the probe, which should be experimentally determined at each frequency of interest.

Obviously the voltage can be fed into any port and the current measured in the same or other port so that any type of immittance and/or phase relationship can be investigated. See, for example, Section 7.7.2.

19.7 Large-Signal Immittance and Phase Measurements

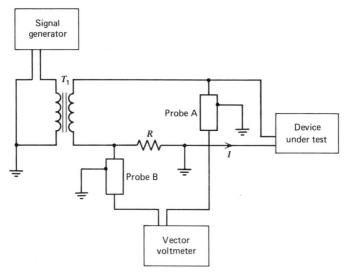

Figure 19.8 Test setup for measuring input impedance.

The above measurement technique has the important advantage that large signals can be impressed on the DUT and the fundamental component of all immittances and phase angles determined since, as previously pointed out, the meter readings are functions of only the fundamental components of the voltages fed to the vector voltmeter.

It should be noted that this setup can also perform the small-signal measurements, described in Section 19.5, but the Vector Impedance Meter is more convenient for that purpose.

19.8 Z_{LL} MEASUREMENT PROCEDURE

Section 19.7 and Fig. 19.8 describe a method of measuring the input impedance of a device (DUT). This device can be the llator of an oscillator. In that case the measurement is performed at the value of $I = I_x$, and Z_{LL} can thus be determined as a function of the frequency f and crystal current I_x.

The procedure described above is practical only when one terminal of the llator is at ac ground potential, as in the Colpitts oscillators. When both terminals are off ground, the indirect procedure described below may be used. This procedure is based upon Section 1.3.1.

The measurement setup is shown in Fig. 19.9. Y_1 is a crystal having a low value of R_L at its rated frequency f_L, which is the oscillator frequency when the crystal sees a llator reactance equal to the reactance of its rated load capacitance, C_L. R_v is a variable resistance and X_v is a variable reactance, which may be either inductive or capacitive depending upon the llator. R_v and X_v are adjusted until the oscillator frequency is f_L and the crystal current is the

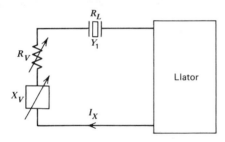

Figure 19.9 Test setup for measuring Z_{LL}.

desired current, I_x. Then, from Eqs. (1.47) and (1.48),

$$Z_{LL} = -(R_L + R_v) - j(-X_{C_L} + X_v) \qquad (19.15)$$

Z_{LL} is thus obtained as a function of I_x. The crystal motional capacitance, C_1, should be relatively high so as to make $\partial f/\partial X$ relatively large.

Since it is very difficult to realize a suitable variable R_v at high frequencies, the procedure can be simplified if the crystal properties are well known. R_v can thus be a set of fixed resistors and f can be allowed to vary slightly since Z_{LL} is not a very sharp function of f, while Z_{Y_1} is a relatively sharp function.

If X_v is capacitive, then the llator has an L_{eq}.

$$L_{eq} = \frac{X_v}{2\pi f} \qquad (19.16)$$

L_{eq} is a useful concept for most of the integrated circuit oscillators in Chapters 15 and 16.

If X_v is inductive, then the llator has a C_{eq}

$$C_{eq} = \frac{159{,}000}{f_L X_v} \qquad (19.17)$$

C_{eq} is a useful concept for the *Pierce family* of oscillators described in Chapter 5.

An alternative modification of the procedure described above is to eliminate Y_1. R_v and X_v are then adjusted until the desired f and I_x are obtained. Equation (19.13) becomes

$$Z_{LL} = -(R_v + jX_v) \qquad (19.18)$$

The latter procedure appears preferable because of its relative simplicity. However, its use may be impractical in many crystal oscillators, particularly of the self-limiting type, because of the possible spurious oscillations and squegging as described in Section 17.8.

References and Bibliography

The reference number includes the chapter in which the reference is first cited or to which the reference is applicable; for example, Ref. 15.6 signifies Reference 6 in Chapter 15.

CHAPTER 1

1.1 Cote, A. J., Jr., "Matrix Analysis of Oscillators and Transistor Applications," *IRE Trans. Circuit Theory*, **CT-5**, 181–188 (Sept. 1958).
1.2 Kuo, Franklin F., *Network Analysis and Synthesis*, Wiley, New York, 1962.
1.3 Uzunoglu, Vasil, *Semiconductor Network Analysis and Design*, McGraw-Hill, New York, 1964.
1.4 Vander Pol, A., "Nonlinear Theory of Electric Oscillations," *Proc. IRE*, **22**, 1051–1086 (1934).
1.5 Edson, W. A., *Vacuum Tube Oscillators*, Wiley, New York, 1955.
1.6 Reich, Herbert J., *Functional Circuits and Oscillators*, Van Nostrand, Princeton, 1961.
1.7 Groszkowski, J., *Frequency of Self-Oscillations*, MacMillan, New York, 1964.
1.8 Hafner, E., *Analysis and Design of Crystal Oscillators*, Part I, Technical Report ECOM-274, U.S. Army Electronics Laboratories, U.S. Army Electronics Command, Fort Monmouth, N.J., May 1964.
1.9 Firth, D., *Quartz Crystal Oscillator Circuits Design Handbook*, Publication AD460-377, U.S. Army Electronics Command, Fort Monmouth, N.J., March 1965.
1.10 Frerking, M. E., *Crystal Oscillator Design and Temperature Compensation*, Van Nostrand Reinhold, New York, 1978.

References on Oscillator Frequency Stability and Noise

1.11 Kartaschoff, P., *Frequency and Time*, Academic, New York, 1978. Contains an extensive bibliography.
1.12 Blair, Byron B., *Time and Frequency: Theory and Fundamentals*, N.B.S. Monograph 140, U.S. Department of Commerce, Washington, D.C., 1974. Contains an extensive bibliography.

1.13 *Proceedings of the Annual Symposiums on Frequency Control*, U.S. Army Electronics Command, Fort Monmouth, N.J. The issues of 1976, 1979, and 1980 contain extensive bibliographies.

1.14 *Proceedings of Annual Precise Time and Time Interval (PTTI) Applications and Planning Meeting*, NASA Goddard Space Flight Center, Greenbelt, Md.

1.15 *Hewlett Packard Application Notes*, Hewlett Packard Co., Palo Alto, Calif. Application Notes AN174-6, AN174-7, AN207, AN225, and AN240-0 are particularly useful.

CHAPTER 2

2.1 Holford, K., "Transistor LC Oscillator Circuits for Low Frequency Low Power Operation," *Mullard Tech. Commun.*, **41**, 17–25 (Dec. 1959).

2.2 Holford, K., "Transistor LC Oscillator Circuits Giving Moderate Values of Power Output," *Mullard Tech. Commun.* **42**, 60–70 (Feb. 1960).

2.3 Clarke, K. K. and Hess, D. T, *Communication Circuits: Analysis and Design*, Addison-Wesley, Reading, Mass., 1971.

2.4 Gray, P. E. and Searle, L. C., *Electronic Principles, Physics, Models, and Circuits*, Wiley, New York, 1969.

2.5 Hunter, L. P., *Handbook of Semiconductor Electronics*, third edition, McGraw-Hill, New York, 1971.

CHAPTER 3

3.1 Cady, W. G., *Piezoelectricity*, Dover, New York, 1964.

3.2 Heising, R. A., ed., *Quartz Crystals for Electrical Circuits*, Van Nostrand, New York, 1946.

3.3 Mason, W. P., "Use of Piezoelectric Crystals and Mechanical Resonators in Filters and Oscillators," *Physical Acoustics*, (W. P. Mason, ed.), Vol. 1A, pp. 335–416, Academic, New York, 1964.

3.4 Gerber, E. A. and Sykes, R. A., "State of the Art—Quartz Crystal Units and Oscillators," *Proc. IEEE*, **54**, 103–116 (Feb. 1966).

3.5 Hafner, E., "Crystal Resonators," *IEEE Trans. Sonics Ultrason.* **SU-21**, 220–237 (Oct. 1974).

3.6 Ballato, A., "The Future of the Quartz Crystal Industry—Worldwide," *Proc. 35th Annual Symposium on Frequency Control*, U.S. Army Electronics Command, Fort Monmouth, N.J., pp. 576–582. Copies available from Electronic Industries Association, 2001 Eye St., NW, Washington, D.C. 20006.

3.7 Ballato, A., "Doubly Rotated Thickness Mode Plate Vibrators," *Physical Acoustics* (W. P. Mason and R. N. Thurston, eds.), Vol. 13, pp. 115–181, Academic, New York, 1977.

3.8 van Randeraat, J., ed., *Piezoelectric Ceramics*, N. V. Philips, Eindhoven, 1968.

3.9 Hafner, E., "The Piezoelectric Crystal Unit-Definitions and Methods of Measurements," *Proc. IEEE*, **57**, 179–201 (Feb. 1969).

3.10 Berlincourt, D. A., Curran, D. R., and Jaffe, H., "Piezoelectric and Piezomagnetic Materials and Their Function in Transducers," *Physical Acoustics* (W. P. Mason, ed.), Vol. 1A, pp. 169–270, Academic, New York, 1964.

3.11 Ballato, A., "Resonance in Piezoelectric Vibrators," *Proc. IEEE*, **58**, 149–151 (Jan. 1970).

3.12 Hafner, E., "Theory of Oscillator Design," *Proc. 17th Annual Symposium on Frequency Control*, U.S. Army Electronics Laboratory, Fort Monmouth, N.J., pp. 508–536, 1963. Copies available from National Technical Information Service, U.S. Department of Commerce, 5285 Port Royal Road, Springfield, Va. 22161.

3.13 Warner, A. W., "Design and Performance of Ultraprecise 2.5-mc Quartz Crystal Units," *Bell System Tech. J.*, **39**, 1193–1217 (Sept. 1960).

3.14 Colorado Crystal Company, Loveland, Colo. 80537 (May 1980), private communication.

3.15 Kusters, J. A., "The SC Cut Crystal—An Overview," *Proc. IEEE Ultrasonics Symposium*, pp. 402–409, IEEE, New York, 1981.

3.16 White, R. M., "Surface Elastic Waves," *Proc. IEEE*, **58**, 1238–1276 (Aug. 1970).

3.17 Matthews, H., ed., *Surface Wave Filters*, Wiley, New York, 1977.

3.18 Oliner, A. A., ed., *Acoustic Surface Waves*, Springer, New York, 1978.

3.19 Cross, P. S. and Elliott, S. S., "Surface-Acoustic-Wave Resonators," *Hewlett-Packard J.*, **32**, 9–17 (Dec. 1981).

3.20 Lukaszek, T. and Ballato, A., "What SAW Can Learn from BAW: Implications for Future Frequency Control, Selection & Signal Processing," *Proc. IEEE Ultrasonics Symposium*, pp. 173–183, IEEE, New York, 1980.

3.21 Shreve, W. R., Bray, R. C., Elliott, S., and Chu, Y. C., "Power Dependence of Aging in SAW Resonators," *Proc. IEEE Ultrasonics Symposium*, pp. 94–99, IEEE, New York, 1981.

3.22 Chuang, S. S., "Quartz Tuning Fork Crystal Using Overtone Flexure Modes," *Proc. 35th Annual Symposium on Frequency Control*, U.S. Army Electronics Command, Fort Monmouth, N.J., pp. 130–143, 1981. Copies available from Electronic Industries Association, 2001 Eye St., NW, Washington, D.C. 20006.

3.23 Dinger, R. J., "A Miniature Quartz Resonator Vibrating at 1 MHz," *Proc. 35th Annual Symposium on Frequency Control*, U.S. Army Electronics Command, Fort Monmouth, N.J., pp. 144–148, 1981. Copies available from Electronics Industries Association, 2001 Eye St., NW, Washington, D.C. 20006.

3.24 Adachi, T., Tsuzuki, Y., and Takeuchi, C., "Investigation of Spurious Modes of Convex DT-Cut Quartz Crystal Resonators," *Proc. 35th Annual Symposium on Frequency Control*, U.S. Army Electronics Command, Fort Monmouth, N.J., pp. 149–156, 1981. Copies available from Electronics Industries Association, 2001 Eye St., NW, Washington, D.C. 20006.

3.25 Vangheluwe, D. C. L. and Fletcher, E. D., "The Edge Mode Resonator," *Proc. 35th Annual Symposium on Frequency Control*, U.S. Army Electronics Command, Fort Monmouth, N.J., pp. 157–165, 1981. Copies available from Electronic Industries Association, 2001 Eye St., NW, Washington, D.C., 20006.

3.26 Okano, S., Kodama, T., Yamazaki, K., and Kotake, H., "4.19 MHz Cylindrical AT-Cut Miniature Resonator," *Proc. 35th Annual Symposium on Frequency Control*, U.S. Army Electronics Command, Fort Monmouth, N.J., pp. 166–173, 1981. Copies available from Electronic Industries Association, 2001 Eye St., NW, Washington, D.C. 20006.

3.27 Kawashima, H., Sato, H., and Ochiai, O., "New Frequency Temperature Characteristics of Miniaturized GT-Cut Quartz Resonators," *Proc. 34th Annual Symposium on Frequency Control*, U.S. Army Electronics Command, Fort Monmouth, N.J., pp. 131–139, 1980. Copies available from Electronic Industries Association, 2001 Eye St., NW, Washington, D.C. 20006.

3.28 Kogure, S., Momosaki, E., and Sonoda, T., "New Type Twin Mode Resonator," *Proc. 34th Annual Symposium on Frequency Control*, U.S. Army Electronics Command, Fort Monmouth, N.J., pp. 160–166, 1980. Copies available from Electronic Industries Association, 2001 Eye St., NW, Washington, D.C. 20006.

3.29 Momosaki, E., Kogure, S., Inoue, M., and Sonoda, T., "New Quartz Tuning Fork with Very

Low Temperature Coefficient," *Proc. 33rd Annual Symposium on Frequency Control*, U.S. Army Electronics Command, Fort Monmouth, N.J., pp. 247–254, 1979. Copies available from Electronic Industries Association, 2001 Eye St., NW, Washington, D.C. 20006.

3.30 Hermann, J. and Bourgeois, C., "A New Quartz Crystal Cut for Contour Mode Resonators," *Proc. 33rd Annual Symposium on Frequency Control*, U.S. Army Electronics Command, Fort Monmouth, N.J., pp. 255–262, 1979. Copies available from Electronic Industries Association, 2001 Eye St., NW, Washington, D.C. 20006.

3.31 Oguchi, K. and Momosaki, E., "+5° X Micro Quartz Resonator by Lithographic Process," *Proc. 32nd Annual Symposium on Frequency Control*, U.S. Army Electronics Command, Fort Monmouth, N.J., pp. 277–281, 1978. Copies available from Electronic Industries Association, 2001 Eye St., NW, Washington, D.C. 20006.

3.32 Guttwein, G. K., Ballato, A., and Lukaszek, T. J., *VHF-UHF Piezoelectric Resonators*, U.S. Patent 3,694,677 (Sept. 1972).

3.33 Bidart, L. and Chauvin, J., "Direct Frequency Crystal Oscillators," *Proc. 35th Annual Symposium on Frequency Control*, U.S. Army Electronics Command, Fort Monmouth, N.J., pp. 365–375, 1981. Copies available from Electronic Industries Association, 2001 Eye St., NW, Washington, D.C. 20006.

3.34 Shockley, W., Curran, D. R,. and Koneval, D. J., "Trapped-Energy Modes in Quartz Filter Crystals," *J. Acoust. Soc. Am.*, **41**, 981–993 (1967).

3.35 Stoddard, W. G., "Design Equations for Plano-Convex AT Filter Crystals," *Frequency*, **1**, 47–50 (July–Aug. 1963); *Proc. 17th Annual Symposium on Frequency Control*, U.S. Army Electronics Laboratory, Fort Monmouth, N.J., pp. 272–282, 1963. Copies available from National Technical Information Service, U.S. Department of Commerce, 5285 Port Royal Road, Springfield, Va. 22161.

3.36 Ballato, A., "Frequency–Temperature–Load Capacitance Behavior of Resonators for TCXO Application," *IEEE Trans. Sonics Ultrason.*, **SU-25**, 185–191 (July 1978).

3.37 Bechmann, R., *Piezoelectric Quartz Vibrators—Properties of Cuts and Modes of Vibration*, Technical Report 2003, U.S. Army Signal R & D Laboratory, Fort Monmouth, N.J., Dec. 1958.

3.38 Gerber, E. A. and Sykes, R. A., "Quartz Frequency Standards," *Proc. IEEE*, **55**, 783–791 (June 1967).

3.39 Vig. J. R. and Le Bus, J. W., "UV/Ozone Cleaning of Surfaces," *IEEE Trans. Parts, Hybrids, Packag.*, **PHP-12**, 365–370 (Dec. 1976).

3.40 Hafner, E., "Quartz Crystal Oscillators," *National Bureau of Standards Seminar* (Aug. 1975), unpublished.

3.41 Ballato, A. and Vig, J. R., "Advances in the Stability of High Precision Crystal Resonators," *Proc. 11th Annual Precise Time and Time Interval Applications and Planning Meeting*, NASA Conference Publication 2129, Goddard Space Flight Center, Greenbelt, Md., pp. 403–438, 1979.

3.42 Hammond, D. L., Adams, C. A., and Benjaminson, A., "Hysteresis Effects in Quartz Resonators," *Proc. 22nd Annual Symposium on Frequency Control*, U.S. Army Electronics Command, Fort Monmouth, N.J., pp. 55–66, 1968. Copies available from National Technical Information Service, U.S. Department of Commerce, 5285 Port Royal Road, Springfield, Va. 22161.

3.43 Stratemeyer, H. P., "The Stability of Standard-Frequency Oscillators," *Gen. Radio Experimenter*, **38**, 1–16 (June 1964).

3.44 Gagnepain, J.-J., "Fundamental Noise Studies of Quartz Crystal Resonators," *Proc. 30th Annual Symposium on Frequency Control*, U.S. Army Electronics Command, Fort Monmouth, N.J., pp. 84–91, 1976. Copies available from National Technical Information Service, U.S. Department of Commerce, 5285 Port Royal Road, Springfield, Va. 22161.

3.45 Ballato, A., "Static and Dynamic Behavior of Quartz Resonators," *IEEE Trans. Sonics Ultrason.* **SU-26**, 299–306 (July 1979).

3.46 Knowles, J. E., "On the Origin of the 'Second Level of Drive' Effect in Quartz Oscillators" *Proc. 29th Annual Symposium on Frequency Control*, U.S. Army Electronics Command, Fort Monmouth, N.J., pp. 230–236, 1975. Copies available from National Technical Information Service, U.S. Department of Commerce, 5285 Port Royal Road, Springfield, Va. 22161.

3.47 Hammond, D., Adams, C., and Cutler, L., "Precision Crystal Units," *Proc. 17th Annual Symposium on Frequency Control*, U.S. Army Electronics R & D Laboratory, Fort Monmouth, N.J., pp. 215–232, 1963. Copies available from National Technical Information Service, U.S. Department of Commerce, 5285 Port Royal Road, Springfield, Va. 22161.

3.48 Bernstein, M., "Increased Resistance of Crystal Units at Oscillator Noise Levels," *Proc. IEEE*, **55**, 1239–1241 (July 1967).

3.49 Gagnepain, J.-J., "Nonlinear Properties of Quartz Crystal and Quartz Resonators: A Review," *Proc. 35th Annual Symposium on Frequency Control*, U.S. Army Electronics Command, Fort Monmouth, N.J., pp. 14–30, 1981. Copies available from Electronic Industries Association, 2001 Eye St., NW, Washington, D.C. 20006.

3.50 Gagnepain, J.-J., Ponçot, J.-C., and Pegeot, C., "Amplitude–Frequency Behavior of Doubly Rotated Quartz Resonators," *Proc. 31st Annual Symposium on Frequency Control*, U.S. Army Electronics R & D Command, Fort Monmouth, N.J., pp. 17–22, 1977. Copies available from National Technical Information Service, U.S. Department of Commerce, 5285 Port Royal Road, Springfield, Va. 22161.

3.51 Ballato, A. and Tilton, R., "Electronic Activity Dip Measurement," *IEEE Trans. Instrum. Meas.*, **IM-27**, 59–65 (March 1978).

3.52 King, J. C. and Sander, H. H., "Rapid Annealing of Frequency Changes in Crystal Resonators Following Pulsed X-Irradiation," *IEEE Trans. Nucl. Sci.*, **NS-19**, 23–32 (Dec. 1972).

3.53 King, J. C. and Sander, H. H., "Transient Change in Q and Frequency of AT-Cut Quartz Resonators Following Exposure to Pulse X-Rays," *IEEE Trans. Nucl. Sci.*, **NS-20**, 117–125 (Dec. 1973).

3.54 Koehler, D. R., "Radiation-Induced Frequency Transients in AT, BT, and SC Cut Quartz Resonators," *Proc. 33rd Annual Symposium on Frequency Control*, U.S. Army Electronics Command, Fort Monmouth, N.J., pp. 311–321, 1979. Copies available from Electronic Industries Association, 2001 Eye St., NW, Washington, D.C. 20006.

3.55 Ballato, A., Lukaszek, T. J., and Iafrate, G. J., "Subtle Effects in High-Stability Quartz Resonators," *Ferroelectrics*, **43**, 25–41 (1982).

3.56 Cook, R. K., Greenspan, M., and Weissler, P. G., "Thermal Voltages of a Quartz Crystal," *Phys. Rev.*, **74**, 1714–1719 (Dec. 1948).

3.57 Hafner, E., "Stability of Crystal Oscillators," *Proc. 14th Annual Symposium on Frequency Control*, U.S. Army Signal R & D Laboratory, Fort Monmouth, N.J., pp. 192–199, 1960. Copies available from National Technical Information Service, U.S. Department of Commerce, 5285 Port Royal Road, Springfield, Va. 22161.

3.58 Riley, W. J., "Frequency Stability in Precision Oscillators," *Electro-Technology*, **79**, 42–44 (Apr. 1967).

3.59 Walls, F. L. and Wainwright, A. E., "Measurement of the Short-Term Stability of Quartz Crystal Resonators and the Implications for Crystal Oscillator Design and Applications," *IEEE Trans. Instrum. Meas.*, **IM-24**, 15–20 (Mar. 1975).

3.60 Burgoon, R. and Wilson, R. L., "Performance Results of an Oscillator Using the SC Cut Crystal," *Proc. 33rd Annual Symposium on Frequency Control*, U.S. Army Electronics Command, Fort Monmouth, N.J., pp. 406–410, 1979. Copies available from Electronic Industries Association, 2001 Eye St., NW, Washington, D.C. 20006.

3.61 "Military Standard. Crystal Units (Quartz) and Crystal Holders (Enclosures), Selection of," MIL-STD-683F, U.S. Department of Defense, 1978.

3.62 "Military Specification. Crystal Units, Quartz, General Specification for," MIL-C-3098G, U.S. Department of Defense, 1979.

3.63 "Military Specification. Holders (Enclosures), Crystal, General Specification for," MIL-H-10056E, U.S. Department of Defense, 1975.

3.64 "IEEE Standard on Piezoelectricity," IEEE, New York, 1978. IEEE Standard 176-1978.

3.65 "IEEE Standard Definitions and Methods of Measurement for Piezoelectric Vibrators," IEEE, New York, 1966. IEEE Standard 177-1966; ANSI Standard C83.17-1970.

3.66 "Quartz Crystal Units for Frequency Control and Selection. Part 1: Standard Values and Test Conditions," second edition, IEC Publication 122-1, 1976. Copies available from American National Standards Institute, 1430 Broadway, New York, N.Y. 10018.

3.67 "Guide to the Use of Quartz Oscillator Crystals," IEC Publication 122-2, Section 3 (1962; Amendment 1, 1969). Copies available from American National Standards Institute, 1430 Broadway, New York, N.Y. 10018.

3.68 "Basic Method for the Measurement of Resonance Frequency and Equivalent Series Resistance of Quartz Crystal Units by Zero Phase Technique in a π-Network," IEC Publication 444, 1973. Copies available from Americal National Standards Institute, 1430 Broadway, New York, N.Y. 10018.

3.69 Parzen, B., "Theoretical and Practical Effects of the Resonator Specifications and Characteristics upon Practical Crystal Oscillator Design and Performance," *Proc. 36th Annual Symposium on Frequency Control*, U.S. Army Electronics R & D Command, Fort Monmouth, N.J., in press.

3.70 Bottom, V. E., *Introduction to Quartz Crystal Unit Design*, Van Nostrand Reinhold, New York, 1982.

CHAPTER 13

13.1 Ebert, J. and Kazimierczuk, M., "Class E High-Efficiency Tuned Power Oscillator," *IEEE J. Solid State Circuits*, **SC-16**, 62–65 (Apr. 1981).

13.2 Driscoll, M. M., "Two-Stage Self-Limiting Series Mode Type Quartz-Crystal Oscillator Exhibiting Improved Short-Term Frequency Stability," *IEEE Trans. Instrum. Meas.* **IM-22**, 130–138 (June 1973).

13.3 Healey, D. J., III, "Low-Noise Frequency Source," *Proc. 27th Annual Symposium on Frequency Control*, U.S. Army Electronics Command, Fort Monmouth, N.J., pp. 170–178, 1973. Copies available from Electronic Industries Association, 2001 Eye St., NW, Washington, D.C. 20006.

13.4 Rhode, U. L., "Mathematical Analysis and Design of Ultra-Stable Low Noise 100 MHz Crystal Oscillator with Differential Limiter and its Possibilities in Frequency Standards," *Proc. 32nd Annual Symposium on Frequency Control*, U.S. Army Electronics Command, Fort Monmouth, N.J., pp. 409–425, 1978. Copies available from Electronic Industries Association, 2001 Eye St., NW, Washington, D.C. 20006.

13.5 Baugh, R. A., "Low Noise Frequency Multiplication," *Proc. 26h Annual Symposium on Frequency Control*, U.S. Army Electronics Command, Fort Monmouth, N.J., pp. 50–54, 1972. Copies available from the Electronic Industries Association, 2001 Eye St., NW, Washington, D.C. 20006.

13.6 Groslambert, J., Marianneau, G., Olivier, M., and Ubersfeld, J. "The Design and Performance of a Crystal Oscillator Exhibiting Improved Short-Term Frequency Stability," *Proc.*

28th Annual Symposium on Frequency Control, U.S. Army Electronics Command, Fort Monmouth, N.J., pp. 181–183, 1974. Copies available from the Electronic Industries Association, 2001 Eye St., NW, Washington, D.C. 20006.

13.7 Neubig, B., "Design of Crystal Oscillator Circuits," *VHF Commun.* (Mar.–Apr. 1979).

13.8 Military Specification, Oscillators, Crystal, General Specification for, MIL-0-55310, U.S. Department of Defense, Dec. 1970. (Latest issue should be used.)

CHAPTER 14

14.1 Truxal, J. G., *Control Engineers Handbook*, McGraw-Hill, New York, 1955.

14.2 Chestnut, H. and Mayer, R. W., *Servomechanisms and Regulating Systems Design*, Wiley, New York, 1959.

14.3 Gardner, F. M., *Phaselock Techniques*, second edition, Wiley, New York, 1979.

14.4 Kulagin, E. V., Pikhtelev, A. I., Sokolov, V. P., and Fateev, B. P. "Natural Fluctuations in a Quartz Crystal Oscillator with Automatic Gain Control," *Izv. Vysshikh Vchebnykh Zavedenti, Radiofiz*, **21**(11), 1618–1626 (Nov. 1976). English translation (1979) available from Plenum, New York.

14.5 Healey, D. J., III, "SC-Cut Quartz Crystal Unit in Low-Noise Application at VHF," *Proc. 35th Annual Symposium on Frequency Control*, U.S. Army Electronics Command, Fort Monmouth, N.J., pp. 440–454, 1981. Copies available from Electronic Industries Association, 2001 Eye St., NW, Washington, D.C. 20006.

14.6 Burgoon, R. and Wilson, H. L., "Design Aspects of an Oscillator Using the 8C-Cut Crystal," *Proc. 32nd Annual Symposium on Frequency Control*, U.S. Army Electronics Command, Fort Monmouth, N.J., pp. 411–416, 1979. Copies available from Electronic Industries Association, 2001 Eye St., NW, Washington, D.C. 20006.

14.7 Babbitt, H. S., III, "Precision Oscillators in the LES-8/9 Spacecraft," *Proc. 31st Annual Symposium on Frequency Control*, U.S. Army Electronics Command, Fort Monmouth, N.J., pp. 412–420, 1977. Copies available from Electronic Industries Association, 2001 Eye St., NW, Washington, D.C. 20006.

14.8 Pustarfi, H. S., "An Improved 5 MHz Reference Oscillator for Time and Frequency Standard Applications," *IEEE Trans. Instrum. Meas.*, **IM-15**, 196–198 (Dec. 1966).

14.9 Felch, E. P. and Israel, J. O., "A Simple Circuit for Frequency Standards Employing Overtone Crystals," *Proc. IRE*, **43**, 596–603 (May 1955).

14.10 Halford, D., Wainwright, A. E., and Barnes, J. A., "Flicker Noise of Phase in RF Amplifiers and Frequency Multipliers: Characterization, Cause and Cure," *Proc. 22nd Annual Symposium on Frequency Control*, U.S. Army Electronics Command, Fort Monmouth, N.J., pp. 340–346, 1968. National Technical Information Service, Accession NR AD844911.

14.11 Leeson, D. B., "A Simple Model of Feedback Oscillator Noise Spectrum," *Proc. IEEE*, **54**, 329–330 (Feb. 1966).

14.12 Sauvage, G., "Phase Noise in Oscillators: A Mathematical Analysis of Leeson's Model," *IEEE Trans. Instrum. Meas.* **IM-26** (No. 4), 408–410 (Dec. 1977).

CHAPTER 15

15.1 Besson, R. E., Girardet, P. G., and Graf, E. P., "Performance of New Oscillators Designed for 'Electrodeless' Crystals," *Proc. 34th Annual Symposium on Frequency Control*, U.S. Army Electronics Command, Fort Monmouth, N.J., pp. 457–462, 1980. Copies available from Electronic Industries Association, 2001 Eye St., NW, Washington, D.C. 20006.

15.2 Holmbeck, J. D., "Frequency Tolerance Limitations with Logic Gate Oscillators," *Proc. 31st Annual Symposium on Frequency Control*, U.S. Army Electronics Command, Fort Monmouth, N.J., pp. 390–395, 1977. Copies available from Electronic Industries Association, 2001 Eye St., NW, Washington, D.C. 20006.

15.3 Farrell, J. J., "Crystals and NMOS: Frequency Controlled MPUs," *Proc. 32nd Annual Symposium on Frequency Control*, U.S. Army Electronics Command, Fort Monmouth, N.J., pp. 332–336, 1978. Copies available from Electronic Industries Association, 2001 Eye St., NW, Washington, D.C. 20006.

15.4 Luxmore, T. and Newell, D. E., "The MXO-Monolithic Crystal Oscillator," *Proc. 31st Annual Symposium on Frequency Control*, U.S. Army Electronics Command, Fort Monmouth, N.J., pp. 396–399, 1977. Copies available from Electronic Industries Association, 2001 Eye St., NW, Washington, D.C. 20006.

15.5 Blood, B., "IC Crystal Controlled Oscillators," *Motorola Applications Note AN417B*, Motorola Semi-Conductor Products, Inc., Phoenix, Ariz., 1977.

15.6 Lane, M. F., "Crystal Oscillator Design Employing Digital Integrated Circuits as the Active Element," *Telecom Australia Research Laboratories Report Number 6949*, Melbourne, Australia, 1975.

CHAPTER 16

16.1 Hildreth, T. E., "IC Crystal Oscillators," *Interdesign/Tridar Monochip Application Note APN-4*, Sunnyvale, Calif.

16.2 Hatchett, J. and Janikowski, R., "Predict Frequency Accuracy for MC 12060 and MC 12061 Crystal Oscillator Circuits," *Motorola Engineering Bulletin EB-59*, Motorola Semi-conductor Products, Inc., Phoenix, Ariz.

16.3 Arnold, M., "Improved Frequency Stability Circuit for 10 MHz Quartz Crystal," *Funk-Tech*, **33**(23), WS 380-1 (Dec. 1978) (in German).

16.4 Thomas, H. P., Sherman, J. H., Jr., and Early, R. C., "Microminiature Crysal Oscillators," *Frequency*, 17–23 (Sept./Oct. 1967).

16.5 Jackson, H. W., "Tactical Miniature Crysal Oscillator," *Proc. 34th Annual Symposium on Frequency Control*, U.S. Army Electronics Command, Fort Monmouth, N.J., pp. 449–456, 1980. Copies available from Electronic Industries Association, 2001 Eye St., NW, Washington, D.C. 20006.

16.6 Embree, D. M., *et al.*, "Miniature Packaged Crystal Oscillators," *Proc. 34th Annual Symposium on Frequency Control*, U.S. Army Electronics Command, Fort Monmouth, N.J., pp. 475–487, 1980. Copies available from Electronic Industries Association, 2001 Eye St., NW, Washington, D.C. 20006.

16.7 "Oscillators, Crystal, Class 1 (Crystal Oscillator (XO)) 1 kHz through 50 MHz Hermetic Seal," U.S. Department of Defense, *Military Specification Sheet, MIL-O-55310/8A*, 5 March 1975.

16.8 Meyer, R. G. and Soo, D. C. F., "MOS Crystal Oscillator Design," *IEEE J. Solid-State Circuits*, **SC-15**(2), 222–228 (Apr. 1980).

CHAPTER 17

17.1 Parzen, B., "Requirements and Evaluation of the Stability of the Circuitry, Excluding the Crystal, in Crystal Oscillators," *Proc. 34th Annual Symposium on Frequency Control*, U.S.

Army Electronics Command, Fort Monmouth, N.J., pp. 471–474, 1980. Copies available from Electronic Industries Association, 2001 Eye St., NW, Washington, D.C. 20006.

CHAPTER 18

18.1 Kodoma, S. and Sato, Y., "An Analysis of Unwanted Frequency Oscillation in a Crystal Controlled Oscillator," *Proc. 33rd Annual Symposium on Frequency Control*, U.S. Army Electronics Command, Fort Monmouth, N.J., pp. 417–424, 1979. Copies available from Electronic Industries Association, Washington, D.C. 20006.

18.2 Bahadur, H. and Parshad, R., "Use of Transistor Heavy Biasing for a Novel Method of Generation of Quartz Crystal Overtones and Mixed Frequency Oscillations," *Proc. IEEE*, **68**(10), 1345 (Oct. 1980).

18.3 Unkrich, M. A. and Meyer, R. G., "Conditions for Start-Up in Crystal Oscillators," *IEEE J. Solid State Circuits*, **SC-17**(1), 87–90 (Feb. 1982).

List of Frequently Used Symbols*

Symbol	Definition	Page of First Appearance
A_L	Oscillator loop gain	16
A_{L_0}	Small signal oscillator loop gain	16
B	Susceptance in ℧, imaginary component of Y	3
BV_{CE}	Breakdown voltage between transistor terminals c and e in mV	187
C	Capacitance in pF	5
C_b	Series capacitor in mode selection networks	156
C_b	Base bypass capacitor	251
C_{bed}	$Cb'ed$ for $r_{bb'} = 0$	48
C_{bed_0}	Small signal C_{bed}	48
$C_{b'ed}$	Bipolar transistor base emitter diffusion capacitance	38
$C_{b'ed_0}$	Small signal $C_{b'ed}$	38
C_{bet}	$C_{b'et}$ for $r_{bb'} = 0$	53
$C_{b'et}$	Bipolar transistor base emitter transition capacitance	39
C_{cem}	Transistor collector emitter Miller effect contribution	236
C_L	Rated crystal load capacitance	70
C'_L	Llator matching capacitor	191
C'_L	Stabilizing capacitor	165
$C_{L_{df}}$	Crystal load capacitance at $f_L + df$	77
C_M	Miller effect contribution to $(X_1 + X_2)$	235
C_N	Total capacitance in overtone selection network	154
C_n	$X_n/(2\pi f)$ $n = 1, 2$, etc.	191
C'_N	Installed part of C_N	139
C'_n	Installed part of C_n ($n = 1, 2$, etc.)	182
C_{r_2}	r_2 bypass capacitor	182

*Unless otherwise stated, the units are those given for the applicable first or second symbol of each letter.

List of Frequently Used Symbols

Symbol	Definition	Page of First Appearance
C_0	Crystal static capacitance	70
C_0'	Crystal static capacitance including external strays	71
C_1	Crystal motional capacitance	70
C_{1M}	Miller effect contribution to C_1	149
C_{2M}	Miller effect contribution to C_2	149
df	Frequency offset from f_L in Hz	71
F	Function of	371
f	Operating frequency in MHz	5
$\mathbf{f}$	Fourier frequency in Hz	376
f_L	Crystal rated frequency with C_L	71
f_N	Crystal overtone frequency	153
f_s	Antiresonant frequency of mode/overtone networks	153
f_T	Transistor gain bandwidth product	38
G	Conductance in $\mho$, real component of Y	3
g_m	Transistor transconductance	48
g_{me}	Equivalent g_m	210
g_{m_L}	Transistor limiting g_m	174
$g_{m_{L_0}}$	Small signal g_{m_L}	174
g_{m_0}	Small signal g_m	28
h_{FE}	Low frequency $h_{fe} \equiv \beta_0$	28
h_{mn}	h parameter	9
I	Current in mA dc or rms*	7
i	Instantaneous current in mA	29
I_A	dc current in terminal a	47
i_A	Total instantaneous current in terminal a	29
I_a	ac current in terminal a	146
I_{a_M}	Current component at frequency Mf in terminal a	47
I_{a_1}	Fundamental current in terminal a	47
I_{BB}	Power supply dc current	196
I_x	Crystal current	78
I_2	ac current in path or terminal 2	7
L	Inductance in μH	70
L_b	Series inductor in mode selection network	156
L_N	Parallel inductor in selection network	154
L_0	Crystal C_0 neutralizing inductor	251
L_1	Crystal motional inductance	70
M	Multiplication order	47
M_{cb}	$1 + V_{L_1}/V_b$	235
M_{ce}	$1 + V_{L_1}/V_e$	235
M_M	$V_{be'}/V_L$	149
M_M	V_{be}/V_e	235
m_r	R_s/R_L	211
N	Crystal overtone number	82

*Note the use of upper and lower-case symbols and upper and lower-case subscripts to denote the type of current and the terminal or path.

List of Frequently Used Symbols

Symbol	Definition	Page of First Appearance
n	X_2/R_t	212
P	Power in mW	1
P_A	Power dissipated in Z_A	257
P_i	Input power	2
P_L	Load power	185
P_o	Output power	2
P_T	Total power	257
P_x	Crystal drive power	78
Q	Quality factor	78
Q_{op}	Operating Q	150
Q_x	Crystal Q	78
R	Resistance in Ω usually connected in series, real component of Z	3
r	Resistance in Ω, usually connected in parallel	38
r	Capacitance ratio, C_0/C_1	82
R_A	Resistance component of Z_A	253
R_b	Equivalent bias circuit series resistance	144
r_b	Thévenin source resistance of the bias circuit	57
$r_{bb'}$	Base spreading resistance	39
r_{b_1}	Bias resistor between base and $-V_{BB}$	57
r_{b_2}	Bias resistor between base and V_{BB}	57
R_{df}	R_e at $f_L + df$	77
R_E	Unbypassed emitter resistor	139
R_e	Crystal equivalent series resistance	72
r_e	Intrinsic emitter dynamic resistance	38
r_e'	Extrinsic emitter dynamic resistance	39
r_{e_0}	Small signal r_e	38
R_{IN}	Emitter input resistance of the common base transistor	55
R_{in}	Equivalent base emitter series resistance	142
R_{IN_0}	Small signal value of R_{IN}	259
R_L	Crystal resistance at f_L	71
R_L	Tuned load resistance	139
R_{L_1}	Fundamental component of R_L	238
r_{mn}	Resistance between terminals m and n	38
R_N	Negative resistance	254
R_n	Crystal PI-network resistance	209
r_{osc}	V_f/I_X	371
r_{par}	Resistor for suppressing parasitic oscillations	361
R_s	Isolating resistor resistance	208
R_T	Total equivalent series resistance	144
R_t	$R_3 + R_{in}$	212
R_x	Same as R_{df}	253
R_1	Crystal motional resistance	70
R_2	Equivalent Z_2 series resistance	142
r_2	Emitter bias resistor	57
$r_{2\,ac}$	ac value of r_2	239

List of Frequently Used Symbols

Symbol	Definition	Page of First Appearance
R_3	Resonator resistance	144
s	s parameter	25
s	f_s/f_N	153
$S_\phi(f)$	phase spectrum in dbc/Hz	376
$S_{\phi_\delta}(f)$	phase noise of the circuitry	376
$S_{\phi_{V_f}}(f)$	phase spectrum of the output signal V_{f_n}	376
V	Voltage in mV dc or rms*	55
v	Instantaneous voltage in mV	55
V_A	dc voltage between terminal a and the datum (ground)	57
v_A	Total instantaneous voltage between terminal a and the datum (ground)	173
V_a	ac voltage between terminal a and the datum (ground)	63
V_{AB}	dc voltage between terminals a and b	48
v_{AB}	Total instantaneous voltage between terminals a and b	48
V_{ab}	ac voltage between terminals a and b	48
V_{BB}	Power supply voltage	57
V_L	ac voltage across the load R_L	139
V_{L_1}	Fundamental component of V_L	235
X	Reactance in Ω, imaginary component of Z, usually connected in series	3
X_A	Reactance component of Z_A	253
X_e	Crystal equivalent series reactance	72
X_L	Combined reactance of L_3 and C_V	255
X_{L+}	X_L at which X_A is maximum	269
X_{L-}	X_L at which X_A is minimum	270
X_N	Negative reactance	254
X_x	Crystal reactance at $f_L + df$	253
X_2	Total reactance of network 2	141
X_2'	Installed reactance of network 2	140
Y	Admittance in $\mho = G + jB$	3
Y_A	Two-port network described by y parameters	9
Y_A	Effective Pierce oscillator active circuitry	375
y_{mn}	y parameter	8
Z	Impedance in $\Omega = R + jX$, usually connected in series	3
Z_A	$Z_x + Z_{IN}$	253
Z_a	Two-port network described by z parameters	10
Z_c	Impedance seen by the collector	252
Z_{IN}	Emitter input impedance	252
Z_{LL}	Llator impedance	144
z_{m_n}	z parameter	8
Z_s	$Z_1 + Z_2 + Z_3$	253

*Note the use of upper and lower-case symbols and upper and lower-case subscripts to denote the type of voltage and the terminals.

List of Frequently Used Symbols

Symbol	Definition	Page of First Appearance
Z_1	Oscillator series impedance contributing to the negative resistance	129
Z_2	Oscillator series impedance contributing to the negative resistance	129
Z_3	Resonator impedance	129
Z_3	Impedance of R_L and X_L	251
α	h_{fb}	28
α	g_m/g_{m_0}	48
β	h_{fe}	28
β_o	Low frequency $\beta = h_{FE}$	28
γ_M	I_{e_M}/I_E ($M = 1, 2, 3$)	48
γ_p	i_{peak}/I_E	48
γ	$\sqrt{2}\, I_e/I_E$	56
Δf	Frequency shift from series resonance	74
Δf	Change in frequency	75
$\Delta f/f$	Fractional frequency stability	404
Δy	y determinant	24
η	Load power divided by the resonator power	185
θ	Phase angle of an immittance	3
ω	Angular frequency $= 2\pi f$	73

Index

Acceleration, 110, 112, 121
  sensitivity, 113
  shock, 67, 110, 112, 121
  tip-over, 110, 112
  vibration, 92, 110, 112, 113, 119, 121
Activity dips, crystal, 117, 118
Aging:
  component, 418
  crystal, 67, 96, 100, 101, 109, 110, 119, 121
ALC oscillator system:
  components, description, 374–382
    of amplifier K_1, ac, 380
    of amplifier K_3, dc, 381
    of comparator, 381
    of detector, K_2, 380
    general, 370–374
      of G_M loop, 370–372
      noise performance of ALC loop, 373–374
      requirements for K_3 and V_C, 372–373
    of noise characteristics, 381–382
    of oscillator, 374–379
    of output amplifier, 381
    of output filter, 382
    V_{REF}, 381
  noise:
    performancece of ALC loop, 373–374
    phase noise, prediction of, 382
    pierce oscillator in, 375–377
Algorithms, 280–352
  for Butler oscillator, stable design, 345–352
  for Butler oscillator, $X_A = 0$ design, 338–344
  conversion efficiency in, 284
  design examples, 284. *See also* Design examples
  form of, 282–283
  for isolated Pierce oscillator, 301–308
  for normal Colpitts oscillator:
    cutoff limiting, 318–327
    collector limiting, 309–317
  for normal Pierce oscillator:
    cutoff limiting, 293–300
    collector limiting, 285–292
  programming for, 284
  for semi-isolated Colpitts oscillator, 328–338
  use of, 284
Amplifiers:
  cascode, 362–363
  clamped biased JFET, 63–65
  neutralized, isolating, 416
Amplitude-frequency, effect in crystals, 115, 117
Angle, *see* Cuts
  orientation, 80, 100, 106–108, 111
  tolerance 108, 109
Automatic level control limiting, ALC, 169–170, 369

Bipolar transistor, 29–59
  biasing, 57–59
  common-base, 52–57
    two port small signal characteristics, Y_{11b}, Y_{inb}, and $\alpha = h_{21b}$, 52
    hybrid-PI model, 53
    large signal characteristics, 53–56
    for sinusoidal i_E, 53–56
    for sinusoidal V_{BE}, 53
  common emitter, 37–51
    large signal characteristics, 47–51
    small signal hybrid-PI model, 37–47
    β calculation of, 41–43

447

448   Index

f_T, discussion of, 42–43
   intrinsic transistor, 29, 38
   local feedback, effect upon
      performance, 44–46
   variation of parameters with voltage,
      current and temperature, 41
   Y_{fe}, Y_{oe}, and Y_{re}, calculation of, 44
   Y_{ie}, calculation of, 43
comparison of common-base and
   common-emitter hybrid-PI
   models, 54
minimum value for f_T:
   resonator in parallel with base and
      emitter, 163–164
   resonator in series with emitter, 252
minimum value for β_0, 164
relationships for parameters, 28
typical data:
   for 2N918, 30–31
   for 2N2857, 36–37
   for 2N2947, 32–36
Bottoming effect, 172
Butler oscillator, 250–269
   circuit analysis, 251–264
   A_{L_0}, gain, factor, concept of, 259
   approximations and assumptions for,
      251–252
   design procedures, discussion of,
      262–264
      assumption A, 264
   frequency stability relationships, 255
   L_0, calculation of, 260–261
   limiting:
      assumptions for, 252
      collector base voltage limiting, 260
      R_{IN} limiting, 260
      type of limiting used, 260
   oscillatory conditions, derivation of,
      252–255
   Q_{L3}, calculation, effect of, 261
   P_A, P_T, P_X, V_e, V_L, V_2, and Z_c, calculation
      of, 256–257
   R_N, X_n, Z_c, in terms of physical
      components, calculation of, 262
   transistor approximations, 252
   Z_s, in terms of physical components,
      255–256
   description of, 250
   stable Butler oscillator, theory and design
      of, 268–279
   background for, 268
   basis of design, 269
   C_{ce}, effect of, 272–273
   design algorithm, 273, 345–352
   design examples, 274–275, 277–278

   oscillator relationships, 269–271
   Q_{op}, calculation of, 279
   trimming, 276, 279
   X_{L+} and X_{L-}, relationships for, 271
   $X_{A=0}$ design procedure, 257–259, 264–268
      design algorithm, 264, 338–344
      design example, 264–266, 267
      Q_{op}, calculation of, 261–262
      relationships for $X_{A=0}$, 257–259
      trimming, 266, 268

C'_L, calculation of, 191–192
C_{r_2}, calculation of, 191, 223
Capacitance, crystal:
   load, 70, 71, 76, 77, 90–92, 117, 120, 121
   motional, 70, 71, 82–85, 87–89, 91–93, 96,
      97, 100, 102–104, 114, 120
   overtone, 83, 84, 88
   ranges, 86, 88
   ratio, 79, 82, 85, 87, 89, 91–93, 95, 96, 100,
      102, 104
   stray, 71, 72, 76, 89, 91
   trimmer, 77, 82
   wiring, 76
Capacitors, associated with crystal, 66, 74, 94
Circuit, crystal:
   equivalent electrical, 66, 69–71, 95, 97, 98,
      101, 104, 105
   motional arm parameters, 71, 79, 83–85, 87,
      89, 91, 95, 96, 102, 104, 105
Circuitry, oscillator, frequency stability
   requirements of, 402–410
Clamp biased amplifier, 63–65
Clamp biased limiting, 63–65, 179
Clapp Gouriot oscillator, 167
Clapp oscillator, *see* Pierce, Colpitts and
   Clapp oscillator family
Colpitts oscillator, *see* Pierce, Colpitts and
   Clapp oscillator family
   normal, 181–207
   semi-isolated, 233–249
Coupling in crystals:
   mechanical, 102, 104, 105, 117
   piezoelectric, 81, 82, 87, 89, 95, 97, 99,
      109
Crystal parameters, measurement of, 424–425
Crystal physical location, problem due to, 415
Crystal responses, problem due to, 413–414
Crystal resonator, *see* Plates
   CR-numbers, 91–93
Current, crystal, I_x, 78, 89, 114, 115, 117, 146,
   151
   measurement of, 206, 421
Cuts, crystal, 69, 80, 82, 85, 92, 93, 100, 104,
   106, 120

Index   449

doubly rotated, 96, 97, 117
   SC, 83, 87, 88, 92, 97–100, 102, 106–108, 110–115, 117–119
   singly rotated, 97
   AT, 81, 83, 86–89, 92, 96, 97, 99, 100, 104, 106, 107, 111–115, 117, 118
   BT, 83, 86–88, 97, 104, 106
   GT, 86, 87, 102, 106
   GT, 86, 87, 102, 106
   SL, 86, 87, 102
   ST, 100, 117

Darlington pair, 358–363
Delay line, 101
Design examples:
   for Butler oscillator:
      stable design, 274–276, 277–278
      $X_A = 0$ design, 264–266, 267
   for isolated Pierce oscillator, 216, 217–218
   for normal Colpitts oscilator:
      cutoff limiting, 227, 228–231
      collector limiting, 224, 225–226
   for normal Pierce oscillator:
      cutoff limiting, 200, 201–204
      collector limiting, 195, 196–198
   for semi-isolated Colpitts oscillator, 241–248
Design procedure, 194. *See also individual oscillator name*
Dielectric constant, quartz, 87, 91
Dielectric permittivity, quartz, 87, 91.
Diode limiter, *see* Limiting, diode
Drive level, crystal, 20, 110, 114, 115, 117–121. *See also* Activity dips; Amplitude-frequency

Electrode, crystal, 113
   deposition, material, 69
   geometry, 89–91, 98, 100, 104, 105, 109, 117
   stresses in, 98, 100, 109, 119
Enclosures, crystal, 66, 68, 69, 90, 95, 104, 115, 120, 121
   ambient, 69, 80, 85, 109, 110
   geometry, 68, 104
   material, 68
   sealing, 68, 109, 110
   terminals, leads, 66, 68
Environment, crystal, 92, 110. *See also* Acceleration; Radiation; Temperature
External limiter, 169, 364–368.

Filter, crystal, 79, 82, 96, 101, 105, 119, 382

Frequency:
   characterization of, 21
   in crystals:
      accuracy, 67
      antiresonance, 73, 74, 78, 79, 82, 83, 91
      changes, 94, 96, 102, 112, 114, 117–119
      constant, N_o, 81, 97
      load, 71, 75, 76, 91, 93, 95, 109
      modulation, 82, 94, 113
      operating, 71, 74, 76, 78, 79, 82, 89, 92, 93, 96, 100, 102–104, 117, 120
      pulling, 82, 93, 95
      range, 67–69, 85, 95, 103
      resonance, 73, 75, 76, 78, 82, 83, 92–95, 97, 109, 113
      series resonance, 73, 76, 92
      stability, 67, 83, 84, 93, 95, 97, 100, 102
      tolerance, 92, 93, 102, 108, 109, 120
Frequency stability, 18, 21. *See also individual oscillator circuit*
   characterizations of, 20–21
   circuitry, evaluation of, 402–410
   long and short term, 18
Frequency stability analysis:
   in Butler oscillator, 255
   in isolated Pierce oscillator, 193, 194, 219
   in normal Colpitts oscillator, 192–194, 227, 232
   in normal Pierce oscillator, 192–194, 199, 205
   in Pierce, Colpitts and Clapp oscillator family, 164–168, 192–194
   in semi-isolated Colpitts oscillator, 193–194, 241
Frequency stability in LC oscillators, 123, 133
Frequency-temperature characteristics:
   crystal, 82, 96, 97, 100, 109, 118
   dynamic, 110, 111
   static, 105–107, 111
      cubic, 99, 108, 111
      parabolic, 108
      turn-over, 108, 111, 112

Gate oscillators:
   crystal operating in inductive region, 392–393
   crystal operating near series resonance, 394–395
   limiting, type of, 391
Gates:
   biasing of, 386–388
   classification of, 384–385
   conversion into equivalent inverters, 389
   inverter approximate equivalent circuit, 390–391

Harmonic, crystal, *see* Overtone
Holders, crystal, *see* Enclosures
  HC-numbers, 68, 69, 81, 90, 99, 115, 122
Hysteresis, crystal:
  amplitude-frequency, 115
  thermal-frequency, 112

Immittance, 3–6
  approximate relationships for, 4
  polar notation, 3
  rectangular notation, 3
  specialized forms, 4–5
  $Z = R + X$ representation, 5, 6
Inductance, crystal:
  motional, 70, 71, 83, 84, 89, 91, 97
  overtone, 83, 84
  region of positive, 73, 79, 83, 93, 95
Inductors for crystals, 93–95, 102
Integrated circuit gate oscillators, 398–399
  Type 74124, 398
  Type 74324, 398
  Type 8224, 398
  Type SP 705, 399
  wristwatch oscillators, 399
Integrated circuit linear oscillators, 399–400
  Type DS 8907, 400
  Type MC 12060, −1, and MC 12560, −1, 400
  Type SL 680A, 400
Inverter equivalent circuit, 390–391
Isolated Pierce oscillator, 208–219. *See also* Normal Pierce oscillator; Pierce, Colpitts and Clapp oscillator family
  advantages of, 208
  circuit analysis, 208–215
    additional design equations, 211–214
    basic equations, 208–211
    practical values of R_s, 214
    R_L, for minimum power consumption, calculation of, 214–215
  design procedure, 215
    design algorithm, 215, 301–308
    design examples, 216, 217–218
    trimming, 216, 219
  frequency stability, 219
  limiting, 215
Isolation, 18, 208, 234, 235, 282, 416–417

Junction field effect transistors (JFET), 59–65. *See also* Transistors
  clamp biased amplifier circuit, 63–65
  common source large signal characteristics, 62–65
  ideal small signal characteristics, 59–61

  noise properties, 59, 379
  small signal high frequency characteristics, 61–62

Leeson oscillator noise model, 382
Limiting:
  ALC, 169–170, 369
  diode, 179–180, 365, 366
  external, 169, 364–368
  self:
    bipolar transistor, sinusoidal i_E, 178–179
      advantages of, 178–179
      collector base voltage, 179
      R_{IN}, 179
    bipolar transistor, sinusoidal v_{BE}, 170–178
      advantages of, 170–171
      base emitter cutoff, 171–172
      collector base voltage, 172–178
      when $R_E = 0$, 175–176
      when $R_E \gg r_{e_0}$, 176–178
    JFET, sinusoidal, v_{GS}, 179
      clamp biased, 179
      drain gate voltage, 179
Linear two port networks, 7–12, 22–28
  parameters for, 7–12, 22–28
    conversion:
      between h and y, 24
      between h and z, 25
      between s and y, 25
      between y and z, 12
    definition:
      of h, y, z, 7–9
      of s, 24
    relationships for, 9–10
    Y networks, relationships for, 10–11
Lithium niobate, 68
Lithium tantalate, 68, 104
Llator, 79, 92
  concept, 14, 403–404
  measurement of, 423–424, 429–430
Load impedance transformation, 160–163
  capacitance coupling network, 162
  capacitive divider coupling network, 161
  PI coupling network, 162
  transformer coupling network, 161
Local oscillator, 130–131
  narrow range, 130–131
  wide range, 130
Losses, circuit components, 261, 283
  ambient, 80, 85
  crystal, *see* Mount; Trapping
  internal friction, 80
  mounting, 80, 81, 85

Measurements:
  ac current, 206, 421
  ac voltage, 420–421
  dc current, 420
  dc resistance, 420
  dc voltage, 419
  large signal immittance, 428–430
    three terminal networks, 428–429
    Z_{LL}, 429–430
  oscilloscope, with, 422
  Q_{o_p}, 206–207
  small signal impedance, 422–428
    c mode selector, 424
    crystal parameters, 424–425
    linear and non-linear components, 423, 424, 425–428
    overtone selector, 424
    passive components, 423
    transistors, 425–428
    Z_{LL}, 423–424
Merit, figure of, crystal, 74, 81, 82, 90, 92, 93, 95, 100, 102, 104
Miller effects, 149–150, 235–237, 262–263, 272–273, 363
Miller oscillator, 133
Miniature packaged crystal oscillators, 400–401
Mode selection networks, 155–160, 240
  measurement of, 424
Modes of vibration, 69, 80, 82–84, 117
  a, b, c thickness, 69, 97, 107
  separation, 97, 98
  spectrum, 96, 98, 99, 105, 113
  unwanted, 84, 104, 105, 121
Motion, crystal
  distribution of, 89–91, 98, 104
  at edges, 81
  measure of, Ψ, 89, 91, 98
Motional arm, see Circuit, crystal
Mount, crystal
  bonding, 69, 109
  losses, 80, 85, 89
  stresses, 109, 110
  support structure 69, 85, 104, 121

Noise:
  circuit components, 381
  crystal, 110, 119
    amplitude-to-frequency, 120
    equivalent resistance, 119
    flicker, higher-order, 119
    polarization, 114
    voltage, 119
  oscillator, 375–377. See also ALC oscillator system

  oscillator spectrum of, 375–377, 382
  phase prediction of, 382
  transistor, 379
Non-linear two port networks, 12
Normal Colpitts oscillator, 220–232. See also Normal Pierce oscillator; Pierce, Colpitts and Clapp oscillator family
  circuit analysis, 221–223
    C_{r_2}, calculation of, 223
    R_L and V_L, calculation of, 221–222
    V_E, calculation of, 223
  description of, 220
  design procedure, 223–232
    be cutoff limiting, 227–232
      design algorithm, 227, 318–316
      design examples, 227, 228–231
      trimming, 227, 232
    collector base voltage limiting, 223–227
      design algorithm, 224, 309–317
      design examples, 224, 225–226
      trimming, 224, 227
  frequency stability, 192–194, 227, 232
Normal Pierce oscillator, 181–207. See also Pierce, Colpitts and Clapp oscillator family
  amplitude variation due to changes in V_{BB}, 193–194
  C'_L, calculation of, 191–192
  C_{r2}, calculation of, 191
  circuit analysis, 181–192
    fundamental design equations, 185–186
    g_m, calculation of, 190–191
    modified PCCF equations, 182–184
    R_L and $V_{L max}$, calculation of, 186–189
    V_E, calculation of, 189–190
  description of, 181
  design procedure, 194–205
    be cutoff limiting, 199–205
      design algorithm, 199, 293–300
      design examples, 200, 201–204
      trimming, 200, 205
    collector base voltage limiting, 194–199
      design algorithm, 195, 285–292
      design examples, 195, 196–198
      trimming, 198
  frequency variations due to external factors, 192–194, 199, 205
    changes in load R_L, 192–193
    changes in V_{BB}, 193, 194
  measurement techniques
    I_3 or I_x, 206
    Q_{op}, 206–207

Osci:
  concept, 14, 402, 403
  replacement for inductive crystal, 405
  replacement for series resonant crystal, 407
Oscillator:
  circuit:
    names of, 3
    selection, 280–281
  classification of, 2
    abbreviations for, 2
  crystal, other circuit configurations, 353–368
    Colpitts, low harmonic and noise, 356
    Darlington pair Colpitts, 361–362
    high-efficiency, 357
    low frequency, 354–356
    two transistor, 364–368
      diode limiting, 365, 366
      emitter coupled, 365, 367–368
      and limiter, 364–365
      semi-isolated Colpitts with cascode, output circuit, 363
      with separate variable gain control, 365, 367
  definition of, 1, 17
  delay line, 101
  harmonic, definition of, 2
  models for, 13–17
    feedback, 15–17
    negative conductance/resistance, 13–15
  names of, 3
  problems, 411–417
    aging of components, 418
    crystal physical location, 415
    isolation, 416–417
    spurious oscillations, 411–413
    spurious signals due to crystal responses, 413–414
    squegging, 414–415
    starting, 415–516
      hysteresis, 415–416
  starting, 415–416
  temperature compensated crystal, TCXO, 82, 109
  turn-on, turn-off, 109
  variable crystal (VCXO), 82, 93
Oscillator:
  definition of, 17
  vs. oscillator, 18
  specifications, 18
  test of, 417
Oven, 92, 112
  noise performance, effect on, 119
  reference point, 112
  time constant, 111
  turn-on, turn-off, 109–111
Overtone, crystal:
  high, 83, 95
  number, 69, 80, 82–85, 97–100, 104, 114, 117, 120
  units, 80, 86, 91
Overtone selection networks, 153–155, 157–160, 240
  measurement of, 424

Phase angle of g_m, effect of, 137–138
Phase noise:
  curves of, 376–377
  prediction of, 382
Pierce, Colpitts and Clapp oscillator family, 125–180
  idealized oscillator, 128–138
    crystal oscillator, 133–136
      crystal inductive region, 133–136
      crystal in series resonance or capacitive region, 95, 136
    LC oscillator examples, 130–133
    phase angles of g_m, effect of, 136–138
  real oscillator, 139, 160
    assumptions and conditions, 141–142
    circuit design equations, derivation of, 145–151
      C_{bc} contribution of, 149–150
      currents and voltages, 146–149
      oscillator operating Q, calculation of, 150
      power, output, 146
      X_2, 147–148
    crystal oscillator, 151–160
      fundamental wide frequency, range oscillators, 152–155
      mode selection circuits, 155–160
      overtone selection circuits, 153–155
    frequency stability analysis, 164–168
      crystal oscillators, in, 167–168
      stabilizing capacitor C'_L, 165–167
    oscillatory equations, calculation of, 142–145
      amplitude equation, 144
      frequency equation, 145
      negative resistance, 144
Pierce oscillator, *see* ALC oscillator system; Isolated Pierce oscillator; Normal Pierce oscillator; Pierce, Colpitts and Clapp oscillator family
Plates, crystal, *see* Electrode, crystal
  dimensions, 81, 90, 91, 103
  geometry:

bevelled, contoured, 81, 91, 98, 105, 117
flat, 69, 80–82, 84, 87, 88, 98, 104, 117
Polarization, crystal, 113, 114, 120
Pole-zero spacing, crystal, 79, 82, 90, 95, 97, 100
Power level, crystal, 20, 78, 115, 118

Q_{op}, 79–80, 150–151. *See also specific oscillator circuit listed by name*
  measurement of, 206–207
Quality factor, crystal, Q_x, 78–81, 89, 92, 93, 95, 96, 100–104, 118, 119
Quartz, 68, 104
  cultured, natural, 68, 118
  etching of, 115, 119
  impure, swept, 68, 118, 119
  surface preparation, 108, 115
Quartz cuts, *see* Cuts

Radiation, 118, 119
Reactance, crystal:
  change with frequency, 75, 78
  equivalent series, 72, 73, 76, 83, 93
  motional, 77
  positive, 79
  sharpness, 79
  slope, 76, 78, 79, 84, 92–95
  variable, varactor, 94
  zeros, 73, 83
Realizable maximum values:
  of resistance of resistor, 6–7, 214
  of resistance of tuned load, 6
Required value of β_o, 164
Required value of f_T:
  sinusoidal i_E, 252
  sinusoidal v_{BE}, 163–164
Resistance, crystal, 100–104, 114, 115, 117, 119
  apparent, 114, 118
  change, with temperature, 80, 102
  equivalent series, 72, 73, 76, 78, 83, 90, 102
  load, 71, 75, 77, 80, 90, 92, 120, 121
  motional, 70, 71, 73, 83, 84, 88–93, 96, 97
  overtone, 84, 87
  ranges, 87
  shunt, 95, 96, 114, 120
Resonance-antiresonance spacing, crystal, 79, 82, 90, 95, 97, 100
Resonance curve, crystal:
  sharpness, 79
  width, 79
Resonators, crystal, 66, 70, 122
  bulk, BAW, 67, 100, 101, 117
  ceramic, 68, 95, 96

miniatire, 84, 92, 102, 104
one-, two-port, 70, 71, 101, 102
other
  electromechanical, 124
  inductance capacitance (LC), 123
  magnetostriction, 124
  surface, SAW, 67, 92, 100–102, 117
Responses, crystal modal, *see* Modes of vibration

Selection:
  mode, 155–160, 240
  oscillator circuit, 280–281
  overtone, 153–155, 240
Self-limiting, *see* Limiting, self
Semi-isolated Colpitts oscillator, 233–249. *See also* Colpitts oscillator; Normal Colpitts; Pierce, Colpitts and Clapp oscillator family
  circuit analysis, 233–240
    description of, 233
    isolation characteristics, 234, 235
    limiting, 237
    Miller effects, 235–237
    overtone and mode selection, 240
    R_L, calculation of, 238
    R_{2ac}, calculation of, 239–240
    V_E and V_{Lmax}, calculation of, 238
  design procedure, 241
    design algorithm, 241
    design examples, 241–248
    trimming, 248–249
    frequency variation due to external factors, 241
Specifications:
  crystal, 94, 115, 120–122
  oscillator, 18–19
Spurious oscillations, 411–413
Spurious signals due to crystal responses, 413–414
Squegging, 414–415
Stability, frequency, *see* Frequency, stability
Stabilizing capacitor, C'_L, 165–167
Stable Butler oscillator, *see* Butler oscillator
Stable LC oscillator, 132–133
Stiffness, crystal, 76, 82, 100
Stress compensation, crystal, 98, 99

Temperature, crystal:
  compensation, 82, 94
  frequency stability over, 92, 100, 108
  inflection, 100, 107, 108
  operating range, 67, 80, 92, 93, 100, 102, 108, 109, 112, 118, 120, 121

reference, 105, 108, 111, 120
transients, 98, 100, 110, 111, 118, 119
Temperature coefficients, crystal:
  first order, 96, 98, 99, 103–105, 107, 117
  higher order, 103–108
Time constant:
  crystal, motional, 80, 85, 87, 89, 90, 97, 99
  oven, 111
Transistors, 22–46. *See also* Bipolar transistors; Junction field effect transistors
  comparison of JFET and bipolar transistor properties, 60
  Darlington connection, two port small signal parameters:
    conversion among common base/gate, common collector/drain and common emitter/source parameters, 26–28
    conversion among h, s, y, z parameters, 12, 24–25
    notation for, 23–24
Trapping, energy, 90, 104
Trimming:
  for Butler oscillator:
    stable design, 276, 279
    $X_A = 0$ design, 266, 268
  for isolated Pierce oscillator, 216, 219
  for semi-isolated Colpitts oscillator, 248–249
  for normal Colpitts oscillator:
    collector cutoff limiting, 224, 227
    cutoff limiting, 227, 232
  for normal Pierce oscillator:
    collector limiting, 198–199
    cutoff limiting, 200, 205

Vibration, crystal, *see* Acceleration; Modes of vibration
  coupled, 69, 103
  edge, 103
  extension, 69, 103
  flexure, 69, 102, 103, 117
  shear, 69, 102, 103
  types of, 69, 80, 82, 84, 85, 102, 103
Voltage, crystal, 113–115

Wristwatch oscillators, 399

Y admittance, 3–4
Y networks, 9–12
Y parameters, 7–8

Z impedance, 3–4
$Z = R + X$, representation, 5–6
Z networks, 10
Z parameters, 8